The Real Wolf

The Real Wolf

The Science, Politics, and Economics
of Coexisting with Wolves in Modern Times

TED B. LYON AND WILL N. GRAVES

With contributions by Rob Arnaud, Dr. Arthur Bergerud,
Karen Budd-Falen, Jess Carey, Dr. Matthew A. Cronin,
Dr. Valerius Geist, Bruce Mahler, Don Peay,
Laura Schneberger, and Heather Smith-Thomas

Skyhorse Publishing

Skyhorse Publishing books may be purchased in bulk at special discounts for sales promotion, corporate gifts, fund-raising, or educational purposes. Special editions can also be created to specifications. For details, contact the Special Sales Department, Skyhorse Publishing, 307 West 36th Street, 11th Floor, New York, NY 10018 or info@skyhorsepublishing.com.

Skyhorse® and Skyhorse Publishing® are registered trademarks of Skyhorse Publishing, Inc.®, a Delaware corporation.

Visit our website at www.skyhorsepublishing.com.

10 9 8 7 6 5 4 3 2 1

Library of Congress Cataloging-in-Publication Data

Names: Lyon, Ted B., author. | Graves, Will N., author.
Title: The real wolf: the science, politics, and economics of coexisting with wolves in modern times / Ted B. Lyon and Will N. Graves; with contributions by Rob Arnaud, Dr. Arthur Bergerud, Karen Budd-Falen, Jess Carey, Dr. Matthew A. Cronin, Dr. Valerius Geist, Bruce Mahler, Don Peay, Laura Schneberger, and Heather Smith-Thomas.
Description: Second edition. | New York, NY: Skyhorse Publishing, 2018. | Includes bibliographical references.
Identifiers: LCCN 2017038432 | ISBN 9781510719613 (pbk.)
Subjects: LCSH: Human-wolf encounters—United States. | Wolves—Reintroduction—United States. | Wolves—Control—United States.
Classification: LCC QL737.C22 L96 2018 | DDC 599.773—dc23
LC record available at https://lccn.loc.gov/2017038432

Cover design by Tom Lau
Front cover photo credit: iStockphoto

Print ISBN: 978-1-5107-1961-3
Ebook ISBN: 978-1-5107-1963-7

Printed in the United States of America

Contents

Foreword

by Idaho Governor C. L. "Butch" Otter

Photo credit Karl Umbriaco/Shutterstock.com

For more than twenty years now I have had a front-row seat to one of the worst natural resources policies ever inflicted upon the West.

I was Idaho's lieutenant governor in the mid-1990s when—like a shotgun wedding—Canadian gray wolves were "reintroduced" to the Idaho back-country by the US Fish and Wildlife Service, expressly against our wishes as a state. I was a member of Congress in the early years of the twenty-first century as those transplanted predators grew into voracious packs that began ravaging our elk herds and terrorizing our livestock. So when I became Idaho's governor in 2007, one of my top priorities was working with our congressional delegation, sportsmen, ranchers, and many others—including Ted Lyon—to overcome the legal hurdles set up by environmental extremists and activist judges to states wresting management of these big marauders from federal bureaucrats.

Ted's book, *The Real Wolf*, does a compelling job of chronicling that process, and why it remains so important to those of us—no matter our political affiliation—who care deeply about states' rights and responsible stewardship of our public lands, wildlife, and other resources. The hard-won experience of too often being at the mercy of that federal fiat we call the Endangered Species Act (ESA) has taught us the value of on-the-ground collaboration, consensus building, and counting on the real-world perspectives of those who live with the outcomes of our public policies.

There have been few issues during my forty years in public life that have provoked the raw passions of so many people from around the world as the debate over wolves.

As Idaho sought to at least control the carnage, I was deluged with some of the nastiest, most disparaging, and truly hateful letters, emails, and phone calls from well-meaning but badly misinformed folks. Most saw wolves only as big, beautiful dogs—harmlessly pursuing their majestic lives in the trackless wild. They argued that wolves are an essential and misunderstood part of the Rocky Mountain ecosystem, and we owe it to our western heritage to enable them to once again roam freely in the Idaho wilderness.

The problem is that wolves don't stay put. Their enormous range, high reproductive rate, and insatiable hunger inevitably draw them out of the back-country into the areas of men. As their numbers in Idaho spiraled far beyond expectations, so did the conflicts, and so did my determination to manage wolves as we do any other species—with an eye toward the bigger picture of a balanced ecosystem that includes man.

Getting to that point was challenging. While my state officially met the federal ESA recovery objectives for wolves in Idaho in 2002, we would not see final delisting for nearly a decade. And the conflicts continued to increase. Depredation grew, and state wildlife officials began noticing significant declines in Idaho deer and elk herds. In many of those areas, wolves were found to be a primary limiting factor for failure to achieve big-game population goals. That was simply unacceptable, so my staff and I continued working to wrench more management flexibility from the federal government's clutches.

We eventually won a few concessions under the government's "non-essential, experimental population" rules, allowing us to move or kill offending wolves. That was welcome, but full delisting was the real goal. However, frivolous legal actions by activist groups delayed those efforts despite the ESA's stated goal of transferring management authority to states once a species has been deemed biologically "recovered."

Ultimately—and after consistent pestering from many others and me—Congress got fed up with the delay tactics and took the unprecedented step of legislatively delisting wolves in Idaho and Montana. If not for that action, wolf management in Idaho still would be up to a pack of federal bureaucrats.

I'm grateful to Ted and the many good people who feel a strong affinity for Idaho, Montana, and the other states where wolves are yet another government-imposed challenge to overcome. But our friends in Wyoming continue struggling to gain state control over wolves, and wildlife managers in Arizona and New Mexico cope with deep mistrust of federal wolf "experts."

Officials in the state of Washington, where some Idaho wolves have migrated, are dealing with angry public outcry and even death threats for merely testing a plan for removing problem wolves—a plan which was agreed upon by a diverse collaborative group of local stakeholders.

Even folks in the Great Lakes states now are looking to Idaho for help breaking through the gridlock. My advice to them: when bureaucratic delays and environmentalist roadblocks prevent the ESA from working as it was intended, get Congress involved.

It's not a perfect solution, but seeking congressional relief can be an effective response to the efforts of ersatz conservationists who speak floridly about the primal necessity of having wolves in our midst.

For them, the real goal is raising money and disrupting or shutting down such traditional multiple uses of public lands as grazing, logging, mining, and especially hunting.

It's a problem created by wolf advocates who repeatedly move recovery targets, forum-shop for sympathetic judges, collect millions of taxpayer dollars to pay their lawyers, and look for any opportunity to abandon their commitment to pay for losses to ranchers and sportsmen.

Ted, and many others who recognize that reality, fought tough odds to turn the tide on the wolf issue. Now Idaho and Montana are managing wolves—wolves that never should have been here in the first place. But since they are, the good news is that the people most impacted by their presence now are managing them in a way that's far more balanced and reflective of the realities of today's West.

They will never be "our wolves," but at least now we have a primary role in controlling their population and impacts.

In some ways, the painful process of wolf introduction has been the canary in the smoldering coalmine that is the Endangered Species Act. Transplanting wolves to Idaho was precursor to a myriad of other ESA-related dustups throughout the Northern Rockies.

Take the greater sage-grouse. Western governors were invited by the Obama administration in 2011 to develop our own grassroots plans for conserving this iconic western species on federal lands. My goal was to develop a strategy that protected the bird without an ESA listing, while maintaining traditional land-use activities. But after crafting a plan that was endorsed by our local federal partners, bureaucrats at the Interior Department in Washington, DC, disregarded our local efforts and opted instead for an unnecessary, more restrictive one-size-fits-all approach across multiple western states.

While the greater sage-grouse was not listed, what we ultimately got was an overly restrictive federal protection plan that I continue to challenge in court.

The Yellowstone grizzly bear population is another example of bureaucracy run amok. Due to the efforts of Idaho, Montana, and Wyoming, Yellowstone's grizzlies were biologically recovered more than a decade ago. But lawsuits heard by activist judges have delayed the handover of management to the states.

In 2011, the 9th US Circuit Court of Appeals found that the states had adequate plans and safeguards in place to assume full management of the bear population. But despite that decision, the Fish and Wildlife Service is holding final delisting hostage unless states adhere to new and added delisting requirements that undermine state sovereignty.

Many other governors share my frustrations. In fact, Wyoming Governor Matt Mead and the Western Governors' Association recently spearheaded an initiative to explore areas of the ESA that are in desperate need of overhaul. The focus is on areas where states can become fully vested partners in conservation and recovery of threatened or endangered species.

After all, it is the states, by way of the Constitution, that have authority over all fish and wildlife within their borders. It's the states that have the biological expertise and the working relationships with those most impacted by conservation decisions that enable us to be the better stewards of our fish and wildlife.

It's my sincere hope that *The Real Wolf* will help open some eyes to the bigger problems with the Endangered Species Act—a once well-intentioned but incredibly flawed law that undermines the real interests and values of conservation by placing the well-being of humans and their livelihoods far down the food chain.

It's time we claw our way back up! Reading Ted's outstanding book is a great start.

Author's Preface

by Ted B. Lyon

By profession, I'm a trial attorney. I try complex cases that involve death, horrible injuries, toxic torts, and environmental litigation. Over the past forty years, I've represented clients in more than 150 jury trials as well as in settlements and arbitrations. I've broken nationwide verdict records and received verdicts that ranked in the "Top 10" on three different occasions. I win because I stick to the facts.

I came to this approach based on my earlier experiences as a law enforcement officer, as a graduate assistant in college where I taught political science, and as a State Senator and State Representative—all in Texas. While serving as a state legislator, my passions were both law enforcement and wildlife policy based on science. Groups as diverse as the Sierra Club, Texas Farmers Union, Texas Black Bass Unlimited, Greater Dallas Crime Commission, World Wildlife Federation, Sportsmen for Fish and Wildlife, and Texas Outdoor Writers Association have given me awards, of which I'm very proud.

A second driving force in my life is my love for nature and outdoor sports, especially fishing and hunting. Once upon a time, I was a licensed fishing and hunting guide. Now, while I live in Texas, my wife and I own a home in Montana where we retreat several times a year. I've been on three African safaris and have made a number of trips to Canada and Alaska.

I'd like to express my gratitude to the many people who saw me through the process of writing and compiling this book, including those who provided support, allowed me to interview them and quote their remarks, talked things over, read, wrote, offered comments, and assisted in the editing and proofreading: David Allen, Ray Anderson, Ed Bangs, Benjamin Barmore, Ryan Benson, Toby Bridges, Mark Connell, Linda Grosskopf, Harriet M. Hageman, Richard Lyon, Richard Mann, Tracy Stone-Manning, W. R. McAfee, Kelley Moore, Miles Moretti, Nancy Morrison, Justin O'Hair, Governor Bruce Otter, former Democratic Majority Leader Senator Harry Reid, Bill Schneider, former Governor Brian Schweitzer, Dale Simmons, James Swan PhD, Dr. Shannon Taylor, Senator Jon Tester and his staff, Josh Tolin, Skip Tubbs, Stephanie Yarbrough, and the staff of the *Yellow Pine Times*, a valuable resource.

For understanding my countless hours devoted to this book, I'd like to thank my wife, Donna Lyon. I also want to thank Tommy Sellers, who for three years rode around Montana listening to me interview people about wolves and elk, while we were supposed to be hunting birds. I also want to thank my son, Payton, who drove me around Montana while working on the editing and revisions of this book.

Last and not least, I beg forgiveness from all those who have helped me over the course of researching and writing this book whose names I've failed to mention.

Author's Preface

by Will N. Graves

My first real job started in 1950 when I went to work for the US Department of Agriculture-Bureau of Animal Industry in Mexico. My assignment was to work for the Mexican-American Commission for the Eradication of Foot and Mouth Disease, formed to prevent the spread of the foot and mouth disease (FMD) from Mexico into the United States, as over fifteen million head of Mexican cattle had been infected. I became the chief of a livestock inspecting and vaccinating brigade in a horseback-only area. My brigade constantly traveled by horseback inspecting and vaccinating all cloven-footed, domestic livestock to prevent them from catching this dreaded disease. Fortunately, there were no active cases of FMD in my sector while I was there. FMD is a highly contagious viral disease. If one animal in a herd catches the disease, within twenty-four hours, every animal in the herd can be infected. FMD is considered the most costly of all animal diseases, as it is often necessary to conduct a wholesale slaughter of animals whenever there is an outbreak. In 1924, there was an outbreak of FMD in California, and the USDA reported that the probable vector for infection of some of the cattle was dogs. An American veterinarian told me that one reason FMD was so difficult to stamp out in Mexico was that dogs and coyotes were spreading the disease. This statement was etched into my mind and had a profound effect on my future interest in livestock and diseases.

In November 1993, I took the opportunity to comment by letter on the Draft Environmental Impact Study about reintroducing wolves into Yellowstone National Park. I wrote that, in my opinion, more research was needed on the potential negative impact wolves would have on bringing and spreading parasites and diseases into the Park and the area in general. Wide-ranging wolves carry and spread many types of dangerous parasites and diseases. The parasites that wolves carry to wild animals may then be passed on to domestic animals, and then pets may pass them to humans. I believed more research needed to be done in regards to the fact that wolves may cause serious harm by spreading dangerous parasites and diseases over large areas.

I believe that wolves have a legitimate role and place in the ecosystem. I support that their numbers be carefully controlled as the result of scientific research on their impact on given areas. However, after all my years researching livestock and Russian wolf behavior, I concluded that, as a general rule, many Western writers and supporters of wolves often over-emphasize the positive role of the wolf and tend to ignore or overlook the negative aspects of wolves in nature.

Photo credit Denis Donahue/Shutterstock.com

Minimum Estimated Wild Gray Wolf Populations

Wolf Populations in the United States as of December 2015*	
STATE	WOLF POPULATION
Alaska	10,000
California	10
Idaho	786
Michigan	636 (and 3 on Isle Royale)
Minnesota	2,221
Montana	536
Oregon	110
Washington	90
Wisconsin	886 to 897
Wyoming	382

*Some scientists believe there are at least twice this many wolves in the United States.

There are also forty-five to sixty red wolves in North Carolina, about a hundred Mexican wolves in the wild in New Mexico and Arizona; many more coy-wolf and wolf-dog hybrids; and there are at least three hundred thousand and maybe as many as five hundred thousand wolves and wolf-dogs in captivity in the United States and an unknown number that are feral. There are also at least sixty thousand gray wolves in Canada.

Photo credit: Critterbiz/Shutterstock.com

Opening Statement

By Ted B. Lyon

"May it please the court, ladies and gentlemen of the jury. In this case I will prove the following . . ." Those two sentences are the way I've started every case for over forty years as a trial lawyer who has tried hundreds of cases.

So, may it please the court, ladies and gentlemen of the court of public opinion, the case that I am about to make to you about wolves and how they have been managed over the last thirty years is a story about misinformation and myths. This will shock the reader into understanding that wolves truly do need to be managed, especially as they get closer to people. Furthermore, they need to be managed by the states and not the US Fish and Wildlife Service. That agency is inefficient and monolithic, so that it is almost impossible to try to change policies for wildlife management at the national level when it comes to wolves, especially when one has to deal with the Endangered Species Act and the endless lawsuits that the act brings in. In reading this book, you will find that a massive campaign has perpetuated misinformation about wolves. Here are some examples:

1. Wolves do not kill or attack people.
   Fact: They do, and regularly.
2. Wolves are the sanitarians of nature and only kill the weak and the sick.
   Fact: Wolves kill any and all forms of animals, both the weak and the strong.
3. Wolves do not destroy game herds.
   Fact: Research has shown time and again that wolves do destroy game herds in most cases, and then move on.
4. Wolves are an economic boom to the economies of Idaho, Montana, and Wyoming.
   Fact: The research in chapter 11 shows that claim to be totally false.
5. Wolves do not carry diseases that are harmful to man.
   Fact: In chapters 8 and 9 you will be shown the truth about wolves and the diseases they carry, backed up by worldwide scientific data that show how dangerous these diseases can be to man.

6. All wild wolves are 100 percent pure wolves.
   Fact: Actually as wolf populations grow in the lower forty-eight states, especially in and around communities, wolves interact with dogs and coyotes, either killing them or breeding with them to result in hybrids, which are not protected by the Endangered Species Act.
7. Wolves are an endangered species that is threatened by extinction.
   Fact: There are at least one hundred thousand wolves in the wild in North America and another three hundred thousand or more wolves and wolf-dog hybrids that are in sanctuaries, education centers, and zoos, as well as pets in North America. There are at least this many wolves in Europe and Asia.

This book will also demonstrate the enormous amount of taxpayer dollars that the US government and states have spent to introduce wolves back into the western United States, well over two hundred million dollars, and the many considerable additional economic costs these states have endured by this introduction.

We will also prove by a preponderance of the evidence that the wolves introduced into Montana and Idaho were not the same wolf that existed there before. These massive animals are much larger than the wolves that roamed the western states over one hundred years ago.

We will also cast serious doubt over the validity of the Mexican wolf recovery effort. There is a concern of whether or not the millions of dollars spent by the US government (over twenty-six million dollars and counting) have been spent on wolf-coyote hybrids. We ask for a congressional investigation into this program to determine whether or not this is true (see chapter "Canis Stew").

We will also demonstrate the different methodology that surrounds the US Fish and Wildlife Service and the United States Department of Agriculture's counts of how many head of livestock are killed by wolves. A disconnect of several thousand exists between the two.

This book will also demonstrate the impact on local residents of areas where wolves have been introduced and how they have effected serious lifestyle changes and cost millions of dollars of losses to ranchers and stockmen across the west.

We will also show the biological, sociological, and political consequences of the present wolf management program nationwide and make suggestions about how to avoid wasting taxpayer dollars on frivolous lawsuits that are currently consuming millions of taxpayer dollars; working against state and federal agencies producing quality wildlife management programs; and causing purebred wolves to disappear into hybrids of wolf, dog, and coyotes, which are increasingly habituated, resulting in more damages to wildlife, livestock, and pets, as well as wolves themselves.

We will make specific recommendations for changes to the Endangered Species Act and the Equal Access to Justice Act so that other states and citizens are not subjected to the same mistakes that were made in Montana, Idaho, and Wyoming and are now being made in New Mexico, Washington, and Oregon.

Further, this book will demonstrate the impact that wolves have had on local residents and communities where they have been reintroduced and these animals have forced serious lifestyle changes that have cost ranchers, farmers, outfitters, businesses, and states millions of dollars in lost revenue.

In chapter 12, we will document how the government has paid out millions and millions of dollars under the Equal Access to Justice Act in recent years. How federal agencies cannot, even today, document how much money has been paid out under that act to environmental and animal rights groups, and how wolf litigation has been the proverbial cash cow that shows the way. We will also show how state agencies that are forced to protect themselves from litigation are being forced to spend considerable money on legal protection that could have been spent on wildlife conservation.

This book will, in the end, strip the skin from the skull and call for changes to the Equal Access to Justice Act and the Endangered Species Act, and call for a serious look at the millions of dollars that have been spent on wolf recovery in the USA, as well as who is responsible for many of the problems this book documents.

Photo credit: S. R. Maglione/Shutterstock.com

CHAPTER 1

. . .

The Real Wolf Story

By Ted B. Lyon

It became clear to me that the issue of how the wolves could be controlled was not science at all; it was pure, unadulterated politics.

—Ted B. Lyon

My first introduction to the wolf issue came in 1999 while my wife and I were staying at a small resort called Chico Hot Springs, located just north of Yellowstone National Park. I was soaking in the hot spring pool when a big guy with a beard slipped into the pool. Since he and I were the only two people in the pool, we started talking. I asked him what he did and he told me that he used to be a big-game outfitter and had worked and lived in the area his entire life. He'd been a licensed outfitter for over fifteen years and had employed over fifteen people for his operation during the hunting season. He also told me that his business had been booming before the wolves were introduced in 1995 into Yellowstone National Park. However, after the wolves were introduced, the elk herd became smaller and smaller each year until eventually he had to shut his business down.

I listened to his story with a good bit of skepticism because I could not believe that just a few wolves could cause that much destruction to an incredibly large elk herd—over nineteen thousand in 1995.

Montana Real Estate

In 2001, my wife and I bought a beautiful piece of property just north of Bozeman, Montana. As we drove onto the property that crisp October morning, a whitetail buck ran across the road and shortly thereafter, as we continued to drive down the road, two ruffed grouse flew off to the side. We stopped the truck and just as we got out of the vehicle an elk bugled off to the south. I told my wife that the place was speaking to us. We eventually bought the property and built a home there. At the time, the area, which is about fifty miles north of Yellowstone, was full of elk, deer, and moose. Just to the west of us between Bozeman and Big Sky Mountain there was the Gallatin Canyon elk herd with between a thousand and fifteen hundred elk.

Horror Stories or Isolated Cases

In the late 1980s and early 1990s, I spent several weeks in the Salmon River Wilderness riding horses, camping out, and hunting elk and mule deer. It was a wild game paradise. There were a number of huge bull elk and mule deer bucks in the area. There were also moose and bighorn sheep. I hunted there with Brent Hill, a long time outfitter in the area. Each year before 1995 he would take around sixty hunters into the area by horseback. It is one of the wildest places in the Lower 48. Brent and some of his wranglers were there in the wilderness in 1995 when Tom Brokaw, Ted Turner, and Bruce Babbitt, then the Secretary of Interior, watched as US Fish and Wildlife officers released the wolves onto an airfield.

Each year about fifty to fifty-five elk were taken by hunters guided by Brent or his guides in the Salmon River Wilderness; this was before Canadian wolves were released into that area in 1995–96. Brent told me that the wolves began killing sheep, mule deer, and elk that winter by the river. The next year, only twelve bull elk were taken by hunters, and the success rate went down each year. Eventually, his outfitting business was closed.

Even though I knew about those stories, I believed that they were isolated cases because the US Fish and Wildlife Service, Idaho Fish and Game, and Montana's Fish, Wildlife and Parks were putting out the same basic story that wolves would not and did not affect the wild game populations to any great extent.[1]

I simply couldn't believe that trained biologists could be so wrong about the wolf and its destructive effect upon wild game. I trusted them because as a State Senator and House member in Texas, I spent fourteen years on committees that dealt with the Texas Parks and Wildlife Department. That agency would continually appear before my committees and would always advocate for the preservation of our wild game and fish resources. I always used the biologists and their staff to support the bills that I pushed, and I developed a tremendous amount of respect for their scientific knowledge and their desire to manage our wild game and fish so that it was abundant, and to better utilize the resource for the public.

In 2007, I was on an annual pheasant-hunting trip with a number of good friends in Choteau, Montana, on the farm owned by my good friend Skip Tubbs. Skip is an avid sportsman and conservationist who owns an art gallery in Bozeman, Montana. He also raises English setters and is a falconer. All of the people invited to Skip's for opening day had one thing in common: we were, as Southerners say, "dog men." Everyone had hunting dogs, from Labradors to setters to Brittany Spaniels, and even one Cocker Spaniel. We all love to hunt with our dogs and live to see dogs that we have trained perform.

On Saturday night, after bagging our limits of pheasants, we started cooking steaks and drinking a little wine. It was at this event when I was first exposed to the strong reaction that Montana hunters as a group had toward wolves and the US Fish and Wildlife Service. That night I defended the decisions of those who put the sixty-six wolves into Yellowstone National Park and the Salmon River Wilderness area in Idaho in 1995 and 1996; ignorantly, I must say. I also defended the statements made by Idaho and Montana state wildlife biologists who parroted the same statements made by the US Fish and Wildlife Service.

The men crowded around the fire that night were adamant that under no circumstances had the introduction of Canadian wolves been a good thing for Montana, Idaho, or anywhere else. Statements such as "wolves are not impacting the elk herds" and "hunters only need to work harder to find the elk," were considered "pure BS."

I just could not bring myself to believe that a US Fish and Wildlife Service official, or an Idaho or Montana state agency wildlife biologist, knowing the economic impact that elk and deer hunting have on Montana or Idaho, would knowingly make misrepresentations about the effects that wolves could have on Montana and Idaho's elk herds or could be that wrong.

That night, as I drove back to where I was staying, I thought that my hunting friends were surely over-reacting. The next thing I expected to hear from them was about black helicopters. But, that evening stayed with me, so I decided to look into what they were saying.

Research and Enlightenment

In 1995 and 1996 the US Fish and Wildlife Service introduced thirty-two Northern gray wolves from Alberta, Canada, into Yellowstone National Park. As you will learn in later chapters, the cost of introducing each wolf has been between two hundred thousand and one million dollars per wolf.

An additional thirty-four Canadian wolves were introduced into the Salmon River Wilderness area in Idaho at the same time. The Salmon River Wilderness area is located some five hundred miles to the north of Yellowstone. To get there from Yellowstone, which is located at the southern end of Montana and the northern end of Wyoming, follow Interstate Highway 90 up the eastern side of the Rocky Mountains to Missoula, Montana. From there you can travel the Lolo pass, made famous by Lewis and Clark in their exploration of the Missouri River, all the way to Idaho. It is a beautiful trip through a place where you expect to see a lot of wildlife.

Fast Forward Two Years

At the time these foreign wolves were introduced, Yellowstone National Park was home to some of the healthiest elk, mule deer, and Shiras moose populations in the world. The slopes of the Rockies on the western side of Montana and the Lolo National Forest were also home to thousands of these ungulates. These vibrant populations were the result of decades of conservation work by sportsmen.

When the scenario repeated itself at Skip's annual hunt two years later in 2009, I was better armed, having read a number of articles that said the wolves were not impacting the moose or elk herds. There was even an "official" scientific study funded by some groups I had never heard of that said this was true.[2]

There was also an economic study that showed that wolves were a positive thirty-five-million-dollar benefit to the Yellowstone area that you will learn more about in a later chapter.

That opening night of the 2009 pheasant season, when we all gathered at Skip's house, a new guy was there, Ray Anderson, who had retired to Montana after a successful career as a businessman. Ray and Dale Simmons, a website designer, were vocal and articulate in their feelings about the wolves. The clincher came when Dr. Shannon Taylor, a professor at Montana State University, and Terry Thomas, a heating contractor, both insisted that the moose had all but disappeared from Yellowstone National Park. Each fall, Terry spends the entire elk season camped out next to Yellowstone, and has done so for years. He said the elk herds were severely depleted and that the moose were gone. These guys were adamant, adding their voices to the chorus.

That night I resolved to thoroughly research the issue. I had to look widely, as very little of what these men were saying was available in popular print. Frankly, as I got into it, I was shocked. The tangible result of that shock is this book. It is about the true story of the greatest destruction of wild game in the United States since the decimation of the bison herds and the elimination of the passenger pigeon in late 1800s, and how people are trying to reverse it.

When my wife and I first bought land in the mountains of Montana in 2001, I thought wolves were harmless. The wildlife biologists from the United States Fish and Wildlife Service and the states of Idaho and Montana said that wolves brought a natural balance to nature; that wolves are not now and have never been a threat to man; that wolves can be trained and educated so that they do not attack livestock; that wolves are not sport killers and only eat what they kill; that wolves do not carry deadly diseases; that wolves were the sanitarians of nature; and that wolves were good for the economy.

The more I looked into this situation, the more I realized that much of the flood of positive information about wolves was just plain wrong. I believed these statements because they were put out by officials at every level of government. The people that told these myths are not evil, but they were wrong. They either failed to research the issue adequately or simply believed the misstatements that had been perpetrated by many people who had either fabricated the scientific data about wolves or ignored data that had been accumulated since the turn of the century.

Then I learned that environmental and animal rights advocacy groups that supported the wolf reintroduction program were making millions from contributions to save the wolves, which meant that they were not interested in telling any other story, even if it was true.

Finally, in January of 2010, Montana Fish, Wildlife and Parks released a report called "Monitoring and Assessment of Wolf-Ungulate Interactions and Population Trends within the Greater Yellowstone Area, Southwestern Montana, and Montana Statewide Final Report 2009" written by Kenneth L. Hamlin, a senior wildlife researcher, and Julie A. Cunningham, a wildlife biologist. This eighty-three-page report detailed the amazing decline of the Northern Yellowstone herd. It showed that in 1995, when wolves were first introduced into Yellowstone National Park, the elk herd numbered over nineteen thousand animals. The count in 2009 was just a little over six thousand elk. It also showed that there had been a precipitous decline in moose. They were almost gone from the same area.

At first, I simply could not understand how an initial population of thirty-two wolves could take an elk herd of over nineteen thousand down to a little over 6,200 in the space of fourteen years. The harsh reality is that the elk herd today in Yellowstone was down to three thousand in 2015, and the Yellowstone moose population has dropped from one thousand to less than two hundred.[3]

This was not an isolated case. Similar devastation has happened to other elk populations throughout the northern Rocky Mountains since the new wolves arrived. In the Lolo National Forest where wolves migrated from Idaho in 1995 the elk herd has dropped from twelve thousand in 1995 to around two thousand in 2011.

More Irrefutable Data

The early "studies" had totally misjudged the rapid rate that the wolf population would grow and spread, from sixty-six in 1996 to conservatively seventeen hundred in 2012, and many scientists believe there are at least five times that many. Unlike other predators like mountain lions or bears, wolves have large litters and they can begin breeding by age two.

On top of that, the introduced wolves were a larger subspecies than the native subspecies, which had voracious appetites. Gray wolves can survive on about two-and-a-half pounds of food per wolf per day, but they require about seven pounds per wolf per day to reproduce successfully. A large gray wolf can eat between twenty-two and twenty-three pounds at one time, and these intro-duced Canadian wolves are much bigger than the ones that used to live in the Northern Rockies.[4] Put larger wolves together with an abundance of prey, and you get a lot of wolves quickly.

Despite what some people were saying about wolves being nature's san-itarians, they do not seem to care what they eat, healthy or not. On occa-sion, wolves simply go on killing sprees, killing and wounding many times the number of animals that they could ever eat, leaving without feeding on their victims.

Side Effects

Predation by wolves is significant, but their impact on herds goes far beyond that. Research by Professor Scott Creel at Montana State University (funded by the National Institute of Science) determined that the cow elk in and around Yellowstone were not getting pregnant as a result of the stress caused by wolves. Think about being the fattest animal in the herd. They run the slowest and therefore become the easiest victim of the predators. Creel's later research showed that the elk that were hunted by wolves were actually starv-ing to death in the winter, as well as not calving, or having many fewer calves.[5]

More Damning Evidence

The damning evidence does not stop there. Before the introduction of the wolves into Montana and Idaho, there was no known incidence of hydatid disease in either of those states or in Wyoming. Hydatid disease, also known as hydatidosis or echinococcosis, is a parasitic infection of various animals, and can infect humans. The disease is caused by a small tapeworm that lives in canids, especially wolves. Tapeworm eggs pass out in the feces of infected wolves. If eaten by a suitable host—ungulates, livestock, and man—these eggs may develop into hydatid cysts in the internal organs of the host, especially the liver, heart, and lung. The disease didn't exist in the moose, elk, mule deer, whitetail deer, mountain goats, or sheep herds before the wolf introduction. It's there now though, and is a serious threat to animals and man, as you will see in a later chapter.

Like most people, I was not aware of this disease before I began my research. It was while I was researching wildlife diseases that I found my co-author, former National Security Agency Security Officer Will Graves, an incredibly interesting man who has spent a good deal of his life researching wolves. Will had written a letter in 1993 to Ed Bangs, the US Fish and Wildlife biologist in charge of transporting the wolves to Yellowstone, about his concerns about hydatid disease in the wolves from Canada. (See Appendix.)

I interviewed Ed Bangs in Helena Montana in June of 2012 and he confirmed that the wolves were wormed twice before they were released. The circumstantial evidence is strong, almost overwhelming, that either the wrong type of wormer was used or that the parasite existed in Montana and Idaho and was simply unknown. Since the wolves were introduced, the parasite that carries hydatidosis has been transported by wolves all across the western states. Humans can become infected with this disease simply by petting a dog that has rolled in an area where a wolf has defecated.

Contrary to what some spokesman for wildlife agencies in the United States have reported, hydatidosis is a deadly disease to humans with reported deaths all around the world where the tapeworm exists.[6] Over 68 percent of the wolves in Montana, Idaho, and Wyoming that have been tested are infected with *Echinococcus granulosus* tape worms. In some areas the infection rate is as high as 84 percent. There's also strong evidence that the tapeworm weakens the ungulate, which is an intermediate host, making it more susceptible to being preyed upon by predators.[7]

Spin Doctors

After realizing that my friends' anecdotal stories about what havoc the wolves had done to the elk were right, I resolved to find out why. One reason that quickly became apparent was that although the US Fish and Wildlife Service had declared that wolves were recovered in 2000—easily passing the goal of ten breeding pairs and a hundred wolves in Montana, Wyoming, and Idaho—neither states, hunters, livestock producers, nor the US Fish and Wildlife Service, was allowed to manage the wolves.

I was shocked to find that respected magazines like *Outdoor Life* and *National Geographic* had published stories that did not look into the research that had been done over the years, detailing how wolves had destroyed elk and caribou herds in Canada. Instead, these respected magazines simply repeated the claims of the pro-wolf side and quoted anecdotal stories from hunters who said they could not find any elk. In one case the chief wolf biologist for Montana Fish, Wildlife and Parks, Carolyn Sime, simply said hunters would have to work harder, that the elk had not disappeared but had simply retreated to the woods, implying that hunters were lazy.

As someone who had been involved in running political campaigns since I was twenty-one years old, as well as serving in office and campaigning myself, I began to realize that the wolf issue had been framed by extremely smart, well-funded spin doctors who were Machiavellian in their approach to the issue of the wolf—masters at manipulating the facts to raise money for their sponsors.

The people on the other side of the issue, hunters, sportsmen, and hunting and fishing groups, were hopelessly outmatched—not from a political power standpoint, but from a political strategic point of view. The pro-wolf advocates spent their money pumping out propaganda while the sportsmen conservation groups were conserving habitat and sponsoring research: work that was not well-known to the general public, but had produced monumental success in restoring big-game populations. It was as if the wildlife conservationists were high school baseball players going up against major leaguers; they did not know what to do politically, or how to handle the mainstream media, and were, as a group, howling at the moon.

The pro-wolfers had outflanked the wildlife conservation groups and live-stock producers at every step along the way.

One of the issues that I first began researching was how to overcome the court losses that US Fish and Wildlife Service had suffered, and continued to suffer, in their efforts to delist the wolves so that states could begin managing wolves in not only the states where northern gray wolves had been introduced like Montana, Idaho, and Wyoming, but also in states like Minnesota, Wisconsin, and Michigan.

Lawsuits and Legal Challenges

A lawsuit was filed in 1993 to keep Canadian wolves from being introduced into the west by Cat Urbigkit and her husband. The couple claimed there were already wolves in the area, and that the Canadian wolves were a different, larger subspecies that would displace or hybridize with the native wolves.

In Minnesota, legal challenges to allow delisting the wolf had been going on since the 1970s. Additional lawsuits by the pro-wolf groups were also filed against the US Fish and Wildlife Service who wanted to delist the wolf in the western states in 2001 when their numbers had reached what was called for in the initial agreement. In each and every case, the pro-wolf forces won, postponing, delaying, and stopping state wildlife agencies from being able to manage wolves.

In the beginning of 2010, the wolf in the United States enjoyed exalted status over all other species. As a private citizen you could be walking down a city road in any of the lower forty-eight states and a pack of wolves could attack your dog, horse, or cow and you would be committing a felony if you shot them. Wolves could attack at will sheep, cattle, and horses, and private citizens were powerless to stop them. Only in defense of human life could a public citizen defend oneself from a wolf attack.

The pro-wolf forces were also manipulating reporting of how many live-stock that were actually killed by wolves. Typically the US Fish and Wildlife

Service or Montana, Idaho, Wyoming, or New Mexico state wildlife agencies would release at the end of each year the "confirmed" kills (and I emphasize the word "confirmed") caused by wolves. In 2010 for the states of Montana, Idaho, Wyoming, Wisconsin, Minnesota, Michigan, and New Mexico, reports of those kills totaled up to less than a thousand.

Real Names and Faces

In an interview I conducted with Montana rancher Justin O'Hair, he advised me that on one occasion he spotted a young Black Angus calf with his entrails hanging out, and a wolf a hundred yards away. Initially the US Fish and Wildlife Service officer would not confirm that the calf had been attacked by a wolf because they said that Justin was "not qualified" to confirm the difference between a wolf and a coyote.

Justin and his family own the eighty-thousand-acre O'Hair ranch outside of Livingston, Montana, where they run eleven hundred head of cattle. He has lived on the ranch his entire life and his family homesteaded the ranch in 1878. They are out in the field checking cattle on horseback almost every day. To say that they do not know the difference between a wolf and coyote is beyond ignorant.

Justin also related that often cattlemen or ranchers will find just an ear tag laying on the ground or a dead cow, but they could not tell what killed it. Cattle that suffer attacks from wolves almost always die since no amount of antibiotics is able to overcome the infection that comes with a wolf bite. And wolves often just maim their victims and move on without eating anything.

USDA Report

The United States Department of Agriculture (USDA) releases a report every five years as a cooperative effort between the National Agriculture Statistics Service and the Animal Plant and Inspection Service-Wildlife Services and Veterinary Service. The report is, and has been, a scientifically validated survey based on producer reports. There's no pro-wolf bias involved in the compilation of these reports. The USDA report published in May of 2011 showed that wolves killed 8,100 head of cattle in 2010.

Photo credit: Debbie Steinhausser/Shutterstock.com

That's thousands more than the few hundred head that US Fish and Wildlife Service reports each year.

If wolf numbers aren't controlled, their population will only continue to climb. With the rapidly declining herds of wild ungulates, wolves will end up either preying on livestock and pets, frequenting garbage dumps, traveling along roads seeking roadkills, or dispersing to other areas, which is already happening. This is not the normal behavior of wild wolves.

Handpicked Judges and Other Deck-Stacking Techniques

I came to the conclusion that something had to be done about the wolves in North America. They had to be managed by people who truly understood what was going on in the field, and those people needed to be respected and supported by the government.

At first, I approached the problem as a trial attorney. In the past, I have enjoyed tremendous success in using the courtroom as a venue for enforcing justice. I thought that as a non-governmental lawyer, my fresh perspective and skill as a trial attorney would allow me to succeed where the government's attorneys had failed for the past decade.

I wondered about all of the lawsuits that had been filed over the years by groups trying to stop the introduction of the wolves into Montana and Idaho. I wondered if they did not have good lawyers. Or, perhaps the lawyers representing the government in their attempts to delist the wolf since 2001 just were not good lawyers. So, I hired two extremely bright young law students: Ben Barmore, who was at the top of his class at Southern Methodist University, and Richard Mann, a Canadian, who attended the University of Texas Law School. I asked them to research what could be done from the perspective of a lawsuit to give the states the power to control wolves. Each day we would talk and come up with legal theories to pursue. The next day they would give me their findings.

What we found is that almost every legal theory that we could come up with to attack the continued protected status of the wolf had already been tried and that the side representing the people who wanted to control the wolves lost at every turn. The US government lost, the states lost, private citizens lost, and non-governmental 501C-3 conservation organizations lost every time. The winners were groups like Defenders of Wildlife, the Center for Biological Diversity, Earthjustice, the Natural Resources Defense Council, and others who purported to represent the environmental movement.

The delisting supporters lost in federal courts from Missoula, Montana, to Albuquerque, New Mexico, to Duluth, Minnesota. And in some of the losses, if the pro-wolf people could prove an error by the federal government, even a technicality, they got their legal expenses paid for by the US government through the Equal Access to Justice Act.

After reading all the cases, we concluded that the lawyers representing clients who wanted to control the wolves had generally done a very good job.

I also found, however, that in some cases the federal judges who heard these cases were handpicked by the pro-wolf groups because their political philosophy was more in line with the groups, who wanted no controls exerted over the wolves. In retrospect, I concluded that it would be almost impossible under the Endangered Species Act to win a legal victory to delist wolves.

Conclusion

In April of 2010, it became clear to me that the issue of how the wolves could be controlled was not science at all, it was pure unadulterated politics. The wolves had been placed in Montana, Idaho, and New Mexico because of politics and they could only be controlled and removed by politics. It seemed obvious that the only way to control wolves was by amending the Endangered Species Act—something that had never been done before. I believed in my heart that if we could just get the truth out to members of the US House and Senate that we would be able to get the act amended.

Later on in the book I will detail the amazing story of how the Endangered Species Act was amended for the first time in history.

"The misinformation promulgated by wolf advocacy groups ranges from minor technical errors to major deception and fraud. Technical biological misinformation, though bothersome to professionals working with wolves, is not as serious as deception about such issues as the status and trends in wolf populations. This latter type of misinformation tends to motivate well-meaning wolf advocates to press their causes through letter-writing campaigns, public meetings, lobbying, and lawsuits . . . These misrepresentations have even made it into conference proceedings. In the non-peer-reviewed proceedings of a nonprofit citizen organization, 'Defenders of Wildlife's Restoring the Wolf Conference,' undocumented claims were made the wolf has been eliminated from '95 percent of its former range' and '95 percent of its historic range in North America.' The actual figures are closer to 30 percent of its global range and 40 percent of its North American range."

Dr. David L. Mech[9]

Endnotes:

1. http://wdfw.wa.gov/conservation/gray_wolf/big_game/faq.html#19; http://fwp.mt.gov /mtoutdoors/HTML/articles/2002/wolvesvselk.htm; https://www.fws.gov/mountain-prairie/species/mammals/wolf/EIS_1994.pdf
2. http://*Bioscience*.oxfordjournals.org/content/53/4/330.full; http://westernwildlife. org/gray-wolf-outreach-project/biology-behavior-4/
3. http://wolf.org/wolves/learn/basic/faqs/faq.asp#19

4. https://www.nps.gov/yell/learn/nature/elk.htm; http://www.bozemandailychronicle
   .com/news/environment/survey-indicates-northern-yellowstone-elk-herd-stable/
   article_5dd2027c-7885-5403-888c-34cc6c10b032.html
5. Creel, S., D. Christianson, and J.A. Winnie, "A Survey of the Effects of Wolf
   Predation Risk on Pregnancy Rates and Calf Recruitment of Elk," *Ecological
   Adaptations* 21:2847–2853, 2010.
6. www.dpsi.nsw.gov.au; "Hydatids You, too, Can Be Affected," NSW DPI, February
   2007, Australian Government Prime Facts.
7. http://www.fao.org/docrep/t1300t/t1300t0m.htm
8. http://articles.chicagotribune.com/1986-06-29/news/8602160259_1_gray-
   wolves-wolf-packs-reintroduction; http://www.usu.edu/today/pdf/2008/august/itn
   0806083.pdf
9. Dr. L. David Mech, Wolf Restoration to the Adirondacks: The Advantages and
   Disadvantages of Public Participation in the Decision, (2001). https://pubs.er.usgs.gov
   /publication/93795

Credit: Karl Umbriaco/Shutterstock.com

Endangered Species Act (ESA)

Congress passed the Endangered Species Preservation Act (ESPA) in 1966, providing a means for listing native animal species as "endangered" and giving them limited protection. The Departments of Interior, Agriculture, and Defense were to seek to protect listed species and, insofar as consistent with their primary purposes, preserve the habitats of such species. The ESPA also authorized the US Fish and Wildlife Service to acquire land as habitat for endangered species.

In 1969, Congress amended the ESPA to provide additional protection to species in danger of "worldwide extinction" by prohibiting their importation and subsequent sale in the United States. One amendment to the ESPA changed its title to the Endangered Species Conservation Act (ESCA).

A 1973 conference in Washington, DC, led eighty nations to sign a treaty called the Convention on International Trade in Endangered Species of Wild Fauna and Flora (CITES), which monitors and, in some cases, restricts international commerce in plant and animal species believed to be harmed by trade.

Later in 1973, Congress passed the Endangered Species Act of 1973 (ESA). It defined the terms "endangered" and "threatened"; made plants and all invertebrates eligible for protection; applied broad "take" prohibitions to all endangered animal species and allowed the prohibitions to apply to threatened animal species by special regulation; required federal agencies to use their authorities to conserve listed species and consult on "may affect" actions; prohibited federal agencies from authorizing, funding, or carrying out any action that would jeopardize a listed species or destroy or modify its "critical habitat"; made matching funds available to states with cooperative agreements; provided funding authority for land acquisition for foreign species; and implemented CITES protection in the United States.

Congress enacted significant amendments in 1978, 1982, and 1988, while keeping the overall framework of the 1973 ESA essentially unchanged. The funding levels in the present ESA were authorized through Fiscal Year 1992. Congress has annually appropriated funds since that time.

CHAPTER 2

...

Selling the Wolf: The Massive Sales Campaign and Its Fallacies

By Ted B. Lyon

"Environmental battles are not between good guys and bad guys but between beliefs, and the real villain is ignorance."

—Alston Chase[1]

Photo credit: Stayer/Shutterstock.com

The Wolf: From Bad Guy to Poster Child

In 1985, Yale sociologist Dr. Stephen Kellert conducted a national survey of public opinion about wildlife. He found that wolves were the least liked of all animals in North America. Fifty-five percent of the people said they were neutral toward wolves or disliked them.[2] Since then, wolves have been reintroduced into the Northern Rockies, the Pacific Northwest, the Southwest, and the Southeast. The wolf populations in the Upper Midwest and New England have grown; wolf populations in Alaska and Canada have increased. Some wolf advocates have set a goal of wild wolves thriving in all fifty states. Similar programs are underway in Europe and Russia. The wolf has gone from bad guy to a poster child for conservation in less than thirty years.

The unprecedented wolf repopulation program brought sixty-six wolves from Canada to the Northern Rockies in 1995 and 1996 and has since sheltered them, allowing the population to skyrocket to at least ten times the number called for in the original plan. This could only have been accomplished with a massive, multi-faceted promotional sales campaign, for as you will learn, one introducing wolves into a modern social landscape is like "Jurassic Park"—the intentions may be honorable, but the results can be catastrophic.

The purpose of this book is two-fold: first, to expose the myths about wolves that have been sold to people in North America and abroad, falsehoods that have resulted in a war of words and seemingly endless courtroom battles, as well as a war in the woods; and, second, to set the record straight so people on all levels can understand the real issues about living with wolves in modern times, and make responsible decisions about the future of our uneasy relationship with *Canis lupus*, the gray wolf.

In Sun Zu's masterful treatise on winning in conflict, *The Art of War*, he insists that to win you must understand your enemy. The sad truth is that the "Save the Wolf" campaign is largely based on romantic half-truths, exaggerations, and distortions, mixed with negative stereotyping, stigmatizing, and even intimidation of anyone who questions the wolf restoration program. But it has been extremely successful. So, let's see how and why this is so.

A Brief History of Public Opinion about Wolves in the United States

The ancestors of the modern gray wolf, the largest living member of the wild dog family Canidae, trace back to the Pleistocene era, perhaps as far back as 4.75 million years ago. The gray wolf was once the most widely distributed large mammal on Earth. Everywhere where wolves and people are found together, there is a history of respect, distrust, and mutual predation. This is a primary reason why wolves are not as common today as they once were.

When European settlers arrived in the United States, they found wolves, as their ancestors had known for thousands of years. Native Americans lived with wolves, which were integrated into their spirituality, mythology, and rituals, but Native Americans also trapped and killed wolves, using their skins for clothing and costumes. Eating them was considered a delicacy. While there were no newspapers or written records of wolves in those days, there are many tales of people being attacked, killed, and eaten by wolves. In the 1800s, as the buffalo were nearly exterminated by market hunters and a planned military strategy to drive Indians onto reservations, elk and deer were killed in large numbers by market hunters and the natural habitat declined dramatically due to logging and farming. In response to the lack of prey, wolves switched their predation to livestock. This triggered a war on wolves—bounties, trapping, hunting, and poisons—that was supported by the US government. This was the first wolf educational campaign—get rid of them—and Congress supported it.

In 1914, the US Congress passed legislation calling for the elimination of predators from all public lands, including National Parks, as wolves and other predators kept down the numbers of elk, deer, moose, and antelope, which were major attractions for tourists as well as game for hunters.[3]

Aided by modern weapons, traps, and poisons, by 1930 wolves were all but gone from the Lower 48, except for small numbers in the Northern Rockies and northern Minnesota, and a handful of Mexican wolves in Arizona, New Mexico, and Mexico. Remaining wolves in Canada and Alaska became very wary of man, and were seldom seen, except in the far north. This was the second wolf educational campaign, again backed by the US Congress.

The use of poison baits (which were heavily used on coyotes after the wolves were nearly eliminated) was not banned until 1972, in large part due to Earth Day 1970, when banning the 1080 poison (sodium fluoroacetate) was a hot issue at teach-ins across the United States.

With an absence of wolves and diminished numbers of bears and mountain lions, as well as habitat conservation programs supported by many conservation groups, by the 1960s elk, deer, moose, and antelope numbers in Yellowstone National Park and elsewhere across the United States mushroomed to record high numbers. In some cases they exceeded the carrying capacity of the land. The wild game restoration campaign was spearheaded by conservation and sportsmen organizations with support from state and federal resource agencies. Increased hunting was considered to be the most popular way to control game animals.

The concept of restoring wolves to the lower forty-eight as a way to control big-game herds was first introduced to Congress in 1966 by biologists.[4] Support for this strategy came from years of study of wolves on Mount McKinley in Alaska by Adolph Murie, and studies of wolves and moose on Isle Royale in Lake Superior by Purdue University wildlife biologist Durward Allen and his students, including David Mech and Rolf Peterson.[5] This research concluded that wolves are shy creatures of the wilderness that do not attack people or seriously reduce large ungulate populations; and that wolves are nature's sanitarians, attacking only the old, the lame, and diseased animals. That perspective became the gospel in wildlife management for decades. The problem is, as you will soon learn, wolves are very adaptable, and in other situations they behave very differently. The research kicked off another wave of wolf education, for the first time in favor of wolves.

The "harmless wolf" research was woven into the 1963 "Leopold Report," otherwise known as "Wildlife Management in the National Parks," written by Aldo Leopold's son, Starker, a renowned wildlife biologist in his own right.[6] The Leopold Report called for active management of wildlife to ensure that "a reasonable illusion of primitive America (what things looked like when white men first arrived there) . . . should be the objective of every national park and monument."[7]

Following Earth Day 1970, support for restoring wolves began rising. "Wolfism" joined racism, sexism, ageism, and pollution as another form of oppression. Riding on the wave of the first Earth Day, the Endangered Species Act was passed in 1973. One year later the gray wolf was added to the list of endangered species in the lower forty-eight states. Saving the wolf became a growing rallying cause for environmentalists, who were joined by animal rights groups, resulting in a "Save the Wolf" movement. But, as the Kellert study found, even by 1985, the general public was still not too keen on wolves. To bring back the wolf, an unprecedented massive public education program was needed to change the prevailing negative opinions of wolves.

With the only wolves found in zoos or remote areas, media became the new sense organs of urban Americans, as a swarm of books, articles, lecture tours, exhibits, public meetings, films, toys, and TV shows in support of wolf restoration exploded. "Wolf experts" were suddenly everywhere. The wolf became a symbol of green ecological action, along with stopping pollution, recycling, sustainability, and fighting global warming. Wolf restoration was also supported by animal rights groups: wolves not only were a species to restore, but a way to reduce big-game herds that supported hunting.

The new wild wolf emerged as a romantic mythic image of wilderness that urbanized Americans, clustered in concrete, steel, plastic, and wood canyons, longed for in their soul. Reviewing thirty-eight quantitative surveys conducted between 1972 and 2000, Williams, Ericsson, and Heberlein find that attitudes toward wolves consistently show that the farther one lives from wolves, the more likely public opinion is in favor of wolf restoration.[8]

Williams, Ericsson, and Heberlein also found, as did Kellert, that people who have the most first-hand contact with wild wolves—ranchers, farmers, outfitters, and hunters—held the most negative views of wolves, and despite the pro-wolf campaign, positive attitudes about wolf restoration have not continued to increase over time. In the United States, they found that 55.3 percent overall were favorable to wolf restoration. In Europe, where wolves have a history of contact with people, attitudes about wolves are less favorable—37 percent are favorable to wolves in Western Europe and 43 percent are favorable in Scandinavia.

While one result of the "Save the Wolf"" movement has been wolf restoration programs, a second consequence is growing antagonism between pro- and anti-wolf groups and advocates. Unfortunately, in the flood of wolf media, there has been very little accurate information about the problems associated with wolf restoration. Setting the record straight is a major goal of this book.

Owning the Truth about Wolves

There are at least four major problems with the "Save the Wolf" movement's educational campaign. The first is that wolf behavior around people is heavily influenced by human behavior. Wolves are intelligent and adaptable, as well as unpredictable. In localities in Europe and Asia where people are commonly armed, as they are in North America, wolves are shy and reclusive. Where the populace is not heavily armed, wolves adapt, become habituated, and act much more boldly, preying on livestock, venturing into towns to attack pets and feed on garbage and attack people. The chapter by ethologist Dr. Valerius Geist shows a predictable behavior pattern of habituation that happens when wolves contact people and meet little or no opposition.

A second major problem is that the "Save the Wolf" campaign also has largely avoided reporting that in addition to rabies, wolves may carry over fifty diseases, some of which can be fatal to humans and livestock, such as hydatidosis. That we have little record of these diseases in the United States is simply due to the previous absence of wolves, and in some cases a lack of reporting of wolf-borne diseases. Warnings about such diseases are at best a footnote in the many "Save the Wolf" messages. It is bad for business. You will learn more about this in a later chapter.

A third major problem is that the economic benefits of a wolf restoration on a large scale are far outweighed by the costs, but the costs are not given anywhere near full coverage.

A fourth major problem is that the pro-wolf media has not only sold us a harmless wolf, but for the first time ever, it has sought to discredit as pure superstition the rich legacy of myths, fables, folklore, and fairy tales about wolves that originates from Europe and Asia. This campaign fails to understand how and why these tales came about, for they represent the earliest wolf educational campaign.

A fifth major problem is that wolves in the wild can and do interbreed with dogs and coyotes. This is already happening, especially in areas where wolf numbers are still small, and as it does the question of what is a "real wolf" to protect becomes more difficult. And, as canid hybridization increases, the behavior of these new hybrids will change.

Fairy Tales, Mythology, and Folklore about Wolves

Fables, folklore, and mythology of Europe and Asia were the first wolf educational campaign; most all teach that the wolf is dangerous. Far from being wrong, in Europe and Asia for thousands of years wolves have attacked and killed big game, livestock, pets, and people. From centuries of study in Europe and Asia, it's known that wolves are adaptable and intelligent predators, both mysterious and unpredictable. Unlike most other predators, occasionally wolves run amok and engage in mass spree killings for sheer joy, such as the pack of wolves that killed 120 sheep on one August 2009 night in Dillon, Montana, or another pack that killed nineteen elk in March of 2016 in Wyoming, eating little or nothing. Put those qualities together with distinctive haunting vocalizations, and the possibility of a rabid wolf, you have an animal that in the right situations people should fear, and with good reason.

The Moral of the Story

After reviewing the history of man-wolf relations, in his award-winning book, *Of Wolves and Men*, Barry Holston Lopez arrives at the conclusion that: "No one—not biologists, not Eskimos, not backwoods hunters, not naturalist writers—knows why wolves do what they do."[9] That is a very good reason why folklore about wolves carries warnings. If an animal is unpredictable and carnivorous, you have a suspicious demon. This is why a terrorist acting alone is often called "a lone wolf."

It's understandable then that in cultures where a significant number of people do not own firearms, and where children may venture into areas where wolves are present, fairy tales and folklore such as Aesop's fables, "Little Red Riding Hood," "The Three Little Pigs," Shakespeare, Grimm's fairy tales, as well as holy books including the Bible, the *Rig-Veda*, and the like cast *Canis lupus* in a negative light. These stories are warnings, especially to children and shepherds, to keep people alive.

We know the wolf on a subconscious level, too, as it may visit us in our dreams. From a psychological standpoint, animals that appear in our dreams are symbols of instincts in the unconscious roots of the psyche—in other words, archetypes. The wolf is an archetypal symbol of pure wildness, both in nature and human nature; and a reminder of one's own inner wolf-like qualities—positive in terms of being a skillful hunter and a family protector, and the wolf's dark shadow side of lust, violence, greed, killing, unpredictability, etc.—that can make a wolf seem like a sociopath. The reality is that wolves are unpredictable, which is why stories of the danger of wolves were created in the first place.

Wolves reproduce rapidly; little wonder that the wolf is a symbol associated with lust. The call of the rogue male out on the prowl for chicks is the "wolf whistle." This is why calling a person a "wolf" means they are not trustworthy and can be dangerous.

The universal belief in half-human and half-wolf creatures—the werewolf—and the rare mental disease of lycanthropy, where a person goes berserk with almost superhuman strength, howling, making wolf-like sounds, and may attack people as if they are prey, all speak of our fear of raw human instinctual emotions that make people behave like wolves.[10] Among the Navajo, the word *mai-coh* means both wolf and witch, which the Navajo see as a werewolf, a person who is most likely to perform evil acts during twilight or at night while wearing a wolf skin.

Referring to the Wolf

The meanings of "wolf" are many. In medieval times, famine was called "a wolf." Werewolves were a principal target of the Inquisition. In Dante's *Inferno*, the wolf presides over the eighth circle of hell where punishment is meted out to those who have committed the "sins of the wolf" in their lives—religious and political hypocrites, magicians, thieves, and seducers. In the fairy tale "Little Red Riding Hood" the wolf uses trickery to try to lure a young girl into his clutches. Ostensibly he is going to eat her, but implicit sexual connotations are also obvious.

In the myth and magic of earlier times, wolves have a strong association with the supernatural. Wolves prowl at night and twilight, a time for the imagination to grow larger. Latin for "dawn" is *interlupum et canum*, which translates as "the time between the wolf and the dog."

There are thirteen references to wolves in the Bible, almost all as metaphors for destructiveness and greed. It should not be surprising then that the *Book of Beasts*, a medieval bestiary derived from a chain of Christian monks adapting earlier natural histories that date to Pliny and Aristotle, states: "The devil bears the similitude of a wolf: he who is always looking over the human race with his evil eye, and darkly prowling round the sheepfolds of the faithful so that he may afflict and ruin their souls. . . . Because a wolf is never able to turn its neck backward, except with movement of the whole body, it means that the Devil never turns back to lay hold on repentance."[11]

In his early days, Adolph Hitler referred to himself as "Herr Wolf" and referred to his sister as "Frau Wolf." The name "Adolf" itself is a derivative of "Athalwolf," meaning "Nobel Wolf." He called his retreat in Prussia "The Wolf's Lair," and he named three of his military headquarters *Wolfsschanze*, *Wolfsschlucht*, and *Werwolf*. His favorite dogs were wolfshunde, and he referred to his SS as "my pack of wolves." Little wonder then that journalists spoke of groups of German submarines patrolling the North Atlantic as "wolf packs."

There are exceptions to the negative wolf mythology, as in Roman mythology when a she-wolf, or Lupa, raises Romulus and Remus after their mother, Rhea Silvia, was forced to abandon the twins.

In Rudyard Kipling's *The Jungle Book*, the boy Mowgli is adopted by wolves, and there have been a few cases where something like this may have happened in India. Japanese farmers once left offerings to the wolf *kami* (spirit) to ask his help in protecting their fields from deer and wild pigs. However, the two species of wolves that once inhabited Japan have been extinct for over a century. The Honshu wolf (*Canis lupus hodophilax*) is said to have become extinct in 1905 due to an epidemic of rabies. The Ezo wolf (*Canis lupus hattai*) of the island of Hokkaido, died out in the Meiji period (1868–1912) when, with the establishment of American-style horse and cattle ranches in the area, wolves came to be viewed as a serious threat to the livestock and strychnine-poisoned bait was used to reduce wolf numbers. By 1889 the Hokkaido wolf had disappeared.

The Bottom Line

The bottom line is that in Asia wolves attack and kill children far more often than they adopt them, and in European history there are many cases of fatal wolf attacks. This is why myths, folklore, and fairy tales almost always portray wolves in a negative light abroad.

Many Native American tribes respect the wolf's prowess as a mighty hunter, and they too have many legends, rituals, and myths about wolves, but Native Americans and Inuits still kill wolves for their fur, for food, and in self-defense. Author Barry Lopez writes: "It is popularly believed that there is no written record of a healthy wolf ever having killed a person in North America. Those making the claim ignore Eskimos and Indians who have been killed."[12]

Psychologist James Hillman found that in the dreams of most modern people, animals are pursuing us or we are trying to kill them.[13] Hillman interpreted this as the result of suppression of our own primal instincts, which Hillman and many others believe is a primary cause of the epidemic of anxiety that inflicts our age. In the same vein, psychologist Aneila Jaffe observes: "Primitive man must tame the animal in himself and make it his helpful companion; civilized man must heal the animal in himself and make it his friend."[14]

Clarissa Pinkola Estes's bestselling book about the wild woman archetype, *Women Who Run with the Wolves*, is an example of the power of a symbolic association with wolves. People who live far from wild wolves have an unconscious desire to reconnect with nature, more than conserving the actual wild wolf, which few have even seen. Her book is really not about wolves at all but rather the need to restore our psychological connection with nature as a way to increase health.

The point simply is that to discredit the rich legacy of wolf folklore and mythology is denying human nature and nature itself. The modern myth of the "harmless wolf" is not only inaccurate but may have contributed to attacks and deaths by wolves in recent years.

As this book is being written, hungry wolves are starting to show up in broad daylight in the city limits of towns including: Sun Valley, Idaho; Jackson Hole, Wyoming; Anchorage, Alaska; Juneau, Alaska; Ironwood, Michigan; Toronto, Ontario; Reserve, New Mexico; Duluth, Minnesota; and Kalispell, Montana, hunting for garbage, killing pets, and testing humans. We know of three people in North America in the last decade, who were unarmed, that were killed by wolves, and many others have been attacked; some attacks were reported, and others not. We will list some of those attacks shortly.

People need to distinguish fact from fiction, and appreciate wolves for what they really are. Save the fairy tales and you save lives.

"He's mad that trusts in the tameness of a wolf, a horse's health, a boy's love, or a whore's oath."

—William Shakespeare, *King Lear* (III, vi, 19–21)

The Wolf as a Cash Cow

"Every great cause begins as a movement, becomes a business, and ends up as a racket."

—Eric Hoffer, *The Temper of Our Time*

The leader of the pack of environmental and animal rights groups promoting saving the wolf is Defenders of Wildlife, which idolizes wolves so much that the wolf is their logo. Founded in 1947 as Defenders of Furbearers, their initial target was banning steel-jaw leg hold traps and poisons. They began with one staff person and fifteen hundred members. Today, Defenders' mission statement is to promote "science-based, results-oriented wildlife conservation," and "saving imperiled wildlife and championing the Endangered Species Act."

They do this with a staff of 150 and, they say, over one million members.[15] According to Charity Navigator, in fiscal year 2010, Defenders of Wildlife had an annual budget of $32,595,000 and its president received an annual salary of $295,641.[16] Much of this is due to their "Save the Wolf" campaign.

The American Institute of Philanthropy gives Defenders of Wildlife a "D" for the percentage of its budget spent on charitable purposes—43 percent—noting that the organization sends out ten to twelve million pieces of direct mail each year to draw in about $25.6 million.[17]

This USPS tidal wave hardly seems "green." Appeal letters are written by special direct mail and telemarketing firms—who crank out the same kinds of letters for all kinds of causes—using focus groups to determine the most emotionally engaging pitch. Sometimes the science behind such appeals is questionable, or wrong, but what you read is crafted to have the greatest potential for drawing in donations. For example, a common emotional hook is a crisis—fear that if you don't give, something terrible will surely happen. The opening line for the 2012 Defenders "Campaign to Save America's Wolves" on their website is: "America's Wolves Need Our Help!" and it is followed by: "America's wolves were nearly eradicated in the 20th century. Now, after a remarkable recovery in parts of the country, our wolves are once again in serious danger."[18]

Of course, if something bad does occur, then they can make another appeal based on guilt—if you had donated more this would not have happened.

Another popular appeal is sentimentality, such as Defenders' "Won't you please adopt a furry little pup like "Hope"? Hope is cuddly brown wolf . . . Hope was triumphantly born in Yellowstone."

For the record, the US Fish and Wildlife Service does not name wolves. They give them numbers. Nonetheless, the World Wildlife Fund also offers donors the chance to "Adopt a Wolf."[19]

Another popular appeal is to identify a dastardly, cruel enemy, who if not stopped will surely cause great damage or extinction of a species, or already is doing so. The American Farm Bureau has been a favorite target. If a magazine, radio, or TV show does not report full support for uncontrolled wolf restoration, it also may become a target for hate mail. In short, from a psychological standpoint, the organization must operate as a crisis addict to keep itself in business, for the new wolf is a cash cow.

To Defenders' credit, they initially had a Wolf Compensation Fund to pay ranchers for livestock lost to wolves. However, on August 20, 2010, Defenders announced cancellation of their wolf compensation fund so states and tribes could take over the cost while they worked with farmers and ranchers on non-lethal means of wolf control.[20] This has placed a heavy burden on states, diverting funds that could have served more critical wildlife needs. Should environmental groups that have supported wolf restoration in excess of US Fish and Wildlife Service projected population of sustainable numbers of wolves be held responsible for damages that the excess populations of wolves cause?

Ranchers and farmers additionally complain that the compensation was paid only for confirmed kills, not lost animals or kills that several species— bears, coyotes, mountain lions, eagles, ravens, foxes—feed on before it can be determined which killed the cow or sheep in the first place. (See the chapter "Collateral Damage" on why compensation claims so often go unsupported.)

Pro-Wolf Organizations
A Google search for "Save The Wolves" today comes up with 128,000,000 results as many organizations and petitions have joined the pack when they saw

that wolves were cash cows. In addition to Defenders of Wildlife, some of the best-known "Save the Wolf" groups that use both "educational" campaigns and litigation include: The Center for Biological Diversity, EarthJustice, Friends of Animals, Humane Society of the United States, the Natural Resources Defense Council, WildEarth Guardians, World Wildlife Fund, and the Sierra Club.

A major theme running through "Save the Wolves" appeals is that wolves are in danger of extinction. This, of course, is false. There may be as many as one hundred thousand wolves in the wild in North America, and despite USDA Wildlife Services, USFWS, United States Park Service, and state natural resources and agricultural agencies removing problem wolves, roadkill, natural mortality, and legal and illegal hunting, the North American wolf population is growing appreciably and spreading. Add to this the at least three hundred thousand (possibly five hundred thousand) wolves and wolf-dogs that are living in wolf sanctuaries, zoos, and education centers, running loose in the wild with packs of feral hybrid canines, or being kept as pets in North America.

Defenders of Wildlife also uses public opinion polls to support their advocacy, but not the same polls that unbiased researchers conduct. For example, on the Defenders' website, they report that in response to an NBC Dateline segment on wolves, more than fifteen hundred viewers responded, with less than 11 percent saying they are opposed to wolf reintroduction.[21] They go on to selectively draw on a few surveys to show that people everywhere favor wolves, although they acknowledge that people in rural areas are more likely to feel negatively about wolves.

The Natural Resources Defense Council, who funds "wolf advocates" in the field, states that "Persistent intolerance among humans . . . is one of the two greatest threats to wolves, the other being loss of habitat."[22] On their website,[23] NRDC says:

> The howling wolf is the very icon of wilderness in the American West. Once all but extinct, today some 1,700 wolves roam the Northern Rockies. Despite this magnificent comeback, the future of wolves is once again in jeopardy. Congress has stripped them of their endangered species protection, leaving wolves at the mercy of states planning to kill hundreds of them.

And of course they add, "DONATE."

Another online NRDC pitch for wolf donations wants people to be outraged and donate to help them defend Wyoming's wolves:

> I am outraged that Wyoming allows wolves to be shot on sight across some 85 percent of the state. I want to help NRDC fight to end the

slaughter and restore Wyoming's wolves to the endangered species list, where they belong right now. Please use my tax-deductible gift to save the wolves and defend our environment in the most effective way possible.[24]

Wolf advocates often launch attacks in the media to discredit and attack those people who want wolves managed, portraying them as intolerant, fearful, uninformed, naive, somehow inferior and mentally unsound and/or unethical, and even a threat to society. If someone targeted does lose their temper, it only helps the pro-wolf advocate organizations raise money as they can say, "See, I told you so."

This is an example of how wolf advocates can take on the personality of the "big, bad wolf" who may attack anyone who is not part of their pack without warning at any time.

The Humane Society of the United States, the nation's largest animal rights group, says on their website as of November, 2011—"Social, family-oriented, and highly adaptable—wolves have a lot in common with humans. And while there's no record of a healthy, wild wolf ever attacking a person in the United States, old myths and fears plus competition for land and prey threaten the survival of this wild canine."[25] (We will discuss wolf attacks on people in a later chapter.)

The April/May 2012 issue of Charity Watch, Charity Rating Guide and Watchdog Report gave HSUS a "D" unsatisfactory rating for the second year in a row based how much money it spends to raise money. In contrast, PETA gets a "C+" and the American Red Cross and the Wildlife Conservation Society get an "A."[26]

Earthjustice's website's wolf page is entitled "Wolves in Danger," and proclaims: "For the past decade, Earthjustice was instrumental in protecting the gray wolves in court. Our work is now shifting to Congress where there have been legislative attempts to derail wolf recovery and push these animals to the brink of extinction." This is far from accurate.

The Sierra Club uses the slogan "Those faithful shepherds" on their wolf campaign page. The Sierra Club states that they "Educate the public about wolves and their biology to dispel negative stereotypes."[27] However, the Sierra Club calls the wolf a "Species at Risk," when in reality wolves are plentiful in many parts of North America, and abroad, and several times as many wolves and wolf hybrids are in captivity.

In response to the 2012 arrival in California of one wolf from Oregon, the Center for Biological Diversity sent out an email message that begins: "Wolves are smart, fast, curious and strong. It was inevitable that they'd find their way to California. It is not inevitable, though, that they'll survive. The livestock industry has already vowed to kill any wolf it sees and is gearing up its lobbying machine to keep them out of the state . . . Make a generous gift to support our California Wolf Fund," ends the message.[28] A perfect example of negative stereotyping and polarization.

It's also true that a new pack of wolves was sighted in northern California in 2015—a family of two adults and five pups. One of the parents is dark, and most of the pups are dark also. Black or very dark fur is a sign of wolves hybridizing with dogs.

Wild wolves are one thing, but APHIS trappers in northern CA have seen and trapped a number of wolves and/or wolf-dogs for the last decade. Some are raised by people and purposefully released into the wild or escaped, and often wolf-dogs are used by illegal marijuana growers to guard gardens, APHIS agents report.

None of the major pro-wolf groups say much about wolves also representing a danger to people and livestock due to up to fifty diseases they may carry.

"Save the Wolf" messages ultimately are picked up by the general media, which further inflames polarization as advocates are paid to dramatize situations to help raise money for their salaries.[29] "Save the Wolf" in many cases actually means "Save my salary."

Often the messages of the wolf advocates are misleading or simply wrong. For example, a common campaign message of many pro-wolf groups is that browsing elk are destroying aspens in Yellowstone National Park. Introducing wolves, they say, is the best way to restore the aspens, establishing a "landscape of fear," that keeps elk away from aspens, which results in habitat improvement that benefits many other species.

Fifteen years after wolves were released into Yellowstone, in the September of 2010 issue of *Science Daily*, USGS scientist Matthew Kauffman reports that elk are continuing to browse on aspens, regardless of wolves. Kauffman states: "This study not only confirms that elk are responsible for the decline of aspen in Yellowstone beginning in the 1890s, but also that none of the aspen groves studied after wolf restoration appear to be regenerating, even in areas risky to elk."[30]

US Fish and Wildlife Service
There could be no successful campaign to bring back wolves without the support of the US Fish and Wildlife Service, which has jurisdiction as wolves are an endangered species. Since the 1980s USFWS has promoted wolf recovery programs all around the United States through news media, public hearings, interviews, websites, exhibits, and personal appearances.

One of the most visible parts of this program was the widespread public review of Environmental Impact Statement that led up to the 1995–96 relocation of Canadian wolves into the Northern Rockies. Ten years after the relocation took place, the Wyoming Game and Fish Department did a review of the predictions made by the US Fish and Wildlife Service in that EIS. This is what they found:

> Despite research findings in Idaho and the Greater Yellowstone Area, and monitoring evidence in Wyoming that indicate wolf predation is having an impact on ungulate populations that will reduce hunter opportunity if the current impact levels persist, the Service continues to rigidly deny wolf predation is a problem.

The 1994 EIS predicted that presence of wolves would result in a 5 to 10 percent increase in annual visitation to Yellowstone National Park. On this basis, the EIS forecast wolves in the region would generate $20 million in revenue to the states of Idaho, Montana, and Wyoming. WG&F reports that annual park visitation has remained essentially unchanged after wolf introduction. A later chapter will examine in detail the real economics of the wolf reintroduction.

WG&F states: "Wolf presence can be ecologically compatible in the GYA only to the extent that the distribution and numbers of wolves are controlled and maintained at approximately the levels originally predicted by the 1994 EIS –100 wolves and 10 breeding pairs." USFWS . . . "has a permanent, legal obligation to manage wolves at the levels on which the wolf recovery program was originally predicated, the levels described by the impact analysis in the 1994 EIS."[31]

In a September 2010 interview with the *Bozeman Daily Chronicle*, two leading federal wolf biologists, Ed Bangs of the US Fish and Wildlife Service and Doug Smith of the US National Park Service,[32] state that from the beginning their long-term goal has been to delist wolves so they can be managed, which would mean controlled hunting.[33]

A number of wildlife biologists, including Dr. L. David Mech, Chair of the World Conservation Union (IUCN) Wolf Specialist Group, support that goal.[34]

Nonetheless, a lot of the general public believes any hunting of wolves is wrong, largely because of the way that wolves have been sold by wolf advocates. For example, a 1999 poll in Minnesota showed that while people favored wolf management, they preferred non-lethal methods.[35] A 2004 poll in Ontario, which does have a native wolf population but not near any population centers, found that 70 percent of the public opposed hunting wolves, 88 percent oppose sport hunting of wolves, and 82 percent do not support killing wolves for their pelts.[36] Such opposition also holds today for Scandinavia, where a 2003 study found that while a majority supported hunting wolves if livestock were being harmed, or wolves were entering cities, they did not favor a general wolf hunt.[37]

The problem is that nonlethal wolf controls seldom work, and if they do, it is short-term and costly. Issuing wolf sport hunting licenses is one way for state agencies to try to pay for the economic burden of wolf management, which is considerable; however, when this happens it raises the hackles of anti-hunting groups, who attack the agencies, forcing them to pay for their defense in court and in the political area.

Ohio State University researcher Jeremy Bruskotter reported in *Bioscience* in December of 2010, that the US Fish and Wildlife Service is essentially supporting the wolf advocates by suppressing research on the real and potential negative consequences of wolf populations, and this is contributing to the polarization of public opinion about wolves as well as misleading people about the negative consequences of expanding wolf populations.[38]

Selling Wolves to the General Public

Exhibits
In 1986, writer Rene Askins launched a traveling exhibit, the Wolf Fund, whose primary purpose was promoting the releasing of wolves into Yellowstone. To her credit, Atkins closed the Wolf Fund when the first wolf was released into Yellowstone.[39]

Defenders of Wildlife at one time also had a traveling wolf exhibit that was the largest such wildlife exhibit in the United States.

Many parks in the United States and Canada have educational exhibits about wolves, especially Yellowstone National Park and Algonquin Provincial Park in Ontario, where they also feature "howl-ins." The same is true for US Fish and Wildlife refuges where wolves are found. Visiting a park and howling with wolves is a much different experience than living with them day to day.

Wolf Education Centers
In the United States and Canada there are at least seventy "Wolf Education Centers" where people may see wolves in captivity, and be exposed to exhibits and educational programs about wolves. Some of the most popular centers include: Wolf Park in Battle Ground, Indiana[40]; Wolf Education Research Center in Winchester, Idaho[41]; Colorado Wolf and Wildlife Center, Divide, Colorado[42]; Wolf Song of Alaska Education Center, Eagle River, Arkansas[43]; Wolf Conservation Center, South Salem, New York[44]; and Northern Lights Wolf Center-Golden, British Columbia.[45]

One of the most professional and popular wolf education centers is the International Wolf Center in Ely, Minnesota, which is located in the center of the Minnesota wolf population and was launched in 1989 by Dr. L. David Mech, one of the world's most respected wildlife biologists who study wolves.

Mech was once a firm believer in wolves in North America not attacking people, but he has reversed his position and has made Mark McNay's landmark study on attacks available on his website, has sponsored a conference on attacks in Europe and Asia, and supports managing wolf populations with hunting. [46]

The International Wolf Center serves about fifty thousand visitors a year and conducts classes, workshops, and lectures in the United States and Canada. Such a facility located near Yellowstone National Park might enable visitors to see wolves and learn about them without the need to try to keep large packs of wolves running free in the park, especially near the highway where they can become habituated or spill over into nearby areas, thus reducing elk and moose populations, and/or attacking pets and eating roadkill and garbage. It would also increase the chances of preserving the existence of purebred wolves, for as you will learn in later chapters, wolves can and do mate with coyotes and dogs. Larger wolf populations that result in wolves interacting more with people will definitely increase the disappearance of purebred wolves and result in wolf-dog-coyote hybrids becoming the only "wild wolves."

Books

There have been hundreds of books written about wolves since the first Earth Day in 1970. [47] Here we will spotlight a few of the best-known wolf books, which have gained recognition and contain misleading information.

Following Jack London's very successful novel, *Call of the Wild*, about a domestic dog that returns back to a wild state, his 1906 novel *White Fang* is about a wild three-quarters-wolf wolf-dog that becomes domesticated. Following its publication, Theodore Roosevelt declared that London was a "nature faker" and that some of the scenes in *White Fang* were "the sublimity of absurdity."[48] Nonetheless, *White Fang* has been made into several films, including a 1991 adaptation starring Ethan Hawke.

When he was just out of college and working in the Southwest for the US Forest Service, Aldo Leopold believed that predators—bobcats, wolves, cougars, and black and grizzly bears—should be removed from special areas so that big-game populations like deer and elk could build up those reserves and spill over into adjacent lands, increasing recreational opportunities for hunters and wildlife watchers. Teddy Roosevelt supported Leopold, as did ranchers, farmers, and hunters. The Kaibab Plateau, adjacent to the Grand Canyon, was chosen as a place to implement Leopold's plan. When predators were eliminated at the Kaibab Plateau, the resident deer population exploded from less than ten thousand to nearly a hundred thousand. But the deer stayed on the reserve and ultimately ate out all the food, leading to massive starvation, a horrific crash in the deer population, and destruction of habitat that lasted for years after. That incident changed Aldo's attitude toward predators.

One of the most commonly quoted pro-wolf passages appears in Leopold's masterful treatise on man and nature, *A Sand County Almanac* (published in 1949), where he recounts an incident in 1909 when he shot a female wolf, but didn't immediately kill her. Approaching the wounded animal to administer the final shot, he looked at the old she-wolf and watched "a green fire dying in her eyes. I realized then, and have known ever since, that there was something new to me in those eyes—something known to her and the mountain."[49]

This story is moving and fits perfectly with the mindset of "Save the Wolf" people, but it would be a mistake to think that Aldo Leopold believed that predatory animals like bears, cougars, wolves, bobcats, and coyotes shouldn't be managed. In 1933, Aldo Leopold, the nation's first professor of wildlife management, teaching at the University of Wisconsin-Madison, wrote in his textbook, *Game Management* (the first college wildlife management text that remains in use by many colleges today), in the chapter on predator control:

> Predatory animals directly affect four kinds of people: 1. agriculturists; 2. game managers and sportsmen; 3. students of natural history; and 4. the fur industry. There is a certain degree of natural and inevitable conflict of interest among these groups. Each tends to assume that its interest is paramount. Some students of natural history want no predator control at all, while many hunters and farmers want as much as they can get up to complete eradication. Both extremes are biologically unsound and economically impossible. The real question is one of determining and practicing such kind and degree of control as comes nearest to the interests of all four groups in the long run.[50]

Additionally, in 1944, according to his biographer Curt Meine, the same year that Aldo Leopold wrote *Thinking Like a Mountain,* which contained his famous description of the "fierce green fire" in the eyes of a dying she-wolf, Leopold also wrote in a manuscript on predator management in Wisconsin:

> No one seriously advocates more than a small sprinkling of wolves. When they reach a certain level they will certainly have to be held down to it. . . . In thickly settled counties we cannot have wolves, but in parts of the north we can and should.[51]

Aldo Leopold and his entire family were lifelong avid hunters. Estella, his wife, was the Wisconsin state women's archery champion. In the Aldo Leopold Archives at the University of Wisconsin-Madison, there is a photo of Aldo's son, Starker (who became a very prominent wildlife biologist and

When this wolf was legally shot on February 15, 2010, it weighed 127 pounds and was approximately eight years old. The wolf had been collared in 2006 by Idaho Fish and Game, at which time he weighed 127 pounds, and they estimated he was four years old. The largest wolf weighed by Douglas W. Smith, wolf biologist for Yellowstone National Park, weighed 148 pounds. For reference, the woman in the photo is five feet, four inches tall; a signed affidavit from the woman in the photo is on file.

university professor, and who edited *A Sand County Almanac* after his father's death), proudly standing beside a Mexican wolf that he shot in New Mexico in 1948.[52]

Naturalist, artist, writer, and predator bounty hunter Ernest Thompson Seton was an early pioneer of the modern school of animal fiction writing. His most popular work, *Wild Animals I Have Known* (published in 1898), contains the story of his killing of a renegade Mexican wolf named Lobo— "The King of the Currumpaw." Seton later became involved in a literary controversy about writers who fictionalized natural history and distort wildlife biology and behavior—"nature fakers" as Teddy Roosevelt called them. In a 1903 article in the *Atlantic Monthly*, John Burroughs charged Seton with purposefully deceiving people with his writing. The controversy lasted four years involving many important American environmental and political figures of the day, including Teddy Roosevelt, who negotiated a deal with Seton to clean up his act.[53] More about this in the section on the documentary film made about Seton's wolfish tale, "The Wolf That Changed America."[54]

Never Cry Wolf is Farley Mowat's 1963 account of a young government biologist who in 1958 is flown to the tundra plains of Northern Canada to study the area's wolf population and gather proof of the ongoing destruction of caribou herds by wolves. After locating them on the remote tundra, the biologist contacts wolves as he discovers a den with pups and devoted protectors of their young. And in the absence of caribou, the wolves happily feed on mice and lemmings. The biologist meets two Inuit who tell him their own stories about the wolves. As he learns more and more about the wolf, he comes to fear the onslaught of hunters out to kill the wolves for their pelts. Ultimately, he runs naked with the wolves as they chase a herd of caribou. The book was made into a popular Walt Disney feature film called *Never Cry Wolf* in which all the wolves seen on camera were tame. The film was nominated for one Academy Award (Best Sound), and it won several other awards for "Best Cinematography." Posters advertised the movie as "based on a true story."

While the book is supposedly based on a real-life experience, Inuits refer to Mowat as "Hardly Know It." Scientists agree. Writing a review in *Canadian Field-Naturalist* in 1964, Canadian Wildlife Federation officer Alexander William Francis Banfield, who supervised Mowat's field work, accused Mowat of blatantly lying, as Mowat was part of a team of three biologists, and was never alone. Banfield also pointed out that a lot of what was written in *Never Cry Wolf* was not derived from Mowat's first-hand observations, but were lifted from Banfield's own works, as well as those of Adolph Murie's studies of wolves on Mount McKinley. Ultimately, he compared *Never Cry Wolf* to "Little Red Riding Hood," stating that "both stories have about the same factual content."[55]

Wolf biologist L. David Mech writes about the book: "Whereas the other books and articles were based strictly on facts and the experiences of the author, Mowat's seems to be basically fiction founded somewhat on facts."[56] Ethologist, Dr. Geist, calls *Never Cry Wolf* "a brilliant, literary prank."

Written before Mark McNay's documentation on wolf attacks and the deadly wolf attacks of 2005 and 2010, Barry Holston Lopez's *Of Wolves and Men* (published in 1978) explores many aspects of the relationship between people and wolves through history, and clearly states that wolves can and do kill people, but he does not acknowledge any attacks on white people in North America in the twentieth century. He says that while most Native American tribes consider wolves to be very important spiritually, seeing wolves as the ancestors of man, in the majority of Native American tribes, wolves were killed for body parts used in rituals, fur for clothing, to stop them raiding food caches, and wolf pups were considered a delicacy.

After surveying literature, mythology, and folklore from around the world, and trying to raise two red wolves, in the "Epilogue" of *Of Wolves and Men*, Lopez concludes: "Wolves don't belong living with people. It's as simple as that."[57]

An Associate Professor of History at Notre Dame, Jon T. Coleman admits at the outset of his 2004 book *Vicious: Men and Wolves in America* that he is an "animal person." *Vicious* begins with a description of John James Audubon watching an Ohio farmer catch three wolves in a pit trap and then slowly kill them. He explains the farmer's behavior as an example of "theriophobia"—an excessive fear of wild animals, saying, "Oblivious to the actual behavior of wolves, anti-wolf people based their hatred on 'myths, tales, and legends.'"[58] To generalize all people who do not like wolves as being the same is engaging in cheap stereotyping, which is hardly scholarly or accurate. Coleman fails to say if the farmer had lost livestock, pets, or even friends to wolves prior to this time, which would mean the farmer was simply very angry and seeking revenge.

On the same page, Coleman states, "There is no record of a non-rabid wolf killing a human in North America since the arrival of the Europeans." The irony is that Audubon personally investigated and confirmed the death of a man by three wolves in 1830.[60]

Two problems with finding information about wolf attacks is that until very recently there was no reliable recording system, you have to look for newspaper reports. And, when people are lost in the woods and are never found, or their remains are found later, confirming if they were killed by wolves or not is very difficult or impossible. When Young and Goldman looked for records of wolf attacks on humans before 1900 in North America, they found thirty accounts of attacks, and six possible human kills.[60]

Coleman's book came out in 2004, two years after Mark McNay's report on wolf attacks, yet he gives no mention of this research. Could the wolf attack deaths of Kenton Carnegie in 2005, and Candice Berger in 2010 have been averted if people like Coleman were not spreading misinformation about potential dangers of wolves? None of these three young people were carrying firearms or even pepper spray when they were attacked.

On page fourteen, Coleman states, "In 1995 the Fish and Wildlife Service set fourteen wolves from Alberta, Canada, loose in Yellowstone. Nine years later the population had grown to 148 predators, and packs of 'non-essential' gray and Mexican wolves loped in Idaho, Arizona, and New Mexico."

Actually, a total of sixty-six Canadian wolves were released into Yellowstone and Idaho by the US Fish and Wildlife Service, establishing "experimental, non-essential" populations according to article 10(j) of the Endangered Species Act. In 2005, the wolf population of the Greater Yellowstone area alone was estimated at a minimum of 325 wolves, not 148. Adding together the wolves in Idaho with those in the Yellowstone area and those in Montana, some of which were already living in the state before the 1995–96 releases, results in at least fifteen hundred wolves in the Northern Rockies, some say the number is at least three thousand.

In March 1998, the US Fish and Wildlife Service released three packs of Mexican wolves, bred in captivity, into the Apache-Sitgreaves National Forest in Arizona, and eleven wolves into the Blue Range Wilderness Area of western New Mexico. There are currently approximately ninety-seven Mexican wolves in the wild in both Arizona and New Mexico. (More about Mexican wolves later.)

Coleman goes at great length to deride the negative mythology, folklore and fairy tales that originated from Europe, failing to report the numerous well-documented attacks and killings of people by wolves throughout Eurasia, and the importance of folk tales, fables, and fairy tales in educating children to avoid becoming victims.[61]

Despite the bias of the author, the many inaccuracies of the work and his lack of understanding of psychology, folklore, and mythology, *Vicious* won awards from the American Historical Association and the Western Historical Association. A falsehood often repeated does not make it true.

Magazine Articles

There have been thousands of articles about wolves in major national magazines and newspapers, including *National Geographic*, *The New York Times*, *Audubon*, *Sierra*, *Newsweek*, *Wall Street Journal*, etc., as well local newspapers and magazines. The majority of these favor wolf restoration, and most are written by journalists parroting what they are told by wolf advocates from environmental groups and federal and state agencies charged with wolf management.

An article worth noting, "Cry Wolf: How A Campaign of Fear and Intimidation Led to The Gray Wolf's Removal from the Endangered Species List" by James William Gibson that appeared in *Earth Island Journal*, Summer 2011, illustrates one way the "Save the Wolf" campaign has often used self-righteousness to create enemies to hate.[62]

Gibson, a professor of sociology at California State University at Long Beach, claims that "an extreme right-wing culture that celebrates the image of man as 'warrior,' recognizes only local and state governance as legitimate, and advocates resistance—even armed resistance—against the federal government . . . built around some shared myths that focus on the evils of wolves in general and the Rockies' wolves specifically" is engaging in "fear-driven demagoguery" targeting pro-wolf groups and individuals, and this is the real reason why wolves were delisted, and not biology. In a December 8, 2011, op-ed article in the *Los*

Angeles Times, Gibson describes proponents of hunting wolves to manage their population as "paramilitary militia advocates."

"Afraid for their lives and their families, regional wolf advocates stopped participating in public hearings held by fish and game agencies and legislative committees and retreated to the relative safety of the Internet to spread their message," Gibson says. He charges that "a dysfunctional political system in which fear—both irrational fear and fear harnessed for political gain—determines policy."

"In the entire twentieth century, wolves attacked about fifteen people in North America, killing none," Gibson claims. He does admit, "In 2010, wolves did kill a woman jogging on the outskirts of her Alaskan town," but he makes no mention of other attacks and fatalities that you will soon learn more about.

Little or no mention is made of the numerous professional wildlife biologists, and many present and former state and US Fish and Wildlife Service biologists, who support wolf management and controlled hunts—none of which belongs to right-wing paramilitary organizations. (Nor do any of the contributors to this book.)

This is not to say that some ranchers, hunters, and others have not written or said negative things about wolf supporters. Gibson, however, makes no mention of the swarm of threats in electronic media and direct personal attacks that have been made by wolf advocates to anyone who differs with their view. He fails to acknowledge how the whole situation is made worse by "wolf advocates" who are paid to carry on campaigns against anyone who questions their agenda, keep controversies going, and fuel crises that can be used by their employers to raise money. (See the five-part series "Environment Inc." by Pulitzer Prize–winning *Sacramento Bee* investigative journalist Thomas Knudsen, documenting how large environmental groups use inflammatory articles, exaggeration, and even purposeful lying to keep themselves visible and money flowing in.)[63]

An important book about the politics of the wolf wars of the Northern Rockies is *Yellowstone Wolves: A Chronicle of the Animal, the People and the Politics* by Cat Urbigkit.[64] Urbigkit, a Wyoming journalist and sheepherder, traces the history of wolves in the Yellowstone area, clearly showing that while hunting, trapping, and poisoning wolves knocked the population down in the mid-1900s, wolves never were eradicated from that area. She traces the history of groups, especially Defenders of Wildlife, who began sending out Action Alerts as early as 1992, and the federal government, who declared at public meetings about wolf reintroduction in the early 1990s, "As the wolf population grows, support can grow along with it . . ." Urbigkit's story of how she and her husband filed suit to block the USFWS wolf relocation program as wolves were already present in the Northern Rockies is a true example of heroism.

In response to Gibson's inflammatory negative stereotyping, Urbigkit quotes a January 1999 editorial in the national newsletter of American Farm Bureau by AFB President Dean Kleckner that describes what happened when the AFB filed a lawsuit to have introduced Canadian wolves removed from the Northern Rockies and returned to Canada:

> Defenders of Wildlife launched a nationwide campaign against Farm Bureau in the press, television, radio and Internet, falsely describing our organization and our lawsuit. [AFB did not call for killing off wolves. It simply wanted them removed from the area.]
>
> Wolf stocking advocates, incited by Defenders of Wildlife, organized a campaign of harassment and intimidation against the Bureau with the aim of forcing us to abandon our own farmer-written policies and drop the lawsuit. We have received several bomb threats and threats against the lives of Farm Bureau officers and their families. Even a federal judge's life was threatened. [The judge found that wolves could be captured humanely and returned to Canada.]
>
> Working Assets (in support of Defenders of Wildlife) brags about sending us 34,000 letters and calls. . . . Defenders of Wildlife and Working Assets crossed a line when they intended to shut down our phone system and encouraged callers to harass us into dropping our case against the Department of Interior's illegal program.

The American Farm Bureau filed a complaint with the Federal Communications Commission and the FBI regarding the threats they had been subjected to, which is how such pressures should be handled.

Films

There are many documentaries and feature films about wolves.[65] Nearly all support the "Save the Wolf" movement, and many have inaccuracies, such as *Wild Wolves* on PBS *NOVA*, which includes an interview with USFWS wolf biologist Ed Bangs, where Bangs states:

> The studies that we've done and that other people have done indicate that wolves normally kill less than one-tenth of one percent of the livestock available to them. To date, in the past fifteen years in the northern Rocky Mountains, we've lost an average of about five cattle and five sheep per year to wolf depredations.

Those statistics do not represent what is actually happening. For example in 2010, wolves killed sixty-five livestock (thirty-seven cattle and thirty-three sheep), three horses, and one dog in Wyoming alone.[66] Bangs continues:

> You know when you teach your dog to not go out of the yard or not go in the flower bed—and your dog learns that for the rest of its life? It's just something it won't do? That's the same reason that wolves never attack people. Behaviorally, they just don't recognize people as anything they want to screw with. And they live their entire lives without ever trying it.[67]

That statement will also be corrected shortly.

In 1990, naturalist, cinematographer, director, and author Jim Dutcher purchased wolf pups born in captivity and was allowed a permit to set up a twenty-five-acre wolf observation camp in the Sawtooth Mountains of Idaho, where he stayed, later joined by his wife, Jamie. The pups were first raised in captivity and then released in the larger pen. Dutchers lived in the enclosure until 1996, raising and documenting the captive pack of wolves and their socializing behavior. The wolves, which became known as the "Sawtooth Pack," then became property of the Wolf Education and Research Center (a non-profit organization he founded) in conjunction with the Nez Perce and moved to northern Idaho. The Dutchers made a documentary film, *Wolves at Our Door*," which won a Primetime Emmy.

The wolves were allowed to roam freely in the twenty-five-acre enclosure, but these wolves were raised by the Dutchers, making them habituated. While the film may have made people feel less afraid of wolves, it also may have helped promote people buying and raising wolves as pets.

According to Rick Hobson, who once worked for the Dutchers, much of the film is not at all based on what actually happened. Hobson says, "The film gives the impression that the Sawtooth Pack was given to the Nez Pierce in Idaho, but this isn't the case . . . The wolves were sold for tens of thousands of dollars to a non-profit organization hastily created to save the wolves, after Dutcher had commented that euthanizing the wolves was an option he was considering. The wolves came with a restrictive agreement which gave most photographic rights to Dutcher. The non-profit group couldn't even use images of the wolves on merchandise in order to support itself or the wolves."[68]

In feature films, we generally find a more balanced coverage of wolves. A good example is *Never Cry Wolf* based on Farley Mowat's 1963 bestselling book, adapted for the big screen in 1983 by Carroll Ballard. However, in the May 1996 issue of the popular Canadian magazine *Saturday Night*, John Goddard wrote a heavily researched review article entitled "A Real Whopper," in which he poked many holes in Mowat's claim that the book was non-fictional. Goddard also reported that Mowat told him, "I never let the facts get in the way of the truth." While Mowat called Goddard's article "bullshit, pure and simple," he refused to refute Goddard's main claims.[69] This is an example of misrepresenting the facts with emotion. Films often, like books, distort the behavior and tendencies of wolves in order to romanticize them, which do injustice to the perception of wolves, and consequently do injustice to the people who have to live with wolves.

In 2011, a feature film *The Grey*, a wilderness survival story written and produced by Joel Carnahan and starring Liam Neeson, describes a plane loaded with Alaskan oil field workers crashing in the dead of winter in a snowy area. The oilmen were subsequently attacked by hungry wolves. As expected, many wolf advocates howled about this film, nonetheless *The Grey* made more than three times its production costs at the box office, and continues to be popular on streaming services.

Could such a thing happen? Some wolf advocates adamantly assert "NO!," however as you will learn later, there is a long history of wolves attacking people in Europe and Asia during times when food was short.

Hope for the Future

As the federal government seeks to turn wolf management over to states, and habitat conservation organizations seek to have the wolf delisted and managed by hunting, pro-wolf groups counter with a seemingly endless escalating barrage of legal challenges accompanied by huge mass mailings and other propaganda. These legal attacks slow down and impede management of wolves as it forces agencies to spend money on legal costs, rather than put it to use in research and field work.

People do want to hear the truth about wolves. In the following pages, you will learn more about what that is. Between 2010 and 2014, three different lawsuits were filed challenging Wyoming's plan to manage wolves, one in federal court in Colorado and two in federal courts in Washington, DC, and there are legal challenges to nearly every wolf population in the lower forty-eight states. For wolf advocates, wolves clearly are cash cows in wolf fur.

"Environmentalists routinely exaggerate problems
so as to alarm people and get support for their agendas."

—John Naisbitt, *Mind Set!*[70]

Endnotes:

1. Chase, Alston. *In a Dark Wood: The Fight Over Forests and the Rising Tyranny of Ecology.* (NY, NY: Houghton-Mifflin, 1995), xiii.

2. Kellert, R.S. "Public Perceptions of Predators, Particularly the Wolf and Coyote," *Biological Conservation* 31 (1985), 167–189.
3. Urbigkit, Cat. *Yellowstone Wolves: A Chronicle of the Animal, the People, and the Politics.* (Ohio: McDonald and Woodward Publishing Company, 2008), 28.
4. Ibid.
5. Murie, Adolph. *The Wolves of Mount McKinley.* (University of Washington Press, 1985).
6. Allen, Durwood and L. David Mech. "Wolves Versus Moose on Isle Royale," *National Geographic*, (Feb. 1963), 200–219.
7. Leopold, A. Starker, et al. "Wildlife Management in the National Parks." (National Park Service, 1963).
8. Williams, CW, Gordan Ericsson, and Thomas Herbelein. "A Quantitative Summary of Attitudes Toward Wolves and Their Reintroduction (1972–2000)," Wildlife Society Bulletin, 30 (2002), 1–10.
9. Lopez, Barry Holston. *Of Wolves and Men.* (Scribners, 1978), 4.
10. http://www.livescience.com/44875-werewolves-in-psychiatry.html
11. White, T.H. *The Book of Beasts.* (New York, NY: Putnam, 1954), 59.
12. Lopez, Barry. Ibid.
13. White, Johnathan. *Talking on the Water: Conversations About Nature and Creativity* (Sierra Club Books, 1994), 121–136.
14. Jaffe, Aniela. "Symbolism In The Visual Arts," in *Man and His Symbols*, ed. Carl Jung. (New York, NY: Dell, l968), 266.
15. http://www.defenders.org/about_us/history/index.php
16. http://www.charitynavigator.org/index.cfm?bay=search.summary&orgid=3605
17. http://www.charitywatch.org/articles/defendersofwildlife.html
18. http://action.defenders.org/site/PageServer?pagename=sayewolves_homepage
19. http://www.worldwildlife.org/gift-center/gifts/Species-Adoptions/Gray-Wolf.aspx?gid =13&sc=AWY1000WCGP1&searchen=google&gclid=CLzrv6rKw6AC FR6kiQodFmSybA
20. http://www.defenders.org/programs_and_policy/wildlife_conservation/solutions /wolf_comp nsation_trust
21. http://www.defenders.org/programs_and_policy/wildlife_conservation/imperiled _species/ wolves/america_votes_yes!_for_wolves.php
22. http://www.nrdc.org/wildlife/habitat/esa/rockies02.asp
23. https://www.nrdc.org/resources/ensure-thriving-populations-wolves
24. https://secure.nrdconline.org/site/Donation2?df_id=8040&8040.donation=form1
25. http://www.humanesociety.org/animals/wolves
26. http://humanewatch.org/index.php/site/post/hsus_earns_some_detention
27. http://www.sierraclub.org/lewisandclark/species/wolf.asp
28. http://www.biologicaldiversity.org/publications/earthonline/endangered-earth-onlineno605.html
29. http://www.pbs.org/now/shows/609/index.html
30. http://www.sciencedaily.com/releases/2010/09/100901111636.htm
31. http://www.pinedaleonline.com/wolf/wolfimpacts.htm
32. http://www.bozemandailychronicle.com/news/article_32c72a40-3c5b-11df-91e 5-001cc4c002e0.html

33. http://digitalcommons.unl.edu/cgi/viewcontent.cgi?article=1040&context=vpc16
34. http://www.wolf.org/wolves/learn/scientific/challenge_mech.asp
35. http://www.wolf.org/wolves/learn/intermed/inter_human/survey_shows.asp
36. http://www.wolvesontario.org/wolves/Full%20Wolf%20Poll%20 March%2011% 202004%20revised.pdf
37. http://www.wildlifebiology.com/Downloads/Article/477/En/10_4_ericsson.pdf
38. http://www.sciencedaily.com/releases/2010/12/101206093703.html; http://bruskotter .wordpress.com/
39. http://www.people.com/people/archive/article/0,20108666,00.html.38;
40. http://www.wolfpark.org
41. http://www.wolfcenter.org/default.aspx
42. http://www.wolfeducation.org
43. http://www.wolfsongalaska.org/education_center_tour.html
44. http://www.nywolf.org
45. http://www.northernlightswildlife.com
46. http://www.wolf.org/wolves/index.asp
47. http://www.inetdesign.com/wolfdunn/wolfbooks
48. Carson, Gerald. "T.R. and the 'Nature Fakers," *American Heritage* (February 1971), 22.
49. Leopold, Aldo. "Thinking Like a Mountain. A Sand County Almanac. (Oxford University Press, 1949), 130.
50. Leopold, Aldo. Game Management. (University of Wisconsin Press, 1933), 230.
51. Meine, Curt. *Aldo Leopold, His Life and Work*, (1988). University of Wisconsin Press, Madison, WI, pg. 458.
52. A. Starker Leopold with wolf, July 1948, Aldo Leopold Papers, series 3/1, box 85, folder 7. http://digital.library.wisc.edu/1711.dl/AldoLeopold
53. Carson, Gerald. Op. Cit.
54. http://factoidz.com/lobo-king-of-the-currumpaw/
55. Banfield, A.W.F. "Never Cry Wolf." *Canadian Field Naturalist* 78, (January-March 1964), 52–54.
56. Mech, L. David. *The Wolf: The Ecology and Behavior of an Endangered Species.* Natural History Press (Doubleday Publishing Co., N.Y, 1978), 389.
57. Lopez, Barry Holston. *Of Wolves and Men.* (Charles Scribner's Sons, 1978), 280.
58. Coleman, Jon T. *Vicious: Men and Wolves in America* (Yale University Press, 2004), 3.
59. Audubon, J.J., and J. Bachman. The Quadrupeds of North America. (New York: Wellfleet Press, 1851–1854).
60. Young, S. and E. Goldman. *Wolves of North America, Vol. 1 and 2.* (Dover, 1944).
61. Von Franz, Marie-Louise. *Shadow and Evil in Fairy Tales.* (Shambala, Boston: 1995).
62. http://www.earthisland.org/journal/index.php/eij/article/cry_wolf
63. http://www.earthisland.org/journal/index.php/eij/article/cry_wolf
64. Urbigkit, Cat. *Yellowstone Wolves: A Chronicle of the Animal, the People, and the Politics.* (McDonald and Wordwood, 2008).
65. http://www.bullfrogfilms.com/catalog/wolf.html
66. http://wolves.biginterest2u.com/docs.html
67. http://www.pinedaleonline.com/news/2012/02/Wyomingwolfpopulatio.htm

68. http://www.pbs.org/wgbh/nova/wolves
69. http://www.imdb.com/title/tt0326413/reviews
70. http://www.salon.com/1999/05/11/mowat/singleton
71. John Naisbitt, *Mind Set!: Eleven Ways to Change the Way You See—and Create—the Future.* HarperBusiness; Reprint edition (December, 2008).

Photo credit: Dennis Donahue/Shutterstock.com

CHAPTER 3

The Myth of the Harmless Wolf

By Ted B. Lyon

Photo credit: Holly Kuchera/Shutterstock.com

There are no known gray wolf attacks on humans in modern times in North America.

—US Fish and Wildlife Service Website,
North Dakota Field Office, August 2016.[1]

Predators

Predators are part of the web of life. Some are very selective in their prey. Others aren't.

There are an estimated eight hundred thousand to nine hundred thousand black bears in North America. They are found in forty states and all Canadian provinces. According to wildlife biologist and bear expert Dr. Gary Alt, black bears normally operate as lone individuals, attack an average of twenty-five people per year, and kill an average of one to two people a year, primarily with the intent of predation and sows protecting their cubs.

There are less than one thousand grizzly bears in the Lower 48. Grizzlies primarily attack in defense of food and cubs. Attacks on humans average three to five per year, occasionally they are deadly. In June of 2016, Brad Treat, a thirty-eight-year-old Montana Forest Service Law Enforcement Officer, was fatally attacked by a grizzly bear while mountain biking on a trail in Glacier National Park.[2]

There are an estimated fifty thousand mountain lions in the Lower 48, increasing in numbers and spreading eastward to feed on the mushrooming whitetail deer population. Recently a mountain lion was killed by an SUV in Connecticut, the first such sighting in a century.[3] Mountain lions have also been sighted in a number of Midwestern states, including Oklahoma, Indiana, Nebraska, Michigan, and Wisconsin. Cougars are solo hunters that shy away from people, unless they run out of food or have never been hunted. Attacks per year for North America run from two to nine, with up to two deaths resulting from the attacks. These attacks are almost always predatory, with cougars targeting runners or hikers whose movement may seem like deer to mountain lions.

There are approximately seventy-five million dogs in the United States. The CDC "Dog Bite: Fact Sheet" says that each year, 4.7 million Americans are bitten by domestic dogs, and as many as fifty Americans die every year from dog attacks. Feral dogs are often the most likely to attack humans (as well as breed with coyotes and wolves), however, 75 percent of the attacks and over 50 percent of the fatal attacks are by two breeds—pit bull and Rottweiler—all bred for aggressiveness. For perspective, there are approximately 4.5 million registered pit bulls, a least that many that are not registered, and there are about half of that number of Rottweilers. There are only about a hundred thousand wolves in the wild in North American but tame wolves and wolf-dog hybrids (three to five hundred thousand) are considered in the top five breeds that attack people, and as you will learn more later, these attacks can be predatory.

Most importantly, domestic dog attacks are almost always defensive, unless the dogs have been trained to attack or they are feral and have not developed the predatory skills of wild canines to catch wild animals.[4] Feral dogs are a growing problem in the United States, as well as worldwide. Dog ownership in the United States has tripled since the 1960s, and the total number of feral dogs is uncertain. In Detroit, there are estimated to be at least fifty thousand feral dogs.

Once a western species, coyotes are now found in all states but Hawaii. No one is quite sure how many coyotes there are in North America; estimates range from ten million to one hundred million. The number of coyote attacks per year in the United States averages around fourteen, and it's rising as more and more coyotes move into urban areas including Chicago, Washington, DC, Columbus, Toronto, San Francisco, and Los Angeles.[5] Coyotes have recently been spotted in Central Park in New York City.

According to University of California at Davis biologist Dr. Robert Timm, forty-eight coyote attacks on children and adults were verified from 1998–2003, compared to forty-one attacks during the period from 1988–1998.[6]

Coyote attacks have been reported in at least eighteen states in addition to California and from four Canadian provinces, with the majority of attacks occurring since the early 1990s.[7] Just how many of these coyotes have some wolf or dog DNA is uncertain.

Human deaths from coyote attacks are rare, but on October 27, 2009, Canadian folk singer Taylor Mitchell was hiking in Cape Breton Highlands National Park in Nova Scotia when she was attacked by two eastern coyotes and died of her injuries the following day.[8]

Coy-wolves, or eastern coyotes, or wolves, are larger than western coyotes and many scientists believe they are hybrids between western coyotes and gray wolves or red wolves.[9]

As the wolf genetics chapter in this book, "Canis Stew," will show, cross-breeding among wolves, coyotes, and domestic dogs will inevitably increase as the three canines come into contact with each other. In hybridizing with other canid species, especially domestic dogs, the wolf loses some of its instinctual learning and wildness, and seemingly is born semi-habituated, which is the case with the coy-wolves of New England and the Mexican wolves of the Southwest.

We've already established that there are between three and five hundred thousand tame wolves and wolf-dog hybrids in the United States and their numbers are rising, even though such animals are illegal in forty states. Typically, such animals are either pets or living in educational centers, zoos, and sanctuaries, and they are unpredictable as people find that what started out as a cute puppy becomes a dangerous adult. When this happens, people will sometimes turn these wolf-dogs loose, resulting in them interacting with feral dogs and coyotes. All too often, wolf-dogs are responsible for attacks on livestock and people.[10] One might predict then that increasing numbers of attacks on people, their livestock, and pets is inevitable, not only due to habituation, but hybridization.

Wolves are ancestors of dogs, but the pure wolf remains a very different species of wild canid. There are at least six thousand gray wolves in the Northern Rockies and Northern Great Lakes states, another fifty to sixty thousand wolves in Canada, and 7,700 to 11,200 in Alaska—seventy to seventy-five thousand gray wolves for all of North America, possibly as many as one hundred thousand.

There also are two subspecies: around a hundred Mexican wolves in New Mexico and Arizona, and forty-five to sixty red wolves in the Southeastern United States.[11] Both Mexican and red wolves were raised in captivity and released into the wild. There is considerable debate if these animals are truly wild or not and both species have hybridized with other canines.

Wolves are predators—they kill other animals to eat, and for pleasure. As the quote from a USFWS website that opens this chapter shows, for decades we have been told in colleges across North America, and in countless popular books, articles, and films, that wolves do not attack and kill people in North America and that wolves have not attacked people in North America for almost a century.[12] The reality is that this is not correct.

Do North American Wolves Attack People?

Wolves are voracious predators. They prefer large ungulates—elk, deer, caribou, moose, and bison. They also will regularly prey on snowshoe rabbits and beaver if they are common.

A pack of three to five wolves will kill an average of two reindeer or caribou every three days. They can eat six to seven pounds of meat per day, and over ten pounds if they have not eaten for a while. That translates into one and a half tons of meat per wolf per year—twenty-two to twenty-three elk if that's what's on the menu. When wolves run out of wild meat, domestic livestock comes next. Always, wolves first test conditions. If humans do not appear to be a threat, they move closer and closer, driven by hunger and the love of killing, taking pets, eating garbage, and ultimately stalking and attacking people. This is proven by the historical record of Europe and Asia, as well as North America. (See later chapter on habituation by Dr. Geist.)

Let us be clear at the outset that the record of wolf attacks on people does not reflect the total number of attacks for several reasons. The first is that there was no written record of wolf attacks in North America until the 1700s. In the oral tradition of Native Americans, there are many stories of attacks, but there is no way to authenticate them.

The second is that there is no central system for wolf attack reports. In general, newspaper reports are all that researchers have to go on.

And third, a surprising number of people are lost in the wilderness and are either not found for some time or are never found. Just how persons fall into either of those categories is unclear, as there is no central data collection system for missing persons in the wilds. However, the Oregon-based Mountain Rescue Association has ninety search-and-rescue teams in twenty states, which complete about three thousand missions each year.

Missing people in the wilds can be victims of bear, mountain lion, or wolf attacks, but we will never know for sure. Once a person in the wilds dies, their body becomes potential food for predators, omnivores, and scavengers, and so determining what animal killed a person in the wilderness, or even if they were killed by an animal, is a forensic nightmare. In Alaska, where the wolf population is at least ten thousand, there are many reports of missing people who are either never found, or only their skeleton is found.[13]

The Alaskan Troopers have an active list of missing persons that has at least ninety-six names on it.

This latter situation is also true for livestock. Did the animal just wander off, or was it chased away and killed in the bush? If a rancher does not come on the dead animal right after it is killed, a number of different predators can dine on the body and the animal that initially killed the cow, sheep, goat, or pig may never be identified.

Thus, anyone who claims that there has never been a fatal wolf attack in North America is basing this statement on wishful thinking, not realism.

Based on what we can learn from scattered reporting, there have not been that many fatal wolf attacks in North America in the twentieth century. A primary reason why is that wolves in the United States and Canada were relentlessly hunted, trapped, and poisoned beginning in the 1800s and continuing through the 1930s. Estimates of the original wolf population in North America range between a quarter and half a million, or more. A secondary consideration is that when European settlers arrived, they brought firearms with them.

The arrival of Europeans with firearms and livestock set the stage for eradication of wolves, as unchecked hunting, especially market hunting, killed off many big-game species and these animals were replaced by livestock. As wolves switched from preying on wild game to domesticated livestock, trapping, hunting, and poisoning wolves drastically reduced their numbers.

Aside from Alaska and the Artic, those few wolves that survived along the Canadian border in the Northern Rockies and Minnesota learned that they needed to retreat to wilderness areas and stay away from people to survive. Wolves are intelligent and North American wolves became seldom seen out of necessity to survive.

This chapter will demonstrate that wolves do attack humans in North America, and the number of attacks is increasing as the wolf population grows and these largest of the wild dogs increasingly come into contact with people, ranches, and towns, as well as coyotes and dogs they can breed with.

Wolf Attacks around the World

Originally, gray wolves were distributed throughout the northern hemisphere worldwide in every habitat where large ungulates were found. Saturating most of the region between 20°N latitude (mid-Mexico and India) and the North Pole, in temperatures from -40° to +40° C, the wolf inhabited areas as diverse as Israel, North Africa, China, Great Britain, Ireland, and Greenland. In North America, they were found from Mexico City north to the Arctic Ocean. Wolves, aside from man, were once the most widely distributed mammal living north of 15°N latitude in North America and 12°N in Eurasia.[14]

There may have been as many as two million wolves worldwide in earlier times, but today the global population is approximately one-eighth of that, because man has trapped, hunted, and poisoned wolves not just in the United States, but worldwide, for centuries. A primary reason for the campaign against wolves has been fears of attacks on livestock and people. Let us look at what history tells us about the propensity for wolves to attack prey other than wildlife.

Historical Wolf Attacks on People in Europe and Asia

There is a long history of wolf attacks in Europe and Asia. This is where most of our western fables and fairy tales originated, and as you shall see, for good reason. Let us briefly look at some statistics about wolf attacks on people abroad up until 2013.

France

Between 1580 and 1830, 3,069 people were killed by wolves in France, 1,857 of these victims killed by wolves that were non-rabid.[14] In the winter of 1455, forty people were killed by wolves in Paris, France. These wolves came to be known as the "Wolves of Paris."[16]

Between 1763–67 in Gévaudan, Auvergne, and Languedoc, France, ninety-nine people were killed by a wolf known as "the Beast of Gévaudan" and its offspring.[17]

Sweden

From December 30, 1820, to March 27, 1821, around Gysinge, near the border of in central Sweden, a wolf attacked thirty-one people, killing eleven children between three-and-a-half to fifteen years of age and one nineteen-year-old woman, most of whom were partially consumed by the wolf. The wolf had been captured as a pup and raised in captivity for three to four years before it was released prior to the attacks.[18]

Italy

Between the fifteenth and nineteenth centuries, 440 people were killed by wolves in the Po Valley of Italy. In 1704, sixteen people were killed by wolves in what is today the popular tourist region of Varesotto in Northern Italy.[19] Today there are between five hundred and one thousand wolves in Italy. According to Luigi Boitani, who heads the La Sapienza University Animal and Human Biology Department, "However, the problems start when wolves return to an area after decades, and the ability to coexist has been forgotten."[20]

Russia

There are more wolves in Russia than any other country in the world today, and those wolves have evolved in a culture where there are scant few hunters, and the general population traditionally owns very few firearms. That we have not heard more about wolves in Russia since the fall of Czarist Russia, author and scientist Will Graves contends, is a function of the Communist government suppressing such information, because if people knew how really dangerous wolves were, the people would demand to be well-armed, and an armed populace could lead to revolution.[21]

During the last century, wolf numbers in Russia have risen and fallen through five cycles. When wolf populations became large, special hunting efforts were organized to reduce them. When the population was lowered, the hunting efforts were disbanded, and the wolf populations began to rise again. The greatest number of wolves was documented during World War I and World War II, when men were called off to war. Since the general populace of Russia has few firearms, and in the absence of men and without firearms, the wolf population grew quickly, as much as 30 percent per year.

In contrast to the US standard line that wolves do not attack people, Russian wildlife biologist Mikhail Pavlov states in his 1982 book *The Wolf in Game Management*:

> Cases of severe unprovoked aggression by wolves toward humans are numerous. In 1988, the editorial board of the magazine *Hunting and Hunting Economics* sent me information on aggressive wolves in Kaluzhskaya Oblast. Ex-chairman of the hunter's society, S. Semiletkin, informed the magazine that in 1943–1947 there were 60 victims of wolf aggression, including 46 children. . . . In Viatskaya in 1896–1897, 205 people became victims of predation, while there were only 10 in Vologodskaya, 18 in Kostromskaya, 1 in Arkhangel'skaya and 9 in Yaroslavskaya.[22]

Graves's exhaustive study, *Wolves in Russia: Anxiety through the Ages*, builds upon Pavlov's work. Graves reports that from 1804–53 non-rabid wolves killed 111 people in Estonia, of which 108 were children, two men and one woman. Another 266 adults and 110 children were killed by wolves in the period of 1849–51 in the European sector of the Russian Empire. Another 1,445 people were killed by wolves in 1870–87 in the European sector of Russian Empire.[23]

Rabid wolves have long been a problem in Russia, with wolves contracting rabies from foxes, raccoons, dogs, and other mammals. Pavlov reports that in 1975–76 attacks of rabid wolves on humans were also recorded in Ulianovskaya Oblast (fifteen cases), Kaluzhskaya Oblast (seven cases), Orenburgskaya Oblast (six cases), and Orlovskaya Oblast (four cases). In Gorkovskaya Oblast, during the ten years from 1929–39, forty people who were bitten by a rabid wolf were treated in hospitals. In the same area, after a rise in numbers of wolves in 1978, some twenty-four attacks on humans were recorded. In the 1980s, there also were numerous stories in newspapers on battles between humans and rabid wolves. Some of these wolves were rabid. In July 1976, in three days a rabid wolf bit sixteen people in the Lubomil district of Volinskaya Oblast.[24]

India and Pakistan

India typically has two thousand to three thousand wolves in rural areas. Pakistan and neighboring Kazakhstan traditionally have many more wolves. Attacks on people in this area are all too common, especially where there are few weapons available for self-defense and little or no police or military presence. A sample of what has taken place there includes 721 people killed by wolves in 1875 North-Western Province and Bihar State, British India.[25] One of the worst cases ever recorded of wolves attacking people occurred in 1878 in British India when 624 people were killed by man-eating wolves.[26]

These tragic events might seem like a thing of the past, but wolves killing people in India has continued over the years. In 1926, ninety-five people were killed by wolves in the Districts of Bareilly and Pilibhit, United Provinces, India.[27] Fast forward to April 1993–95 in the Bihar State of India when wolves killed sixty children, and another twenty children were rescued from what would surely have been death. All the children were taken from settlements primarily during March to August between 5 p.m. and 9 p.m. There were more female victims (58 percent) than males and 89 percent were three to eleven years old. Of the eighty child casualties, only twenty were rescued.[28]

Yadvendradev Jhala, a scientist at the Wildlife Institute of India, reports that in Uttar Pradesh during a two-year period (1996–97), a wolf or wolves killed or seriously injured seventy-four humans, mostly children under the age of ten years.[29]

Literature provides a decidedly different tale, as a friendly wolf pack raised the boy Mowgli in Rudyard Kipling's classic children's tale, *The Jungle Book*.

Iran

Two recent fatal attacks by wolves in Iran suggest why wolves continue to be feared in the Middle East, even when the land is so heavily developed. January 4, 2005, an Iranian homeless man was eaten alive by wild wolves in the village of Vali-Asr, near the town of Torbat Heydariya in northeast Iran.[30]

In November 2008, a wolf attacked an eighty-seven-year-old woman in the village cemetery of the central Iran town of Kashan. It bit one of her fingers, but she fought back and suffocated the wolf to death.[31]

Norway

In Scandinavia over the past three hundred years, ninety-four people have been killed by wolves. All of those cases were before 1882 and most were children under the age of twelve.[32] Wolves were widely killed off in Scandinavia, Great Britain, and in many areas of Europe beginning in the 1600s. In contrast to the myth that native people don't hunt wolves, for centuries Eurasian wolves have been hunted by the Saami, the indigenous people of northern Scandinavia, for protection, furs for clothing, and to safeguard their herds of reindeer, which are a main source of meat for the Saami.

Historical Accounts of Wolf Attacks in North America

Wolves were largely exterminated from heavily armed North America a hundred years ago when cattle and sheep replaced buffalo and other wild game as primary sources of meat and ranchers responded with traps, guns, and poison. A 1997 CNN article states: "While the wolves pose some threat to humans' domesticated animals, there is little risk to people themselves. And while humans have killed an estimated two million wolves in this century, there is not a single documented case of a human being killed by a healthy wild wolf."[33] That statement is blatantly false.

Let us begin a discussion of wolf attacks in North America by clearing up a common misconception. Often we hear from pro-wolf people that Native Americans and other traditional cultures considered the wolf to be a spiritual being, implying that indigenous people revere wolves and want them protected. Among traditional cultures all around the world, there is a spiritual dimension to all parts of nature—animals, stones, clouds, bodies of water, plants, places, and heavenly bodies. And, each part of nature has a unique spirit and power that man may be able to access. That an animal like a wolf has a spiritual aspect or power does not mean that a native person will not kill it, wear its pelt, or even eat it. In fact, in many cultures killing a wolf and wearing its skin is considered a path to gaining the wolf's powers as a brave hunter, as well as protection from cold weather. This is one reason why the Mongolian Winter Olympic teams wear parkas with wolf fur.

Additionally, one should never assume that all Native American tribes hold the same beliefs about wolves. There are 562 federally recognized tribes in the United States, as well as Aleuts, Polynesians, and Inuits. While some tribes regarded wolves positively, others, such as the Navajo, saw wolves as witches—people taking the shape of wolves, wearing wolf skins, and becoming like werewolves.[34] The Chilcotin tribe of British Columbia believes that contact with wolves can lead to madness and death. The Blackfeet, while respecting wolves as great hunters, believe that wolves can never be trusted.[35] According to Blackfoot guide Alvin Yellow Owl, "The elders teach us to respect the wolf, for he is a good hunter. However, the elders also teach to never trust the wolf (or the coyote), for he can turn on his own kind, as well as anything else." In 2011, the Blackfoot tribe in Montana issued ten wolf tags for hunters on the reservation.

Among the Koyukon tribe of Alaska, the wolf's spirit is seen as third in ranking of animal power after the bear and the wolverine, and according to tribal customs all three mammals may be trapped or shot, providing proper ceremonies are conducted to honor the animals' spirits.[36]

In present times, much has been made of the Nez Perce tribe in Idaho supporting wolf reintroductions, and the tribe has made a considerable contribution to wolf conservation along these lines. However, in 2009, when hunting licenses for wolves were issued in Idaho, thirty-five licenses were sold specifically for killing wolves on Nez Perce lands.

Farther south, the White Mountain Apaches support Mexican wolf restoration, but the San Carlos Apaches oppose the Mexican wolf restoration program in New Mexico and Arizona.[37]

Native Americans and Inuits have co-existed with wolves for thousands of years. They respect wolves as great hunters, but they also feared attacks by both rabid and healthy wolves. According to Barry Lopez, "It is popularly believed that there is no written record of a healthy wolf ever having killed a person in North America. Those making the claim ignore Eskimos and Indians who have been killed and are careful to rule out rabid wolves."[38]

Early Europeans encountered wolves upon landing in the New World. Recalling what they had known about wolves in Europe, the settlers responded with firearms, which drove wolves back into the wilderness, where there was abundant wildlife, especially buffalo, which once numbered sixty million in the United States. Accounts from the Lewis and Clark Expedition reported large packs of wolves following herds of buffalo. Even though wild game was abundant, wolves were considered fearless and might attack people, including the Expedition.[39]

In 1807, Clark's squad was camped along the Yellowstone River near Billings, Montana, when a wolf stole into camp and bit a sleeping Sergeant John Pryor "through the hand." The wolf then turned on Private Richard Windsor before being dispatched by another squad member.[40]

The noted naturalist and artist John James Audubon reported in 1830 that wolves attacked two men who defended themselves with axes. Three wolves were killed, while the wolves killed one man.[41]

In his book *The Great American Wolf*, Bruce Hampton reports that in 1833 at a fur-trapping rendezvous for trappers in Wyoming, a rabid wolf wandered into camp on two evenings and bit thirteen people.[41]

Contracting rabies is possible in any warm-blooded animal, but in Asia, Africa, and parts of North America, dogs are the primary host, especially feral dogs, with foxes, raccoons, and skunks being primary carriers. Wolves and coyotes are dogs, and can also carry rabies.

There are an estimated fifty-five thousand human deaths annually from rabies worldwide, about thirty-one thousand in Asia, and twenty-four thousand in Africa. The disease has been known since 3500 BCE. The first written record of rabies appears in the Codex of Eshnunna (ca. 1930 BCE), which dictates that the owner of a dog showing symptoms of rabies should take preventive measure against bites. If another person was bitten by a rabid dog and later died, the owner was fined heavily. [43]

Japan once had wolves, but the Japanese exterminated their wolves by 1905, when most of the population of wolves had contracted rabies. [44] Currently, Japan is considering reintroducing wolves to control sika deer, which are very abundant and cause considerable crop damage. There are pro-wolf groups currently in Japan including the Japan Wolf Association.

Cases of rabies in wolves are numerous in the eastern Mediterranean, Middle East, and Central Asia. Wolves commonly develop a "furious" phase of rabies, aggressively attacking whatever they encounter. This, coupled with their size and strength, make rabid wolves among the most dangerous of rabid animals. Heptner and Naumov state that bites from rabid wolves are fifteen times more dangerous than those of rabid dogs. [45]

Linnell and associates report cases of rabid wolves biting humans in numerous countries, including Italy, France, Finland, Germany, Poland, Slovakia, Spain, Baltic States, Russia, Iran, Kazakhstan, Afghanistan, China, India, and North America. [46]

The noted naturalist George Bird Grinnell, who was skeptical about wolf attacks on humans, confirmed that in the summer of 1881, an eighteen-year-old girl in Colorado was attacked and bitten on the arms and legs by a wolf. [47]

With the disappearance of the buffalo, wolves had to look elsewhere for food or starve to death, for other species of wildlife were not nearly as common as some claim. As Dr. Charles Kay has pointed out, ". . . early explorers who spent 765 days in the (Yellowstone) ecosystem on foot and on horseback between 1835 and 1876, reported seeing elk only once every eighteen days and bison were only seen three times, but not in the park itself."[48]

When wolves began killing livestock as a substitute for wild game, people began a campaign to exterminate wolves with poison, traps, and hunting.

When Young and Goldman looked for wolf attacks on humans before 1900 in North America, they found thirty accounts of attacks, and six possible human kills. [49]

Wolf Attacks in Recent Times in North America

Today we tend to think of habituated or rabid wolves being the ones most likely to attack. However, the historical record shows that healthy wolves have attacked people in wild settings in North America, and continue to do so.

Again, this is only a sample of what has occurred as formal records of wolf attacks have not been kept until very recently. Let us look at some examples. In February 1915 on the Coppermine River in the Canadian Arctic, a large white female wolf came into the camp of the Canadian Arctic Expedition of 1913–18, as they were eating breakfast. The wolf attacked one man and then bit another man on the arm before one of the party shot the wolf, which was later mounted and is today in the collection of the National Museum in Ottawa. [50]

In 1922, an elderly trapper near Port Arthur, Ontario, disappeared in the woods. Two Native Americans were sent to find him. When they did not come back, a search party was organized and the bodies of all three were found, killed by wolves. [51]

On January 26, 1950, a missionary driving a dog sled from Palmer to the Copper Basin, carrying supplies and gifts for orphans and needy families, was surrounded by wolves in deep snow near Sheep Mountain, near the headwaters of the Matanuska River. All the dogs were killed. The missionary survived by keeping the wolves away by building a fire. [52]

By the late 1930s, wolves had been virtually eliminated from the lower forty-eight states due to hunting, trapping, and poisoning. Wolves were still thriving in Alaska, however, and as more and more people moved to Alaska, encounters with wolves began to increase. In 1974–75 in Fairbanks, Alaska, wolves attacked and killed 165 dogs, a number of which were being walked by owners or were in the yards of their owners. In response, thirteen wolves were shot in and around Fairbanks, and attack incidents dropped dramatically. [53]

On April 18, 1996, five wolves attacked and killed twenty-four-year-old Patricia Wyman in the Haliburton Forest and Park Reserve in Haliburton, Ontario. The wolves had been raised in captivity but had never been trained to socialize with people. [54]

In September of 2006, a wolf attacked six people in Algonquin Provincial Park in Ontario. The wolf had a broken clavicle and tooth when it was shot by park rangers following the attacks, but it did not have rabies. [55]

Dogmatic teaching in colleges and universities, coupled with claims by wolf advocates, pushed the belief that wolf attacks on humans could only be abroad, and surely not in North America. In 2000, two wolf attacks on humans in North America changed things forever.

In April of 2000 in Icy Bay, Alaska, six-year-old John Stenglein and a nine-year-old friend were playing outside his family's trailer at a logging camp, when a wolf came out of the woods towards the boys. The boys ran, but the wolf attacked young Stenglein. It bit him on the back and buttocks. Adults, hearing the boy's screams, came and chased the wolf away. The wolf returned a few moments later and was shot. According to Alaska Department of Fish and Game, the wolf was a healthy wild wolf that apparently attacked without provocation. [56]

This incident moved the Alaska Fish and Game Commission to request that a comprehensive study of wolf attacks on people in recent times in North America be undertaken. In response to this request, in 2002 Alaskan wildlife biologist Mark McNay published a report of a two-year study documenting eighty aggressive encounters between wolves and people in North America in the twentieth century. Forty-one were from Canada, thirty-six were from Alaska, and three were from Northern Minnesota. Thirty-nine were aggressive encounters with healthy wolves. In only twelve of the attacks were the wolves rabid. Since McNay's report came out there have been two fatal attacks by healthy wolves—one in Canada and another in Alaska, and an unknown and increasing number of non-fatal aggressive encounters and attacks on people and their pets in the United States and Canada. [57]

Also in 2000, on Vargas Island, British Colombia, university student Scott Langevin, twenty-three, was on a kayaking trip with friends. They camped out on a beach in an area where other campers had stayed. In the middle of the night, Langevin awoke with something pulling on his sleeping bag. He suddenly came face to face with a wolf. Langevin yelled at the wolf and it bit him on the hand. Langevin attempted to fight back, but the wolf jumped on his back and started biting him on the back of his head. Friends, hearing his yells, came to his aid and scared the wolf away. Fifty stitches were required to close Langevin's wounds. The BC Ministry of Environment speculated the reason for the attack was wolves being habituated at that location as a result of being fed by people who used that campground. Langevin's party did not feed the wolves. [58]

In Canada, wolves have been extensively hunted and trapped for many years with a bounty as an incentive. This resulted in over one thousand wolves a year taken in Alberta alone. Such trapping and hunting keeps the wolf population in check and drives remaining wolves back into remote wilderness areas. The attack on Langevin took place in a campground, where wolves were protected and it is likely they were fed. Farther east in Algonquin National Park in Ontario, which has a viable wolf population and special wolf howling trips for visitors, there is a history of wolf attacks on people, including five attacks over an eleven-year period. [59]

Alaska has seven to eleven thousand wolves, which are also hunted and trapped. However, as more and more people flock to Alaska and its urban areas increase in size, wolves investigate these new enclaves where hunting isn't permitted. [60] For example, in the winter of 2007, the Elmendorf wolf pack around Anchorage attacked a number of people walking dogs, killing some dogs. [61] This made the news in Anchorage and Fairbanks, but not much was heard farther south about wolf attacks until on November 8, 2005, college student Kenton Carnegie was hiking on a road near Points North Landing in northern Saskatchewan when he was attacked and killed.[62] A thorough provincial inquest found that wolves were responsible. It was later determined that the wolves had become habituated to a garbage dump in that area, and that just a few days prior to Carnegie's death, some other people were confronted by habituated wolves in the same area.[63] Dr. Geist was one of the experts called in to testify at a special inquisition held by the government to determine the cause of Carnegie's death.

Another wolf attack in Saskatchewan occurred on New Year's Eve, 2005, when Fred Desjarlais was coming home from his job at Key Lake, about 550 kilometers north of Saskatoon. The wolf suddenly came at him from a ditch and bit him several times on the back, arm, leg, and groin.[64] These incidents caused Canadian media to do some research, which discovered more wolf attacks on people in Canada, including those in 1994, 1995, 1996, 2000, and 2004.[65]

This occurrence illustrates that many wolf attacks do occur but do not make the national news or the scientific literature.[66] Four quick examples:

Resort owner Ken Gangler in Manitoba reports that recently one of his summer resort staff went out jogging on the landing strip for the private airport that serves Langler's Resort. She was suddenly struck from the back by a large wolf that knocked her down. Before the wolf could attack any further, it was driven off by people nearby. [67]

In Idaho, in 2008, a father and son were out hunting coyotes using a predator call. The pair was separated by a hundred yards. Suddenly a pack of wolves surrounded the son. The father came over and both men fired shots to scare the wolves away. The wolves left reluctantly, the men said. This was the year before it became legal to hunt wolves in Idaho.[68]

Then, on March 9, 2010, Candice Berner, a thirty-two-year-old special education teacher working in Chignik Lake, Alaska, went jogging at dusk on a road near town and was attacked and killed by wolves. As this was on US soil, finally the reality of wolf attacks on people in North America became a front-page story in the United States. The attacking wolves were not rabid.[69]

In December of 2012, Alaskan trapper Lance Grangaard was riding his snowmobile checking traps with his father when a black wolf appeared and attacked him. Grangaard raised his arm to protect himself and the wolf bit him through his parka and three layers of clothing to put a three-inch gash on his arm. Grangaard jumped off the snowmobile and wrestled the wolf to the ground, screaming and yelling, and hit the animal's head on some ice. The wolf finally let go. As there was no way to test the wolf for rabies, Grangaard had the series of rabies shots just to be sure.

There have been nineteen confirmed cases of rabies in Alaska wolves since testing began in 1971, all of them in northern or western Alaska, where the disease is prevalent in foxes and occasionally spills over into the wolf population according to Alaska Fish and Game in Fairbanks.[70]

Wolf Attacks in North America 2013–2016

The first edition of this book came out in 2013. As wolves have become more established in North America, especially the lower forty-eight states, reporting of wolf attacks has improved, although there still is no central reporting system, such as is used by the Center for Disease Control for dog bites. Nonetheless, thanks to the Internet, it's been possible to compile a list of attacks on people, pets, and livestock that have occurred between the fall of 2013 and the fall of 2016 in North America. This is a sample of what is happening. The total number of such incidents is much higher.

Alaska
March 2014: Wolf kills, eats dog after battle with owner in daylight attack near Haines.[71]
June 2014: Wolf attacked and killed a dog being taught what to do in such an encounter.[72]
November 2015: Wolf attacks dog in Whitehorse subdivision.[73]
June 2016: Two wolf attacks on Dogs in Ketchikan—one dog killed.[74]

Arizona
November 2013: Four wolves attack man on horseback and his dogs.[75]
April 2015: Three wolves attack couple walking their dog.[76]

Idaho
August 2013: Wolves kill 176 sheep in one night in Idaho.[77]
Fall 2013: Wolf attacks bicyclist.[78]
May 2016: Wolves kill six hunting sounds.[79]
August 2016: Idaho Wildlife Services investigated ninety-one wolf livestock killings during fiscal 2015, down from 107 the year before and 129 in 2013.[80]

Michigan
October 2013: Eight more recent wolf attacks on domestic animals.[81]
November 2013: Wolf attacks nearly double previous years.[82]
August 2014: Upper Peninsula wolf attacks on dogs continue.[83]
January 2015: Wolf attacks on livestock and dogs in 2014 jump to one of the highest levels in more than a decade.[84]

Minnesota

August 2014: The Sheriff's office in Cook County, Minnesota, has issued a wolf warning after five dogs were killed by wolves in the last two weeks.[85]

March 2015: Dog killed by wolves in Duluth.[86]

April 2015: Wolf kills pet dog in Suburb of Duluth.[87]

April 2015: Wolf killings of dogs up dramatically in 2015.[88]

April 2015: Wolves trapped after attacks on dogs in the Duluth-Two Harbors area.[89]

June 2015: Wolves attack and kill livestock.[90]

October 2015: Fifteen Northern Minnesota wolves killed after dog attacks.[91]

December 2015: Minnesota agriculture officials see wolf attacks in new locations.[92]

February 2016: Wolf attacks and kills dog, at Brighton Beach.[93]

February 2016: Hungry wolf kills off-leash dog on a walk with owner in Duluth.[94]

Montana

April 2015: Thirty-six head of cattle, six sheep, and one horse killed by wolves in 2014.[95]

Oregon

March 2014: Cows suffering from PTSD-like sickness after watching wolves kill.[96]

September 2014: Eight sheep and two dogs killed by wolves in Umatilla.[97]

September 2014: Wolves kill two dozen sheep in Northeast WA.[98]

March 2015: Umatilla pack wolves stampedes 250 head of pregnant cows.[99]

Washington

June 2015: Wolves kill three sheep, one dog, and one calf in recent attacks.[100]

August 2015: Huckleberry wolf pack attacks livestock guard dog.[101, 102]

November 2015: Wolf attacks cattle, kills one in southern Oregon.[103]

November 2015: Wolf attacks three calves in Klamath.[104]

March 2016: Imnaha wolf pack adds fourth attack (three cows and one sheep).[105]

March 2016: State officials kill four wolves after attacks on livestock.[106]

Wisconsin

August 2014: Pet dogs killed by wolf packs in Wisconsin.[107]

September 2014: Thirteen dogs in Wisconsin verified as killed by wolves in 2014, including six during the first three weeks of August. [108]

September 2015: Deer hunter defends self with pistol from wolf pack attack.[109]

October 2015: USDA Wildlife Services confirmed thirty-one wolf attacks on dogs. Those include twenty-three hunting dogs and eight pet dogs in 2015.[110]
November 2015: Second wolf incident lends credit to wolf attack account.[111]
May 2016: Wolf kills Bowler woman's dog.[112]
May 2016: Family speaks after DNR confirms wolves attack family pet.[113]
July 2016: More than thirty wolf attacks on hunting dogs since January of this year.[114]

British Columbia
March 2015: Wolf warnings posted after multiple dog attacks on west Vancouver Island.[115]
March 2015: Wolf attacks and drags off domestic dog in daylight on Vancouver Island.[116]
November 2015: People take matters into own hands as wolf attacks on pets escalate.[117]

NW Territory
June 2016: A starving wolf stalks a woman and her dog for twelve hours.[118]

Ontario
March 2015: Wolves attack, kill family dog in woman's front yard.[119]

Saskatchewan
September 2015: Experts say declining deer numbers to blame for wolf attacks.[120]
September 2015: Canadian man captures wolf attacking his dog on camera.[121]
August 2016: A wolf attacks a shift worker at Cameco's Cigar Lake uranium mine, who was walking between buildings shortly after midnight. A security guard interrupted the attack and scared the lone wolf away.[122]

Wolf Attacks in Other Parts of the World 2013–2016

Azerbaijan
October 2015: Wolves attack locals in Azerbaijan's Agshu, two injured.[123]

China
August 2014: Wolf attack in Chinese village leaves six people disfigured and one missing an ear after pack of five wolves surrounded a small farming community and attacks.[124]

France
April 2015: Wolves attack flock of sheep on edge of a town.[125]
June 2015: French farmer's son left terrified after attack by nine wolves.[126]
September 2015: Fifty French farmers take park director hostage over wolf attacks.[127]
October 2015: Team of hunters deployed to hunt wolves in French Alps.[128]
October 2015: Wolves go on killing spree of in French village—more than six thousand sheep were lost to these wolves last year alone.[129]

Iran
March 2015: Between April 2001 and April 2012, there were fifty-three gray wolf attacks on humans in the West of Iran.[130]

Russia
January 2013: "State of emergency" declared over wolf attacks in Siberian region.[131]
December 2013: Locals say wolves threaten their reindeer.[132]
November 2015: Pack of wolves attack man in dogsled in Russian national park.[133]

Sweden
October 2015: Wolves attack hunting dog wearing a Go-Pro.[134]

These articles are just a sample of attacks on people and pets from 2013–16 in North America and Eurasia. There is no central, official, count of wolf attacks. Setting up an impartial counter of wolf attacks, like the CDC who cover dog bites, might help to clarify exactly what is the relationship between wolves and people, and how to minimize attacks.

To keep up with wolf news, check in with the website of Wolf Education International—http://wolfeducationinternational.com/.

Some Perspective
According to Dr. Valerius Geist, wolves are generally not dangerous when they are well fed on wild animals, by virtue of successfully preying on abundant wild prey where they have either very little contact with people, or where they are hunted. Geist says that in general, the evidence indicates that wolves are very careful to choose the most nutritious food source easiest obtained without danger. They tackle dangerous prey only when they run out of non-dangerous prey, and they shift to new prey only very gradually, following a long period of gradual exploration. In a later chapter Geist will describe how and why wolves become habituated.

Geist believes that wolves are most likely to attack humans and/or their pets if: 1. they have become habituated; 2. they have rabies; 3. they have been provoked; 4. people in that area have few or no firearms; and 5. garbage is easily accessible.

In the Northern Rockies and northern Midwest, in areas where wolves have not been hunted, big-game herds—elk, deer, and moose—that are wolves' favorite prey, have declined dramatically due to wolf predation. This then sets the stage for wolves to approach livestock, pets, and people as prey.[135]

In a study of wolf predation worldwide, Linnell and associates found that there are four factors associated with wolf attacks on humans: 1. rabies; 2. habituation; 3. provocation as in people approaching dens with young wolves; and 4. environmental conditions including little or no natural prey, garbage dumps, children who are unattended, poverty, and limited availability of weapons, especially firearms.[136]

Unfortunately, there has not been a detailed follow-up study of wolf attacks on people like Mark McNay's, which concluded in 2002.[137] As wolf populations grow across North America, wolves are testing boundaries and moving into new territory. As wolves come into closer contact with people, the chance of attacks increases, especially if we approach wolves naively. Expanding wolf packs in North America are visiting the suburbs of Minneapolis and walking the streets of Sun Valley, Idaho. There are numerous reports of the four wolf packs living around Anchorage, Alaska, attacking dogs and people walking them.[138]

In Yellowstone National Park, where visitors line the roads to take pictures of wolves, in 2009 rangers had to shoot a wolf because it was chasing bicyclists.[139]

Wolf biologist Dr. David Mech advises people to never feed wolves and allow them to become habituated. He says that if you meet a wolf, do not run away—yell, look as big as you can, throw rocks. Pepper spray helps. The sound of a gun will let them know you mean business.[140] Wolves are smart, prolific, and adapt quickly. Mech says that so long as there is adequate food and habitat it is necessary to kill off between 28 percent and 53 percent in an area just to keep that wolf population stable, which is part of the necessary management program for wolves living in the twenty-first century in areas where so many people now live.[141]

The evidence of history clearly shows that when wolves and people do come together, especially when people are not armed and wolves are hungry, sooner or later, attacks will occur and someone is going to be hurt or killed. And, the historical record also shows that on occasion, both healthy and rabid wolves do go on killing sprees that can include humans, especially children. This reality must be factored into any responsible wolf management program.

Conclusion

Ultimately, by not accepting that wolves can and do attack people and have done so from time immemorial, people then believe that wolves are not dangerous to humans, an approach that leaves them vulnerable to wolf attacks.

"Only wolves and tigers seem to have learned to hunt man for food, and perhaps sharks and crocodiles."

—John Muir[142]

Endnotes:

1. http://www.fws.gov/northdakotafieldoffice/endspecies/species/gray_wolf.htm
2. http://abcnews.go.com/US/family-montana-man-killed-grizzly-bear-attack-speaks/story?id=40263924
3. http://latimesblogs.latimes.com/nationnow/2011/08/mountain-lions-flocking-to-greenwich-ct-them-too.html
4. http://www.dogbitelaw.com/PAGES/statistics.html
5. http://digitalcommons.unl.edu/vpc21/1/
6. http://digitalcommons.unl.edu/cgi/viewcontent.cgi?article=1075&context=icwdm_wdmconfproc
7. http://www.latimes.com/science/la-me-coyotes-20141218-story.html and https://www.researchgate.net/profile/Robert_Timm/citations
8. Note: The authors reported in the first edition of *The Real Wolf* that Ms. Mitchell was killed by coy-wolves. They have since received updated information. Several coyotes suspected of involvement in the attack were killed and genetically tested. The results were reviewed by animal geneticist Dr. Matthew Cronin, the author of chapter 18, who concluded they were eastern coyotes. News story about attack: http://latimes-blogs.latimes.com/outposts/2009/10/musician-taylor-mitchell-dies.html
9. http://www.coywolf.org/coywolf-basic-info/
10. http://en.wikipedia.org/wiki/Wolfdog&http://www.wolf.org/wolves/learn/intermed/inter_human/algonquin.asp
11. http://en.wikipedia.org/wiki/Red_Wolf
12. http://www.forwolves.org/wolves.html
13. http://www.adn.com/we-alaskans/article/missing-alaska-without-trace/2014/07/20/
14. http://www.wolf.org/wolves/learn/scientific/challenge_mech.as
15. Moriceau, Jean-Marc. Histoire du méchant loup : 3 000 attaques sur l'homme en France. (2007), 623.
16. Thompson, Richard H. *Wolf-Hunting in France in the Reign of Louis XV: The Beast of the Gévaudan.* (Edwin Mellen Press, 1991), 367.
17. http://en.wikipedia.org/wiki/Wolf_attacks_on_humans
18. http://westinstenv.org/wibio/2010/02/22/the-danger-of-wolves-to-humans/
19. http://www.physorg.com/news90260221.html
20. Pavlov, Mikhail P. "The Danger of Wolves to Humans" in *The Wolf in Game Management.* Translated by Valentina and Leonid Baskin, and Patrick Valkenburg. Moscow: Agropromizdat, 1982.

21. Graves, Will. *Wolves in Russia: Anxiety Through the Ages*. Detselig Enterprises, 2007.
22. http://westinstenv.org/wibio/2010/02/22/the-danger-of-wolves-to-humans/
23. Knight, John. *Wildlife in Asia: Cultural Perspectives*. Abingdon, UK: Routledge, 2004. 280.
24. http://hindu.com/2001/05/08/stories/1308017f.htm
25. Maclean, Charles. The Wolf Children. (1980): 336.
26. http://www.wolfsongalaska.org/Wolves_South_Asia_child.htm
27. http://www.wolf.org/wolves/learn/intermed/inter_human/india_abstract_003.asp
28. http://www.iranfocus.com/en/iran-general-/homeless-man-eaten-by-wolves-in-iran-01150.html
29. http://en.wikipedia.org/wiki/Indian_Wolf#cite_note-19
30. http://www.cnn.com/EARTH/9711/12/yellowstone.wolves/
31. http://www.wolfsongalaska.org/wolves_in_american_culture.html
32. Boitani, Luigi. Wolves: Behavior, Ecology, and Conservation. University of Chicago Press (2003).
33. Nelson, Richard. Make Prayers To Raven: A Koyukon View of the Northern Forest. University of Chicago Press: (1983).
34. http://www.rangemagazine.com/archives/stories/summer03/ground-hog.htm
35. Lopez, Barry. Of Wolves and Men. Charles Scribner's Sons (1978).
36. http://www.sinauer.com/groom/article.php?id=24
37. http://fwp.mt.gov/mtoutdoors/HTML/articles/2006/LCmisadventures.htm
38. Audubon, J.J., and J. Bachman. The Quadrupeds of North America. New York: Wellfleet Press (1851–1854).
39. Hampton, Bruce. The Great American Wolf. NY, NY: Henry Holt (1977), 94.
40. http://en.wikipedia.org/wiki/Prevalence_of_rabieshttp://en.wikipedia.org/wiki/Prevalence_of_rabies
41. http://bestjapanguide.com/483/
42. Heptner, V.G, N.P. Naumov. Mammals of the Soviet Union. Science Publishers, Inc: 267.
43. http://www.lcie.org/Docs/Damage%20prevention/Linnell20NINA20OP20731%Fear%20of%20wolves%20eng.pdf
44. Grinnell, G.B. Trail and Campfire—Wolves and Wolf Nature. New York (1897). http://www.aws.vcn.com/wolf_attacks_on_humans.html
45. Kay, Charles E., "The Kaibab Deer Incident: Myths, Lies and Scientific Fraud," Muley Crazy: January/February 2010.
46. Young, S. P., and E. A. Goldman. The Wolves of North America: Parts 1 and 2. Dover Publ. Inc., New York (1944), 636.
47. Jenness, Stuart E. "Wolf Attacks Scientist: A Unique Canadian Incident," Arctic: 38 (June 1985), 129–132.
48. http://www.rangemagazine.com/archives/stories/summer03/ground-hog.htm
49. http://www.farnorthscience.com/2007/12/13/news-from-alaska/ravenous-wolves-attack -missionary/
50. http://www.farnorthscience.com/2007/12/13/news-from-alaska/ravenous-wolves-attack -missionary/
51. http://en.wikipedia.org/wiki/Wolf_attacks_on_humans

52. http://www.propertyrightsresearch.org/2006/articles09/six_injured_in_rare_wolf_attack.htm
53. http://juneauempire.com/stories/042700/Loc_wolf.html
54. http://www.adfg.alaska.gov/static/home/library/pdfs/wildlife/research_pdfs/techb13_full.pdf
55. http://www.cbc.ca/sask/features/wolves/attacks.html
56. http://www.wolf.org/wolves/learn/intermed/inter_human/algonquin.asp
57. http://alaska.wikia.com/wiki/Wolves
58. http://www.alaskastar.com/stories/122007/new_20071220007.shtml
59. http://en.wikipedia.org/wiki/Kenton_Joel_Carnegie_wolf_attack
60. http://www.cbc.ca/canada/saskatchewan/story/2007/11/01/wolf-verdict.html
61. http://www.cbc.ca/canada/story/2005/01/04/sasktimberwolf050104.html
62. http://www.cbc.ca/sask/features/wolves/attacks.html
63. http://missoulian.com/news/local/article_a43b34b2-5b16-11df-a302-001cc4c002e0.html
64. Personal conversation, January 2009.
65. Personal conversation, March 2009.
66. http://www.msnbc.msn.com/id/35913715/ns/us_news-life/
67. Personal conversation 2010
68. Personal conservation 2010
69. http://www.nbcnews.com/id/35913715/ns/us_news-life/t/fatal-wolf-attack-unnerves-alaska-village/
70. http://www.newsminer.com/wolf-attacks-trapper-on-snowmachine-near-tok/article_78ddd3dd-a279-5bab-ab4a-3a95db4c6549.html
71. http://www.adn.com/alaska-news/article/wolf-kills-eats-dog-after-battle-owner-daylight-attack-near-haines/2014/03/17/
72. http://www.adn.com/alaska-news/article/wolf-attack-killed-dog-lesson-what-do-during-encounter/2014/06/25/
73. http://www.whitehorsestar.com/News/dog-on-the-mend-following-wolf-attack
74. http://juneauempire.com/state/2016-05-19/ketchikan-sees-two-wolf-attacks
75. http://citizenreviewonline.org/since-wolves-were-reintroduced-some-eastern-arizona-ranchers-claim-the-animals-have-destroyed-their-lives/
76. http://www.mogollonrimnews.com/warning-to-heber-overgaard-human-wolf-encounter-in-section-31/
77. http://www.ktvb.com/story/news/local/2014/07/02/11975053/
78. http://www.montanaoutdoor.com/2013/07/wolf-attacks-bicyclist-in-idaho/
79. https://www.starvalleyindependent.com/2016/05/20/wolves-kill-six-hunting-hounds-in-idahos-madison-county/
80. http://www.mtexpress.com/news/environment/money-to-deter-wolf-attacks-goes-unclaimed/article_ee24c44e-9300-11e5-a103-1f6050ea10ed.html
81. http://www.misenategop.com/casperson-says-eight-more-recent-wolf-attacks-on-domestic-animals-proves-need-for-scientific-wolf-management/
82. http://www.9and10news.com/story/27886521/dnr-wolf-attacks-doubled-in-michigans-upper-peninsula-in-2014
83. http://www.9and10news.com/story/26245392/upper-peninsula-wolf-attacks-continue

84. http://www.mlive.com/news/index.ssf/2015/01/wolf_attacks_on_livestock_and .html

85. http://www.northlandsnewscenter.com/news/iron-range/Warning-Timber-wolves-attacking-dogs-and-what-to-do-in-that-situation-273210521.html

86. http://www.northlandsnewscenter.com/news/local/Woman-believes-Max-the-dog-was-brutally-attacked-by-wolf-in-Duluth-296968741.html

87. http://www.mprnews.org/story/2015/04/27/endangered-wolf-control

88. https://www.minnpost.com/glean/2015/04/wolf-killings-dogs-dramatically-2015

89. http://www.duluthnewstribune.com/news/3726049-wolves-trapped-after-attacks-dogs-duluth-two-harbors-area

90. http://www.offthegridnews.com/current-events/he-watched-wolves-attack-and-kill-his-livestock-but-could-do-nothing-because-of-a-fed-court-ruling/

91. http://www.twincities.com/2015/04/21/northern-minnesota-wolves-killed-after-dog-attacks/

92. http://kstp.com/news/stories/s3998714.shtml

93. http://mix108.com/timberwolf-attacks-and-kills-dog-at-brighton-beach

94. http://www.startribune.com/hungry-wolf-kills-off-leash-dog-on-a-walk-with-owner-near-duluth-lakeshore/368323061/

95. http://www.spokesman.com/blogs/outdoors/2015/apr/03/montana-reports-decrease-wolves-attacks-livestock/

96. http://extension.oregonstate.edu/news/release/2014/03/cows-witnessing-wolf-attacks-suffer-symptoms-similar-ptsd

97. http://www.columbian.com/news/2014/sep/29/wolf-attacks-kill-sheep-dogs-oregon/

98. http://www.spokesman.com/blogs/outdoors/2015/apr/03/montana-reports-decrease-wolves-attacks-livestock/ and http://q13fox.com/2014/09/04/ranchers-cautioned-as-nearly-2-dozen-sheep-killed-by-wolves-in-stevens-county/

99. http://www.wallowa.com/wc/editorials/20150317/guest-column-wolf-attack-a-cow-mans-worst-nightmare

100. http://www.statesmanjournal.com/story/news/2015/06/25/wolves-kill-five-domestic-animals-two-attacks/29282985/

101 https://stevenscountycattlemen.com/2015/08/13/huckleberry-pack-attacks-guard-dog-near-hunters/

102. http://www.capitalpress.com/Livestock/20150814/ne-washington-wolf-pack-injures-guard-dog

103. http://www.kptv.com/story/30456692/wolf-kills-3-cattle-in-first-southern-oregon-attack

104. http://www.capitalpress.com/Oregon/20151106/wolf-attacks-three-calves-in-klamath-county

105. http://www.statesmanjournal.com/story/news/2016/03/28/second-attack-wolf-pack-could-spur-lethal-action/82347508/

106. http://www.oregonlive.com/environment/index.ssf/2016/03/state_officials_will_kill_wolf.html

107. http://www.ammoland.com/2014/08/more-pet-dogs-killed-by-wolf-packs-in-wisconsin/

108. http://www.outdoornews.com/2014/09/04/wolves-blamed-for-13-dogs-deaths-in-state/

109. https://www.nrahlf.org/articles/2015/9/30/worldwide-exclusive-wisconsin-deer-hunter-fends-off-wolves-with-walther-pk-380/

110. http://www.outdoornews.com/2015/10/29/with-wolf-attacks-officials-hands-tied/

111. http://www.ammoland.com/2015/11/wi-second-wolf-incident-lends-credit-to-wolf-attack-account/

112. http://wbay.com/2016/05/13/wolf-kills-bowler-womans-dog/

113. http://abc10up.com/warning-graphic-wolves-attack-dog-dnr-confirms/

114. http://sportingclassicsdaily.com/wolf-attacks-hunting-dogs/

115. http://globalnews.ca/news/1881242/wolf-warnings-posted-after-multiple-dog-attacks-on-west-vancouver-island/

116. http://www.montanaoutdoor.com/2014/03/wolf-attacks-and-drags-off-domestic-dog-in-broad-daylight/

117. http://www.hashilthsa.com/news/2015-11-26/people-take-matters-own-hands-wolf-attacks-pets-escalate

118. https://www.washingtonpost.com/news/morning-mix/wp/2016/06/17/a-starving-wolf-stalked-a-woman-and-her-dog-for-12-hours-then-along-came-a-bear/?tid=pm_national_pop_b

119. http://www.tbnewswatch.com/News/369495/Wolves_attack,_kill_family_dog_in_woman%E2%80%8099s_front_yard

120. http://www.winnipegfreepress.com/local/Declining-deer-numbers-to-blame-for-wolves-attacking-pets-experts-say-328921841.html

121. http://www.outdoorhub.com/news/2015/09/30/video-canadian-man-captures-wolf-attacking-his-dog-on-camera

122. http://www.thenownewspaper.com/national/391765151.html

123. http://www.infoaz.org/new/index.php/en/criminal/22913-dzolor-bladzk-wolves-attadzk-lodzals-in-azerbaizhan-s-azhshu-two-inzhured

124. http://www.dailymail.co.uk/news/article-2723818/Wolf-pack-attacks-Chinese-villagers-tearing-victim-s-ear-leaving-two-seriously-injured.html

125. http://www.connexionfrance.com/Wolves-sheep-Roquebilliere-Alpes-Maritimes-16848-view-article.html; http://www.thelocal.fr/20150415/wolves-go-on-killing-spree-in-french-village

126. http://www.telegraph.co.uk/news/worldnews/europe/france/11659510/French-farmers-son-left-terrified-after-wolf-attack.html

127. http://www.france24.com/en/20150902-french-farmers-take-park-boss-hostage-over-wolf-attacks-shepherds

128. http://www.bbc.com/news/science-environment-34510869

129. http://www.thelocal.fr/20150415/wolves-go-on-killing-spree-in-french-village

130. https://www.researchgate.net/publication/273902595_Characteristics_of_Gray_Wolf_Attacks_on_Humans_in_an_Altered_Landscape_in_the_West_of_Iran

131. http://siberiantimes.com/other/others/news/state-of-emergency-over-wolf-attacks-in-siberian-region/ and http://www.ibtimes.co.uk/russia-wolves-siberia-yakutia-wwf-423480

132. http://canislupus101.blogspot.com/2013/12/stories-from-siberia-locals-say-wolves.html
133. http://www.pravdareport.com/news/society/stories/11-03-2016/133786-wolves_attack-0/
134. http://metro.co.uk/2015/10/23/this-is-what-being-attacked-by-wolves-looks-like-some-readers-may-find-footage-upsetting-5459004/
135. http://westinstenv.org/wp-content/Geist_when-do-wolves-become-dangerous-to-humans.pdf
136. Linnell, et. al. 2002, Op. Cit.
137. http://wolfclash.com/
138. http://seattletimes.nwsource.com/html/localnews/2004087262_wolves22m.html?syndication=rss
139. http://www.nationalparkstraveler.com/2009/05/yellowstone-national-park-rangers-kill habituated-wolf
140. http://www.msnbc.msn.com/id/35913715/ns/us_news-life/
141. http://www.npwrc.usgs.gov/resource/mammals/wpop/results.htm
142. http://www.brainyquote.com/quotes/keywords/wolves_3.html

Photo credit: John L. Absher/Shutterstock.com

CHAPTER 4

. . .

Russian Wolves and American Wolves: No Real Difference

By Will N. Graves

Based on Russia's experience, the wolf is a different animal than we have been led to believe from North American books and movies, and the picture is not rosy.

—Will N. Graves

Recent Russian Experience

August 7, 2005: Authorities in Ukraine say a pack of wolves has attacked more than a dozen people in the country's eastern region.[1]

September 2, 2009: A rabid wolf has bitten six persons on the premises of the Chernobyl atomic power plant, four of them employees of SSE Chernobyl and two employees of the contractor organization UAB.[2]

January 31, 2012: A pack of wolves terrorized locals in the streets of a Karelian town, not returning to the woods until police opened fire, killing two. The wolves were reported by local residents from several locations in the city of Petrozavodsk, some 660 kilometers northwest of the capital Moscow.[3]

January 20, 2012: At least six people were attacked by wolves in southeastern Tajikistan in the last several weeks. A survivor of one of the wolf attacks, Ozodamoh Saidnurulloeva, told RFE/RL, "Before, the wolves only attacked livestock; now they have started attacking humans." Saidnurulloeva, 89, said she walked into her yard early one morning this week to prepare for the Muslim morning prayer when a wolf attacked her, knocked her down, and dragged her for several meters. She said she thought the wolf was the "angel of death." A neighbor, who happened to be outside, saw the attack and started screaming for help. The wolf then left Saidnurulloeva and ran away. Locals in the Sijd and

Rivak villages told RFE/RL they have pitchforks and shovels to fight wolves when necessary.[4]

Russians have lived with wolves for thousands of years. In contrast, contact between European settlers and wolves in North America for the entire US-Canada region has only been slightly more than three centuries and much less than a century and a half for many regions. A good deal of the folk tales, adages, fairy tales, and myths that warn us about wolves that have been passed down through many generations, and which are so often today discounted by pro-wolf advocates, originated from Russia. There is good reason for Russia to place so much interest in wolves. Russia has always had many wolves, but in the last century, the USSR and then the Russian Federation of States have had the largest population of wolves in the world, at least forty-five thousand on average, almost three times the wolf population of the lower forty-eight states.[5] In addition, the Russians have never had the widespread ownership of firearms that citizens of North America have always enjoyed.

Average numbers, however, do not tell the whole story. As we've learned in North America, wolves are intelligent, hardy, skillful hunters, and prolific, their populations quickly increasing when the opportunity presents itself. In Russia since the 1870s, there have been five periods when the wolf population swelled dramatically. During each population upswing, nature alone did not keep the wolf population in check. As their population increased and wild game became scarce, wolves came into more direct contact with people and livestock in their quest for food. Damage and danger quickly escalated. Each time, as the situation became critical, the government was forced to organize wolf control programs to intervene. Overall, approximately 1.5 million wolves have been killed by hunting, trapping, and poisoning during these population surges. Nonetheless, the current wolf population remains well above that which was present when the Bolsheviks overthrew the Czar in 1917 and established the USSR and its fifteen republics, the largest of which was the Russian Soviet Federated Socialist Republic (RSFSR), with its capital in Moscow.

Russia's longtime experience with wolves foreshadows what we can expect in North America as rapidly growing and expanding wolf populations come into contact for the first time with modern settlement patterns and big-game populations that have been carefully built up since the early twentieth century. Based on Russia's experiences, we know that the wolf is a different animal than we have been led to believe from North American books and movies, and the picture is not rosy.

First Period of Wolf Abundance

The first recorded period of wolf abundance in modern times in Russia was the 1870s in Czarist Russia. In 1873 in only forty-five provinces, wolves killed 179,000 adult cattle and 562,000 sheep and goats. During the Czarist period, peasants were not allowed to own weapons. Not only was firearm ownership banned, but also farmers were not allowed to own metal pitchforks; they had to be made from wood. As the wolf population soared, the government instituted a number of management programs, and the wolf population declined, but these programs soon were curtailed as the country was moving toward revolution and civil war, and men and weapons were needed for national security.

Second Period of Wolf Abundance

There was a large increase in the Russian wolf population during World War I (1914–18) and in the years immediately after. This was in part due to men of military service age being conscripted, leaving rural areas with little protection and few arms. In 1924–25, wolves killed about one million head of cattle, or 0.5 percent of all the cattle in the entire USSR. Especially hard-hit areas were the Volga area where 2.2 percent of all cattle were killed by wolves. In Siberia 1.6 percent of all cattle were killed by wolves; Kazakhstan lost 1.5 percent of all its cattle to wolves. As is the case in North America, such predation was not spread evenly across an entire area. Some farms and areas suffered much more than others. As the damage reached crisis proportions, the government mounted a campaign of hunting, trapping, and poisoning, and the wolf population dropped back again.

Third Period of Wolf Abundance

During World War II, which started in Europe in the early 1930s and ended in 1945, once again the Russian wolf population increased dramatically as men left communes to fight and few peasants in rural areas had firearms. Immediately following WWII, it was estimated that there were at least two hundred thousand wolves in every region in the USSR. In the RSFSR in the late 1940s, the wolf population was estimated at one hundred thousand, and these wolves were killing approximately two hundred thousand head of livestock a year. Not only were livestock and wild game severely reduced, but also wolf attacks on humans skyrocketed. In 1945, immediately after the war ended, wolf-hunting brigades using light aircraft, trucks, and jeep-like vehicles were formed, and bounties were offered. In 1946, sixty thousand wolves were killed. From 1947 to 1951, the annual kill was forty to fifty thousand.[6] Between 1947 and 1951, 240,000 wolves total were culled.[7] The wolf population did not significantly decline until the early 1960s when the annual kill dropped below twenty thousand. It continued to remain at this level through the mid-1970s. During this period, wolf damage to livestock dropped by a factor of ten.

As wolf numbers dropped, bounties were removed in many areas. The USSR publication of Farley Mowat's book *Never Cry Wolf* appealed to the growing urban Russian population, which had never had firsthand contact with wolves, and subsequently organized control programs ceased. From 1971–1975, the annual wolf cull throughout the entire USSR dropped to an average of fifteen to sixteen thousand per year.

Fourth Period of Wolf Abundance

Between 1973 and 1976, the wolf population of the USSR increased by fifty to sixty thousand, setting the stage for an explosion of wolves that began about 1978. In just five years, the wolf populations in Ukraine, Belarus, and central areas of the RSFSR doubled. In the April 25, 1978, issue of *Kommunist Tadzhikistana*, A. Chutkov, deputy director of the Tadzik Republic State Forest Committee's Administration on Hunting, Game Preserves, and Nature Preserves, wrote:

> Tadzhikistan's 10,000 wolves are harmful to agriculture, game animals, and privately owned livestock . . . The number of wolves is growing, as is the damage they do to the economy . . . The bounties should be increased to one hundred rubles for a mature wolf and sixty rubles for a cub, which are the going rates in the Russian and Ukraine Republics, and more airplanes and motor vehicles should be made available to hunters.

One consequence of the growing wolf population in this era was that the incidence of rabies in wolves increased six-fold between 1977 and 1979.

In 1978, the popular Soviet outdoor sports magazine *Hunting and Game Management* undertook a campaign to increase the wolf kill by creating a coalition of professional hunters, whose task it was to kill wild game and sell it for food; amateur hunters, of which there are about 3.5 million in Russia; and part-time professional hunters, who are largely farmers that hunt and trap in winter months. The total number of hunters in Russia has grown in recent years, but it still is small compared to the United States. A further limitation is that the amateur hunters, who must belong to a hunting collective like a club, may only possess and use shotguns. In response to growing wolf populations, increased livestock predation, and numerous attacks on people, the number of wolves culled in the USSR doubled in the 1980s as compared to the 1970s. In January 1981, A. Sitsko wrote: "In 1976 in the RSFSR, there were 7,000 wolves culled; in 1980 there were more than 15,900 culled."[8] In 1985,

38,900 wolves were culled throughout the entire USSR. Again, when culling was increased, damage associated with wolves decreased.

Fifth Period of Wolf Abundance

When the USSR collapsed in 1991, the government began a process of reorganization. Wolves slipped into the background of governmental attention, and quietly they began to increase in numbers. From 1987 to 1997, the wolf population in the Russian Federation alone rose from 29,000 to 42,000. One consequence of this latest wolf population explosion is that between 1988 and 1997, the moose population decreased 33 percent—from 904,000 to 604,000. In the same time period, the number of moose bagged by hunters dropped from 66,000 to 22,000.[9] During this era, the reindeer population in Anadyr in northeast Siberia dropped from 500,000 to 150,000, while in the Koryak Autonomous Area of the Kamchatka Peninsula, the wolf population increased from 150 to 800, and in 1998 wolves killed 10,000 reindeer.[10] Most reindeer herders are not authorized to have rifles, so they have little way to control hungry wolves. The native peoples in these areas were heavily impacted by the burgeoning wolf population of this period.

Meanwhile, in Kyrgyzstan during the same period, the wolf population increased three-fold, resulting in increased predation on livestock and attacks on people. In 2003 it was written, "Officials say attacks on people and domestic animals have increased overall, almost doubling in the last nine months. One man died after being bitten."[11]

Russia's long experience with wolves provides us with a wealth of information that is especially valuable as North American people experiment with reintroduction of wolves into areas with both an abundance of big game and human populations that are denser than at any other time in history.

A wolf is not afraid of a dog, but is afraid of a weapon.

—Russian proverb

Let us briefly look at some of what the Russians have found from their lengthy experience of living with wolves and coping with the five population peaks.

Wolves as Disease Vectors

Russian scientists provide ample evidence that wolves carry many diseases including some that can be dangerous to humans and to livestock, such as distemper, brucellosis, hoof and mouth disease, mange, and of course rabies. The wolf is a major host of rabies in Russia, Iran, Afghanistan, Iraq, and India. It is well known that victims bitten by animals carrying rabies face a 100 percent chance of dying unless they receive treatment before symptoms appear. Rabies is primarily transferred by a rabid animal biting another. It causes a dramatic change in the behavior of infected animals. Rabid wolves become extremely agitated and aggressive, deserting their packs and traveling as much as eighty kilometers in a day. On such trips they lose all fear of humans and attack anything in their path. In 1957, a rabid wolf in Belorussia traveled 150 kilometers in one and a half days. During that time, it bit twenty-five people, about twenty-five livestock, and an unknown number of wild animals. Rabid wolves tend to aim for the head and face during their attack. In 1976 to 1980, there were thirty registered cases of wolf rabies and thirty-six attacks on humans by rabid wolves in the Russian Kazakh and Georgian Soviet Republics; there were additional cases reported in the Russian Federation (RF). Between 1972 and 1978 in the Aktyubinsk region of Kazakhstan, wolves attacked fifty people, and thirty-three suffered bites from rabid wolves. More often than not, wolves attacked agricultural workers in the fields.

According to the World Health Organization, there were thirty cases of wolf rabies in Turkey from 1977 to 1985: four in Germany, five in Poland, one in Romania, and one in Poland. Foxes, raccoons, and stray dogs, as well as wolves, are the primary species for transmitting rabies in Eurasia. From 1990 to 1997, there were seventy-six cases of human deaths from rabies in thirty administrative territories of the RF.[12] In 1998, there were 2,868 cases of rabies registered in the RF.[13] In North America, we can also add bats and coyotes, as well as the various hybrid wild canids—wolf-dogs, coydogs, and coy-wolves that are increasingly being found in North America.

In addition to rabies, wolves carry over fifty types of parasites that cause dangerous diseases such as echinococcosis, cysticercosis, coenurosis, trichinosis, tularemia, listeriosis, and ascariasis. They also can carry and spread diseases that affect livestock including anthrax, *neosporum caninum*, and hoof and mouth disease, although they themselves are immune to these diseases. Numerous studies have found that the majority of wolves carry at least one type and often several types of tapeworms. A Soviet scientist researching wolves in the extreme northeast of Siberia found that, in areas with high wolf populations, there was also a high incidence of brucellosis (which can be carried by wolves) in reindeer, which are a primary source of food and income for people in these areas.[14] (For more about wolves as vectors for zoonotic diseases, see chapter 14: The Wolf as a Cash Cow.)

Attacks on People

In his 1982 definitive book, *The Wolf*, M.P. Pavlov writes, "Facts about attacks of wolves on people are not myths but reality . . . This danger is increased by the appearance of wolf-dog hybrids."[15] There are countless documented cases of wolves attacking people in Russia, especially children. The attacks are by both rabid and non-rabid wolves. Following are some examples:

According to the Russian Police Department, from 1849 to 1851, wolves in the Russian Empire killed 376 people, including 266 children.

From 1896 to 1897, there were 205 attacks by wolves on humans in the Vyatsjij Province of Russia, ten attacks on people in Vologodskij, eighteen attacks in Kostromskij, one in Arkhangelskij, and nine in Yaroslavskij.

Between 1886 and 1901, in Odessa, Ukraine, there were more than 150 cases of people being attacked and bitten by rabid wolves.[16]

In 1924, two rabid wolves attacked and bit twenty people in Kirov; ten died.

In the Vladimirskij Region from 1945 to 1947, there were ten attacks by wolves on humans. All wolves were shot, and they weighed at least 60 kilograms each and were healthy.

A 1947 report by Dr. P. A. Mantejfel that summarizes non-rabid wolf attacks on humans in the USSR finds that, in recent years, there were twelve documented cases of non-rabid wolves killing and eating eighty humans.

In 1957, it was reported that, in Czarist Russia from 1849 to 1851, there were 266 adults and 110 children killed by wolves. In 1875, 160 people were reported killed by wolves.

In 1975–76, rabid wolves attacked fifteen people in Ulyanovskij, seven in Kaluzhskij, six in Orenburgskij, and four in the Orlovskij Region.

On August 21, 1978, near the village of Styazhnoe, Bryansk Oblast, a wolf attacked and bit six people. After a hunter killed the wolf, its body was examined, and it was found to be healthy.

In August 2005 in a three-day period, wolves attacked fourteen people in the eastern part of Ukraine in the Zaporizhia Region. One wolf that was killed by a car was found to be rabid.

> "If wolves are howling near a village in winter, it portends hunger and high prices for food."
>
> —Russian saying

Wolf Predation and Russian Damage

Russian wolves prey on three principle wild game species: moose in the forests, elk or wapiti in more open areas, and reindeer on the tundra. Moose and elk are wild game species. A significant number of reindeer have been domesticated and represent a very important source of food for people living in the northern regions. A five-year study in the region north of Moscow found that, between 1970 and 1975, 82.4 percent of the food of wolves was moose. This research also found that bears and wolves were not in competition for food.[17]

Russian research on wolf predation also contradicts the widely touted claim by biologists in the United States and Canada that wolves are "nature's sanitarians." Numerous studies show that wolves do not just attack the weak and the old animals, thus making the herd healthier. Actually, wolves seem to prefer healthy wild ungulates, especially the young and the female. As part of the five-year study, the bodies of sixty-three moose killed by wolves were examined. Only nine of the sixty-three total had any defects or deviations from the normal. Five had defects with their teeth; three had defects with their antlers; and one had defects with its hooves. During this same study in February 1971, it was noticed that wolves would approach some moose but would not attack them. The moose they did not attack were later found to have ringworm. Of the three moose that died from ringworm, wolves would not feed on any of them.

Numerous studies in Russia report that wolves engage in surplus killings. One of the most dramatic incidents was a three-week period in February 1978 when wolves embarked on a mass killing of young Caspian seals on the ice of the Caspian Sea near Astrakhan. It is estimated that wolves killed between 17 percent and 40 percent of all the young seals in the area and left their bodies strewn on the ice with few eaten. Large, adult seals were also killed and not eaten.[18]

It's more difficult to count the exact numbers of wild game killed by wolves, though reindeer herders in various regions have lost thousands of their semi-domesticated stock to wolves over many years. For example, between 1945 and 1954, about 75,000 reindeer were lost to wolves in northern regions. In a number of documented cases, wolves killed as many as thirty reindeer in one attack, leaving most of them uneaten. V. Glusjkny reports that from 1967–72 in the Vyatskij forests, every year wolves killed about 28.6 percent of all moose, while grizzly and black bears accounted for only 4.7 percent.

When wild species decline, then wolves turn to livestock, which occurs when either a food shortage or severe weather causes wildlife population to decline or the wolf population increases. Wolf predation on domestic animals in Russia has at times, especially during the five peak periods, been extreme. Following are some examples:

In 1944 in the Buryatskij region ASSR, wolves killed over 5,300 cattle; in the Penzenskij region, 8,700 cattle; in the Kujbyshevskij region, 4,200 cattle; and in the Tambovskij region, 8,000 cattle.[19]

In 1987, there were approximately sixty-two thousand wolves in Kazakhstan, which were causing damage to both livestock and game animals. In the first half of 1988, wolves killed 38,719 domestic animals. The Central Administration of Hunting organized three hundred teams of professional hunters to cull wolf numbers. This was in addition to other private and professional hunters, who were also culling wolves. However, their combined effort was not sufficient to hold back increases in the number of wolves. In 1992, there were approximately ninety thousand wolves, and there was a tendency for the numbers of wolves to increase. In October 1994, at the International Symposium in Poland titled "Large Predators and Man," a Russian scientist reported that presently there were probably a hundred thousand wolves in Kazakhstan.[20]

By 1995, the damage by wolves in just one region of the Russian Federation, the Altai Region, was estimated to be about ten million rubles per year. It was also reported: "There are sounds of a calamity coming."[21]

In the Sakha Republic, formerly the Yakutsk Republic, it was reported in 2013 that the population of wolves was about 3,500. The Veterinarian Department of the Ministry of Agriculture reported that wolves killed 9,995 reindeer in 2000, 8,869 in 2001, and 9,540 in 2002. In 2012, wolves killed 313 horses and 16,111 reindeer—the wolves make no distinction between a reindeer that is sick or healthy. The estimated value of one reindeer is $328, which adds up to a loss of about five million dollars to reindeer herders; 750 wolves have been recently culled. The governor of Sakha, Yegor Borisov, said in January 2012 that the goal of Sakha is to reduce the wolf population from 3,500 to five hundred. As an incentive for hunters and herders to cull wolves, he has placed a bounty of about $630 per wolf pelt. Vladimir Krever, the Russian World Wildlife Federation's Head of the State Committee Biodiversity Program, added: "When wolves start attacking deer and livestock, they have to be killed and the population controlled. This is the right policy." He added: "Even if they were able to kill 3,000 wolves, the population would recover quickly."[22]

Conclusion

To summarize, since 1870 in Russia, the area of the world with the most wolves, during five historical periods when governments reduced support for wolf culling, the wolf population expanded rapidly. During these population peaks, wolves significantly reduced wild ungulates, and when wild game became scarce, the wolves turned their predatory attention to domestic animals. The wolves grew bolder and bolder, culminating in wolves entering villages and preying on dogs, cats, sheep, goats, geese, and large livestock. In many cases, wolves killed far more animals than they needed for food. The wolf predation overall caused considerable economic damage, as well as loss of food. Attacks of rabid and non-rabid wolves on people also escalated, as did incidence of diseases carried by wolves.

The wolf population was not controlled by some "balance of nature." Population explosions could only be controlled by significant human intervention. As ownership of firearms was and still is primarily limited to the government troops, law enforcement, and a relatively small number of the upper class, when wolves did enter villages and communes, they met with little resistance. The interventions had to be supported by the government, as the general populace did not have the resources to undertake large-scale culling.

The lessons for North Americans to take from the Russian experience with wolves are many. What Russia's experience tells us about wolves is that, while wolf behavior is always unpredictable, the overall patterns of wolf behavior adapting to modern settlements follow Dr. Geist's model of habituation and have consequences that are not pleasant for man or animal, wild or domestic.

"Ivan, didn't the farmers around Gorky tell you it is the wolf you do not see that you must fear."
—Tom Clancy, *The Hunt for Red October*[23]

Endnotes:

1. http://www.freerepublic.com/focus/f-news/1459944/posts
2. http://www.rense.com/general87/rabid.htm
3. http://en.rian.ru/strange/20120131/171045959.html
4. http://www.rferl.org/content/tajikistan_wolf_attacks/24457676.html
5. http://www.bbc.co.uk/nature/animals/conservation/wolves/intro. shtml?oo=15732
6. Biblikov, D., "The Wolf," Science, 1985; Biblikov, D., *Hunting and Game Management*, July 1978.
7. Hunting Management in the USSR. Moscow: Forest Industry, 1973: 196.
8. Sitsko, A., *Hunting and Game Management*, January 1981: 1–2.
9. Utilin, Dr. A., President of the Central Board of the All Russian Association of Hunters and Fishermen, Conservation Force Web Page, 2001.
10. Andrei Ivanov, IPS World News, March 26, 1998, Internet.
11. Davis, Catherine, Sky Wolf News, BBC Central Asia Correspondent, January 27, 2003.
12. Bectimirov, T., and A. Movsesyants, "Epidemic and Epizootic Rabies Situation in the Russian Federation," Moscow: L.A. Tarasevich State Institute, Moscow. February 9, 2003.
13. Vedernikov, V.A., V.A. Sedov, et. al., Moscow: Institute of Experimental Veterinary Medicine. Internet, October 8, 2003.
14. Zheleznov, N.K., Wild Ungulates of the USSR North-East, Vladivostok: Academy of Science 9f of the USSR, 1990: 411.
15. Pavlov, M.P., *The Wolf*. Moscow: Forest Industry, 1982: 109, 129.
16. Milenushkin, Y.I., *Wolves and Their Destruction*. Moscow: Military Press, 1957: 103.
17. Moskvin, N., "The Wolf of Mologo-Sheksninskij Mezhdurech," *Hunting and Game Management*, February 1978: 12–13.
18. Rumyantsev, V., and L. Khuraskin, "Wolves and Caspian Seals," *Hunting and Game Management*, February 1978: 22–24.

19. Makridin, V., "Do Not Cultivate Wolves, But Develop the Economy," *Hunting and Game Management*, August 1968, p. 18–19.

20. International Symposium on "Large Predators and Man," Poland, October 1994, attended personally by Will Graves. Dr. Mech was also there.

21. Internet, Pravda, March 12, 2003.

22. Internet, "Wolves: An Emergency Situation," January 23, 2013; also recent email exchange between Will Graves and a resident of Sakha.

23. Clancy, Tom. *The Hunt for Red October*. New York: Berkeley Books (1984): 6.

CHAPTER 5

· · ·

The Seven Steps
of Wolf Habituation

By Valerius Geist, PhD

*"There is another very important point made by native people
and routinely ignored by officialdom—namely, wolves eat the evidence
as well as distribute it widely or bury it."*

—Valerius Geist, PhD

Four Decades of Field Research

Nothing convinces like personal experience! As an academic, I confess to this
with some distress because—by training, experience, and attitude—I should
be above it. That I am not alone in this habit is of little comfort. And so it is
with wolves.

In my four decades of field research on ungulates in North America—
mountain sheep, goats, elk, deer, moose, etc.—I also observed wolves, and my
experience with continental wolves matches that of my colleagues; namely,
that wolves are intelligent and wary, and keep their distance from people.
Consequently, throughout my academic career and four years into retirement,
I thought of wolves as harmless, echoing the words of more experienced North
American colleagues, while considering the reports to the contrary from Russia,
Europe, and Scandinavia as interesting, but not relevant to an understanding of
North American wolves. I was wrong!

Wolves up Close and Personal

I saw my first North American wolf in the wild early one morning in May
1959 on Pyramid Mountain in Wells Gray Provincial Park, British Columbia.
In the spotting scope, his image was crisp and clear. It was an ash-gray wolf

with a motley coat, sitting and watching me from a quarter mile away with an eager, attentive look about his dark face. His red tongue was protruding, while golden-yellow morning light played on his fur. I do not know if my heart skipped a beat upon sighting that wolf, but it well might have. Whose wouldn't?

About five months prior, in early January, I had had an informative brush with a wolf pack just a few miles from that spot. A friend and I were observing moose. We were in the midst of a migration, and some two dozen moose, mostly bulls who had shed antlers, were dispersed over the huge burn. A few were feeding on the tall willows, but most were resting in the knee-deep snow. Suddenly we heard a low, drawn-out moan. When I glanced at the moose, I saw that all were standing alert, facing down the valley. We were green then and perplexed about this unearthly sound. As if to answer us, a high-pitched voice broke in, and then another and another, and we realized that we were hearing wolves. Within minutes, a chorus was underway, and so were the moose. All were hastily moving up the valley, and ten minutes later, the moose had vanished. I opted to stay at our lookout while my friend, taking my rifle, decided to search for the wolves. He saw them at dusk as they walked across a small lake, a pack of seven. Try as he might to fire the rifle, it would not fire as it had frozen in the severe cold. This may have been kind fortune, for the first wolf I shot with that rifle instantly attacked me, although it collapsed before reaching me. The second screamed, which has triggered pack attacks in the past. Had the pack attacked, I would have been minus a friend in minutes. While a large man can subdue an attacking wolf, even strangle it, there is no defense against an attacking pack.

Two years later during my study of Stone's sheep in northern British Columbia, I had exceptional opportunities to observe wolves in pristine wilderness. My closest neighbors, a trapper family, lived some forty miles to the west, and the closest settlement, Telegraph Creek, was about eighty miles to the north. The Timberlines were low, and the wolves spent much time in the open, plainly visible. I watched them for hours on end. These large, painfully shy wolves on occasion even panicked over my scent. Though they killed a few sheep, their hunts were largely unsuccessful. However, I began to appreciate their strategies and tenacity as hunters. In traversing the valley, one crossed a wolf track about every fifty paces. They were scouring the valley for moose.

On rare occasions, a wolf would follow my tracks, sit, and listen to what I was doing in my cabin at night. (Grizzly bears did that, too.) One evening three wolves began to surround me on a frozen lake. One raced towards me, but scrambled madly to get away once he got downwind of me. Another cut my fresh track, then jumped straight up, and raced back. Thus, my early experiences with mainland wolves indicated that they were inquisitive, but also shy and cautious. Moreover, my contacts with wolves were few compared to the huge number of Osborn caribou. I then thought that this was normal. Years later, a first doubt arose when a student of mine could hardly find a caribou where I had seen hundreds. However, a wolf pack of forty-three individuals was recorded where I had observed for years a pack of only seven.

Evidently, my experiences with wolves were anomalous for, a decade earlier, there had been massively broadcast poisonings of wolves to control rabies. The "pristine wilderness" had been tampered with; it had experienced a "rebound" of ungulate populations after they had been freed from severe predation. When my wife and I tell of forests of antlers as caribou bulls gathered on the Spazisi Plateau for the rut, colleagues look at us as if we came from another age.

Threatening Wolves

Nothing in my previous studies had prepared me for what I was to experience with wolves on Vancouver Island following my retirement in 1995 from the University of Calgary, where I had served as Program Director and Professor of Environmental Sciences for many years. In my student days in the late 1950s, wolves on Vancouver Island were so scarce that some thought they were extinct. In the early 1970s, they reappeared and swept the island. The annual hunter kill of black-tailed deer dropped swiftly from about twenty-five thousand then to less than three thousand today. There were reported incidents of wolves threatening people, and a colleague was treed by a pack. When he spoke of this, nobody believed it, so he clamped up. Wolves threatening people? Ridiculous! According to my colleagues, massive clear-cutting of old-growth forests and the rapid spread and growth of the wolf population caused the carnage to ungulates. Those who witnessed it tell of deer carcasses everywhere—and then no more deer. The loggers left standing small patches of mature timber as deer winter range. However, wolves, cougars, and black bears discovered those patches and cleared out the remaining deer. The clear-cuts also led to a population explosion of black bears, some of whom became experts in killing elk calves and deer fawns.

Deer are still so rare in the mountains of the island that I see about three dozen bears for every deer. However, deer are common in towns and suburbs and about farms, where they are relatively safe, at least from wolves. The elk population is holding its own, but at a low level compared to the vast amounts of food on the clear cuts. The bulls are huge, with massive antlers, but with a predator-induced silence during the rut. Enough calves perish so that there is little recruitment, and hunters are held to one permit per forty to 150 applicants.

I retired to an agricultural area on Vancouver Island. During walks near our home, I explored through all seasons a meadow system associated with dairy, beef, and sheep farming. These meadows and adjacent forests contained, year-round, about 120 black-tailed deer and half a dozen large male black bears. In winter came some sixty to eighty trumpeter swans, as well as large flocks of Canada geese, widgeons, mallards, and green-winged teal. Pheasants and ruffed grouse were also not uncommon. In the fall of 1995, I saw one track of a lone wolf. I cannot recall seeing any wolf tracks in the four years following.

Empty Landscape

Then in January 1999, my oldest son Karl and I tracked a pair of wolves in the snow, suggesting a breeding pair and thus pack formation. A pack did indeed arrive that summer. Within three months, not a deer was to be seen, or tracked, in these meadows—even during the rut. Using powerful lights, we saw deer at night huddling against barns and houses, where deer had not been seen previously. For the first time, deer moved into our garden and around our house, and the damage to our fruit trees and roses skyrocketed. The trumpeter swans left. The geese and ducks avoided the outer meadows and lived only close to the barns. Pheasants and ruffed grouse vanished. The landscape looked empty, as if vacuumed of wildlife.

Wolves Become Aggressive

Wolves attacked and killed or injured dogs, at times right beside their shouting, gesticulating owners. Wolves began following our neighbors when they rode out on horseback. A duck hunter shot one wolf and wounded fatally another as three wolves attacked his dog. Wolves ventured into gardens and under verandas trying to get at dogs and ran after quads, tractors, and motorcycles to attack the accompanying farm dogs. My neighbor warded off three such attacks on his dogs with his boots, and his hired man ran back with a tractor in panic with the wolves in pursuit of the two dogs that tried to stay under the running tractor. One wolf approached within about fifteen paces of my wife and a group of eleven visitors that were taking an evening stroll about half a mile from our house. The wolf howled and barked at the people. Our neighbor then went out armed with his dogs, and the wolf, a small female, promptly attacked the dogs and was shot at fifty feet. Nine days later, my neighbor killed a second wolf that was openly following and barking at him—this wolf may have been defending a sheep it had killed and dragged half a mile. These wolves weighed between sixty and seventy pounds, small for wolves, a sign of poor nutrition.

Wolves of the "misbehaving" pack began to be killed as they persisted in attacking livestock and pets and aggressively approaching people. Wolves had been seen in the neighborhood sitting and observing people; we know from captivity studies that wolves are observation learners. One wolf, a male, approached my wife, my brother-in-law, and me across a quarter mile of open meadow and stood looking us over for a very long minute about ten paces away, before moving on into the forest. Neighbors, myself included, repeatedly saw wolves showing interest in us.

A neighbor raising sheep lost many to wolves, so he acquired five large, sheep-guarding dogs. These dogs and the wolf pack engaged in a "dear enemy" ritual of frequent, nightlong barking and howling duels at the forest edge. I observed subsequently, on the evening of October 19, 2002, how the last of the pack, a male, fraternized successfully with the sheep dogs. He kept it up and was eventually shot March 12, 2003, while sitting amongst these dogs. However, before that, he visited us when our German longhair pointer female, Susu, was in heat and barked at my wife in our doorway; that is, he acted like other male dogs that were attracted to Susu in heat, only bolder.

There was a quiet spell of two years after the first wolf pack had been killed. Migratory waterfowl began to reappear, but only half of the swans returned. Then on March 27, 2007, a second "misbehaving" pack appeared and took up where the first pack had left off.

The worst incident happened about 350 yards from our house. Our neighbors went in the morning to inspect their dairy cattle and pastures. Their old dog ran ahead of them. Just as they entered the forest, five wolves attacked the dog. My neighbor grabbed a cedar branch and advanced on the wolves, who turned towards him, snarling. His wife jumped into the caboose of their excavator, which was parked there. My neighbor's energetic counterattack freed the dog and intimidated all but one wolf, who advanced on him, snarling. However, that wolf too withdrew, albeit reluctantly. While my neighbor ran home to get a gun, his wife ran to us, shouting for me to get a rifle. We did not see the wolves again, though they were sighted briefly in the evening, and a neighbor walking his dog had an encounter with two wolves about a mile away; he was able to chase them away. The following morning, our neighbors took a rifle along during their inspection trip of their property. The wolf pack promptly went for them again, and my neighbor shot the most aggressive one, a male weighing seventy-four pounds.

From Misbehaving to Murderous

I saw the neighbors' cattle, spooked by a wolf, crash through fences while fleeing for the security of their barn. I found two of the three cattle killed and eaten by wolves; the third was severely injured about the genitals, udder, and haunches and had to be put down. I saw the docked tails, slit ears, and wounded hocks on the dairy cows. Our neighbor's hired man saw from a barn a wolf attacking a heifer with a newborn calf. He raced out and put the calf on his quad. As he ran to the barn, the wolf ran alongside, lunging at the calf, right into the barn. A predator control officer was called, and thirteen wolves were removed within a mile of our house from the first "misbehaving" pack and four from the second "misbehaving" pack.

The "tameness," "hanging around," and increasing boldness and inquisitiveness are the wolf's way of exploring its potential prey and assessing the strength of its potential enemies. These incidents with the two "misbehaving" packs moved me to look for other situations where patterns of habituation might be found. I did not have to look far. Two wolves in June 2000 severely injured a camper on Vargas Island just off the coast of Vancouver Island. These wolves were tame before the attack—they nipped at the clothing of campers, licked their exposed skin, and ate wieners from their hands.

Our observations here suggested that wolves, attracted to habitations by the scarcity of prey, shift to dogs and livestock for prey, but also increasingly, though cautiously, explore humans, before mounting a first, clumsy attack. I reported such at a Wildlife Society conference on September 27, 2005, in Madison, Wisconsin, in an invited paper on habituation of wildlife. That was about six weeks before wolves in northern Saskatchewan killed Kenton Carnegie.

On November 8, 2005, four wolves at Points North Landing, Wollaston Lake area in northern Saskatchewan, killed a twenty-two-year-old, third-year geological engineering student at the University of Waterloo by the name of Kenton Joel Carnegie. This case is unique in that it is the first direct human fatality from a wolf attack in North America in recent times to receive a thorough investigation.[1] There have been people bitten by rabid wolves and killed, but such kills "do not count" as it is the rabies virus, not the wolf bite, that kills.

Todd Svarckopf confronting a wolf in the same area where, four days later, Kenton Joel Carnegie was killed by wolves.

I was asked by Kenton's family to investigate matters as were—independently—Marc McNay from Alaska, who ultimately served as an expert witness in the coroner's investigation, and Brent Patterson from Ontario. Our findings—independently conducted and assisted by other colleagues from Alaska and Finland—came to the same conclusions as did the earlier investigations by a native coroner, Mrs. Rosalie Tsannie-Burseth, and a native Royal Canadian Mounted Police, Constable Alfonse Noey, who attended the scene within hours of Kenton's death. Both had been raised in the Saskatchewan Wilderness and were highly experienced with native wildlife.

Mrs. Burseth was not only the coroner, but also chief of the Hatchet Lake band and director of education; she held two university degrees and was working on a doctorate. However, Constable Noey's and her thorough investigations were discounted by the Saskatchewan coroner. The two scientists called in by the Saskatchewan coroner, who examined the photography of Constable Noey, mistook the wolf tracks across an overflow on the frozen lake for bear tracks and proclaimed a bear as a killer. The subsequent scandal led to a coroner's hearing, which concluded that wolves had killed Kenton Carnegie.

Unfortunately, this coroner's inquiry had a narrow focus and did not deal with policy. Consequently, it never became public that Saskatchewan's legislation was, in good part, to blame for Kenton's death and that, had he been in British Columbia, Kenton would have lived. In Saskatchewan, wolves can be killed by trappers or wardens or by special permit only. In British Columbia, any licensed hunter can shoot garbage-habituating wolves virtually throughout the year.

Fresh snow allowed accurate track reading that Mrs. Burseth was eminently qualified in, raised as she was by her father as a hunter well before her formal schooling in her teens. Constable Noey is also an experienced Northern hunter. Kenton was by himself when the wolves, both from behind and from the front, approached him. He fell three times before failing to rise. Constable Noey fired three shots to spook the wolves from Kenton's body and posted an armed guard—the husband of Mrs. Burseth and a work colleague, both experienced hunters and track readers—while he and Mrs. Burseth examined and photographed the kill and feeding sites.

We were aware that the four wolves in question, long observed by others, were garbage fed and were photographed four days earlier while attacking two employees of the camp, who beat back the wolves. Kenton was aware of this encounter. Unfortunately, neither he nor those who discussed the matter with him, as reported by Constable Noey's report and by reporters of the *Saskatoon Star Phoenix* of November 14, 2005, were aware that tame and inquisitive wolves are a signal of danger. Consequently, the first requirement is that the general public and especially outdoorsmen know that when they see tame, inquisitive wolves, they should get out of the area quickly, but without undue haste, while being prepared to defend themselves. Running away invites an attack.

Additional Documented Wolf Attacks

As Mark McNay and others have established, there have been other attacks in Canada, both historical and more recent.[2] Five-year-old Marc Leblond was killed by wolves on September 24, 1963, north of Baie-Comeau, Quebec, as indicated by tracks and signs in the snow and the sighting close by of a wolf reluctant to leave, but his case drew little media attention. On April 18, 1996, twenty-four-year-old Patricia Wyman was attacked and killed by five adult North American gray wolves (*Canis lupus*) at the Haliburton Forest and Wildlife Reserve near Haliburton, Ontario. And on December 31, 2004, Fred Desjarlais was attacked and wounded by a wolf in northern Saskatchewan.[3]

There are also plenty of unreported recent attacks by wolves like this one in Saskatchewan, not all fatal and not reported in any central formal wolf attack research center. A local rancher was attacked by three wolves while deer hunting; he killed two. Another attack happened on July 5, 2007, on Anderson Island off Aristazabal Island, north of Bella Bella, British Columbia. The man was a fit thirty-one years old. He was put into the hospital by an emaciated, old she-wolf with broken teeth, even though the man stabbed her nine times. She was killed by a shotgun blast two hours later; there was no sign of rabies. Candice Berner, a thirty-two-year-old schoolteacher, was killed on March 8, 2010, by wolves in the village of Chignik Lake on the Alaska Peninsula.

The Problem of Evidence

There is another very important point made frequently by native people and routinely ignored by officialdom—namely, wolves eat evidence as well as distribute it widely or bury it. Had the search parties looking for Kenton Carnegie not disrupted twice the wolves feeding on his body, very little of the body would have been left by the next morning. Within five hours, two or three wolves had consumed a very large portion of Kenton's body. Little wonder that only about one in seven calves killed by ranch-visiting wolves are ever attributed to wolves.

It is important to note that wolves learn differently than dogs. Wolves learn by observing, and are insight learners; that is, they can solve problems by observing, such as how to unlock a gate. In some studies of captive wolves, researchers have found that wolves, and coyotes for that matter, not only learn to open their own cages, but also those of others. With these intelligence traits, wolves also develop an ability to assess the vulnerability of prey. For example, the sight of a human—walking boldly and carrying a firearm—will give them enough information to know that the potential prey is not vulnerable.

Seven Stages of Habituation

By studying the two "misbehaving" packs of Vancouver Island wolves and comparing this data with data obtained from scientists in Russia, Scandinavia, Europe, and the Middle East, a seven-stage model of habituation has emerged that progressively leads wolves from shy, wild animals to those that begin targeting people as prey. What we know is that, when wolves run out of wild prey, they first begin targeting livestock and then pets and finally exploring people as prey. The presence of garbage also helps draw wolves closer to people. It should be noted here that the considerable number of attacks by wolves in Europe and Asia is at least partially due to the absence of firearms in the hands of the citizenry. Distilled from my own observations and reviews of research in North America and abroad, these are the seven stages of habituation leading to an attack on people by wolves.

- Wolves move in closer to people. Within the pack's territory, prey is becoming scarce—not only due to increased predation on prey animals, but also by the prey evacuating home ranges en masse, leading to a virtual absence of prey. Alternatively, wolves increasingly visit garbage dumps at night. We observed the former in the summer and fall of 1999. Deer left the meadow systems occupied by wolves and entered boldly into suburbs and farm, causing—for the first time—much damage to gardens and sleeping at night close to barns and houses, which they had not done in the previous four years. The wintering grounds of trumpeter swans, Canada geese, and several species of ducks were vacated. The virtual absence of wildlife in the landscape was striking.
- Wolves begin to approach human habitations at night in search of food. In our experience, their presence was announced by frequent and loud barking of farm dogs. A pack of sheep-guarding dogs raced out each evening to confront the wolf pack, resulting in extended barking duels at night. The wolves were heard howling even during the day.
- Wolves appear in daylight and, at some distance, observe people doing their daily chores—wolves excel at learning by close, steady observation. Wolves approach buildings during daylight.
- Wolves act distinctly bolder in their actions. Small-bodied livestock and pets are attacked close to buildings even during the day. Wolves preferentially pick on dogs and follow them right up to the verandas. People out with dogs find themselves defending their dogs against a

wolf or several wolves. Such attacks are hesitant, and people save some dogs. At this stage, wolves do not focus on humans, but attack pets and smaller livestock with determination. However, they may threaten humans with teeth exposed and growl when humans are defending dogs or show up close to a female dog in heat or close to a kill or carrion defended by wolves. The wolves are still establishing territory at this stage.

- Wolves explore large livestock, leading to docked tails, slit ears, and damaged hocks. Livestock may bolt through fences, running for the safety of barns. The first seriously wounded cattle are found; they tend to have severe injuries to the udders, groin, and sexual organs and need to be put down. The actions of wolves become more brazen, and cattle or horses may be killed close to houses and barns where the cattle or horses were trying to find refuge. Wolves may follow riders and surround them. They may mount verandas and look into windows.
- Wolves turn their attention to people and approach closely, initially merely examining them carefully for several minutes on end. This is a switch from establishing territory to targeting people as prey. The wolves may make hesitant, almost playful attacks—biting and tearing clothing and nipping at limbs and torso. They withdraw when confronted. They defend kills by moving towards people, growling and barking at them from ten to twenty paces away.
- Wolves attack people. Initial attacks are clumsy, as the wolves have not yet learned how to take down efficiently the new prey. Persons attacked can often escape because of this clumsiness. A mature, courageous man may beat off or strangulate an attacking wolf. However, against a wolf pack, there is no defense. Even two able and armed men may be killed. Wolves as a pack are hunters so capable a predator that they may take down black bears and even grizzly bears. Wolves may defend kills.[4] Attacks on people may not be motivated by predation, but rather may be a matter of more detailed exploration unmotivated by hunger. This explains why wolves on occasion carry away living, resisting children; why they do not invariably feed on the humans they killed, but may abandon such, just as they may kill foxes and just leave them; and why injuries to an attacked person may at times be surprisingly light, granted the strength of a wolf's jaw and its potential shearing power.

Upon presenting this model at professional meetings and sharing it with other wildlife biologists, I met Dr. Robert Timm at the University of California at Davis, who has been studying coyote attacks. It turned out that coyotes targeting children in urban parks act in virtually the same manner.[5]

Peer Review

Peer review is essential to science. My work has been translated into Swedish, Finnish, and German. Then a review of the Russian wolf experiences by Professor Christian Stubbe in Germany and Will Graves's book *Wolves in Russia* added more support. Italian and French historians have also published papers and books detailing how thousands of people had died in earlier centuries from wolf attacks.

Some historians rightly asked the question, "How did North American scientists ever conclude that wolves were harmless and no threat to people?" We now know the answer: in the absence of personal experience or sound language competence, they chose to disregard, even ridicule, the accumulated experience of others from Russia, France, Italy, Germany, Finland, Sweden, Iran, Kazakhstan, India, Afghanistan, South Korea, and Japan. They were also unaware that, during most of the twentieth century, tens of thousands of trappers in the heartland of wolf distribution in Canada and Alaska were killing every wolf they could, legally and illegally. Moreover they were encouraged to do so by bounties, while concomitantly predator control officers removed wolves, aerial poisoning and shooting campaigns were carried out, and wolves were free to be killed by anybody wishing to do so. Little wonder that wolves were scarce and very shy, attacks on people unheard of, livestock losses minimal, and wolf-borne diseases virtually unnoticed.

Conclusion

The argument—that there is little danger from wolves because they have rarely attacked humans in North America—is fallacious. There are very good reasons why wolves in North America, as opposed to Europe, have attacked people rarely. In North America in the 1700s and 1800s, wolves were killed off and driven into wild places. In their absence in the past decades, we have experienced in North America a unique situation: a land with a recovery of wildlife. Few North Americans are aware today that a century ago North America's wildlife was largely decimated and that it took a lot of effort to bring wildlife back. This restoration of North America's wildlife, and thus this continent's biodiversity, is probably the greatest environmental success story of the twentieth century. Such a recovery begins with an increase in herbivores. It is followed after a lag-time by an increase in predators. While predators are scarce and herbivores are abundant, wolves are well-fed. Consequently, they are very large, but also very shy of people. We expect to see then no tame or inquisitive wolves. Wolves are seen rarely under such conditions, fostering the romantic image of wolves so prevalent in North America today. However, when herbivore numbers decline while wolf numbers rise, we expect wolves to disperse and begin exploring for new prey. That is when trouble begins.

Historical and current evidence indicates that one can co-exist with wolves where such are severely limited in numbers on an ongoing basis, so that there is continually a buffer of wild prey and livestock between wolves and humans. The current notion that wolves can be made to co-exist with people in settled landscapes—in multi-use landscapes surrounding houses, farms, villages, and cities—is not tenable. Under such conditions, wolves—becoming territorial—will confront people when they walk dogs or approach wolf-killed livestock. In addition, even well-fed, habituated wolves will test people by approaching such, initially nipping at their clothing and licking exposed skin, before mounting a clumsy first attack that may leave victims alive but injured, followed by serious attacks. While a healthy man can fight off a lone wolf with some chances of success, a lone person cannot defeat a pack.

One cannot defend the current romantic notions about harmless, friendly, cuddly wolves. It is necessary that the public be informed that there exists a large amount of experience and information to the contrary. And the public should know the signs of danger before heading into the wilds. Tame, inquisitive wolves are one such sign.

"If the wolf had stayed in the wood there would have been no hue and cry after him."

—German folk saying

Endnotes:

1. Geist, V., "Death by Wolves and the Power of Myths: The Kenton Carnegie Tragedy," *Fair Chase*, Winter 2008, 23.
2. McNay, Mark, "A Case History of Wolf-Human Encounters in Alaska and Canada," Alaska Department of Fish and Game, 2002.
3. http://www.cbc.ca/news/canada/story/2005/01/04/sask- timberwolf050104.html
4. Geist, V., "Let's Get Real: Beyond Wolf Advocacy, Towards Realistic Policies for Carnivore Conservation," *Fair Chase*, 2009, 24.
5. Baker, R.O. and R.M. Timm, "Management of Conflict between Urban Coyotes and Humans in Southern California," 18th Vertebrate Pest Conference, University of California-Davis, 1998.

CHAPTER 6

· · ·

Mathematical Error or Deliberate Misrepresentation?

By Don Peay

*"I knew the answer to the problem was not biological.
It was not legal. I knew the answer was a political answer."*

—Don Peay

Red Flags and Failed Experiments

From the beginning, the Foundation for North American Wild Sheep (FNAWS) strongly opposed the introduction of wolves, one of the few sportsmen-based conservation groups to do so. The rest of the groups drank the proverbial Kool Aid—stating that wolves would only kill the sick and the weak, thereby helping the health of the herds. The FNAWS members, most of whom have hunted in Canada, along with many Canadian outfitter members knew the wolf restoration effort in the Northern Rockies would be a disaster for abundant ungulate populations and the hunting and ranching industries. They should know, as they live with wolves.

I started the fledgling Utah chapter of FNAWS in 1991. That group has raised more than five million dollars to support an amazing wild sheep restoration effort in Utah, going from some five hundred bighorns in just a few isolated herds in the 1990s to about five thousand desert and Rocky Mountain bighorn sheep in over twenty-eight different places in Utah today.

From the beginning, I was personally opposed to the wolf restoration and went on record to that effect with both the Utah chapter of FNAWS and Sportsmen for Fish and Wildlife, which I founded in 1994.[1] The red flag went up when I saw that the mathematical models and projections being proposed by the "professional biologists" were completely flawed—as this failed experiment has proven.

I wasn't a professional biologist; I had a degree in chemical engineering, an MBA, and some successful business experience, but I did have another credential. I was a "rocket scientist" and had worked on some very complex mathematical equations while working at an aerospace firm that manufactured nitroglycerin-based propellants for strategic rocket motors. One very critical part of our jobs was to find out when processes went from linear to exponential transitions—from smooth flight to catastrophic explosions. Similarly, in my MBA training, I spent a fair amount of time looking at financial modeling and the growth of investment portfolios. The rate of return, with small variations over a ten- to fifteen-year period of time, had tremendously different outcomes.

Either Wrong or Lying

Many of my college professors told us that math is a universal language and that if you understood math well, you could work in any profession. My experience has proved those thoughts to be true. So, from the very beginning, when some "professional biologists" told us that we didn't have a biology degree and therefore we had no idea what we were talking about, I was very leery of what the wolf reintroduction experts were telling us. I will give two specific examples, early on, that proved to me that the biologists pushing for wolf introduction and others were either wrong or they had an agenda and were not telling the truth.

Besides the mathematical modeling, engineering, and business training, I had spent my life in the outdoors of the western United States, and I have a pretty good understanding of basic biology. When I looked at the models that were being used to forecast wolf populations back in the 1990s, I could see a similar mathematical process with wolves and elk about to happen—a catastrophic failure in ten to twenty years.

Biggest Myth

The biggest myth perpetrated on the public with wolf reintroduction is that nature perfectly balances itself and that wolves only eat the sick and the weak, thus actually helping the herds become healthier. The federal and state biologists stated in their own documents that wolves would only have a 7 to 13 percent impact on elk and very little, if any impact, on moose. As we now know, there has been an 80 percent reduction in the greater Yellowstone elk herds, moose are for all practical purposes gone from Yellowstone, and now bison, the final prey, are declining as well.

Second Major Myth

The second major myth was that the wolf populations in each state would only grow at a 3 to 5 percent growth rate, based upon wolf growth rates in Canada and Alaska where a "steady state" has been reached. It doesn't take a rocket scientist to understand that when you dump a few predators in the middle of the "Garden of Eden" of wildlife, the wolves are going to multiply at very high rates—25 to 35 percent a year.

Mathematical Implications

Let's take a look at the mathematical implications of these errors. Let's assume that you have a two-million-dollar portfolio and that your goal is to never touch the principal. You are, on average, getting an 8 percent return, so you develop your plan to spend $160,000 a year, and things are going along fine.

Before the wolf introduction, excess big-game animals were taken by hunters, with hunter harvest very carefully regulated through permits for both the male and female species, and for years, the populations of elk (or your two-million-dollar investment portfolio) remained fairly constant. Your base principal is there, and you can spend $160,000 every year for a long time to come.

Now, let's assume two things happen. Number one, your interest rate of return drops to 4 percent (the beginning effect of wolves eating calf elk). Your investment manager tells you that you have to cut back spending to eighty thousand dollars a year or you will be eating into your principal. Translating this to the elk situation, state agencies cut back the number of hunting permits, but they knew that wolves were not cutting back but were growing exponentially. The net effect of wolves killing calves and then adults is like doing this to your investment portfolio: in just five short years, your principal is now one million, not two million dollars, as the growing pack of wolves ate half of it, and compounding the problem, with less cows (your principal), your interest rate is now dropped to a 2 percent return. Without dipping into your principal, the most you can spend is twenty thousand dollars a year. In just a short time, say five to seven years, you have gone from being able to spend $160,000 a year and never touching your principal to only being able to spend twenty thousand a year, and even with reduced spending, you know your principal is dropping dramatically. Again translating this to wolves and elk, the wolves are rapidly gobbling up the one-million-dollar portion.

The Yellowstone elk herd dropped from over nineteen thousand head in 1995 to just over four thousand in 2012. During this time period, wolf numbers were growing exponentially. Anyone who understands math at all could see this biological train wreck coming. Your consumption is growing exponentially, your interest rate (calf survival) is dropping exponentially, and your principal (total elk population) is dropping rapidly. Even though hunting permits were ended—we told biologists to end them sooner—elk populations still dropped precipitously.

The second point that concerned me is that these biologists had a political agenda. Ed Bangs, Wolf Recovery Coordinator for the US Fish and Wildlife Service, presented a talk about the wolf recovery effort at the University of Utah three or so years after wolf reintroduction began in 1995–96. In his analysis, he presented some data that painted the fairy tale picture—wolves were eating the old, the sick, and the weak, and hunters were killing the healthy elk. Bangs's data showed that, at that time, wolves were killing elk that averaged eight years old and that hunters took elk that averaged four years old—i.e., wolves were good, and hunters were bad. I raised my hand and asked one question: "Did you count in your average age of harvest the baby calf elk component into the wolf take?" Bangs replied, "Uh, well, no."

A similar analogy would be like talking to five hundred people that make fifty thousand dollars a year, then throwing in a few billionaires, and then telling everyone that, on average, they each make a quarter million dollars a year so everyone should be happy! WRONG! If the biologists had included the total number of calf elk—ranging from two days to six months—killed by wolves with the adult elk killed, the average age of elk killed by wolves would have been dramatically younger than the age of elk taken by hunters. This discussion proved to me beyond a shadow of a doubt that these wolf biologists had an agenda to let wolves run free and destroy game herds and that they were using phony math to buy time.

Out of Hand in a Hurry

Let me put one other issue in mathematical context. Assume you have one hundred wolves in 1995, and you predict that they are going to grow at 5 percent per year. Compounding annually, by 2010 you will have 208 wolves. However, if that same population of wolves grew at 25 percent compounding annually, by 2010, there would be 2,842 wolves—an ERROR almost fourteen times larger than the promised number. And unfortunately, this is exactly what has been happening. We were promised the goal of ten breeding pairs and a total population of one hundred wolves in Montana, Idaho, and Wyoming. However, in 2012, we had at least three thousand wolves, and some experts believe twice that many in the three states. Don't you wish your $100,000 investment grew not to $208,000 from 1995 to 2010, but to $2.84 million? This is the magnitude of the error, and with some very quick and easy mathematical modeling, I knew this wolf situation was a complete fiasco that would unfold in a period of ten to fifteen years.

Let me give you one last example of putting numbers in terms of what they really mean. With five hundred wolves in the woods (two hundred more than the agreed-upon minimum population for the three states) and with an average kill per wolf of twenty-three elk per year, the wolves would be killing around 11,500 elk a year. With just 2,500 wolves, they would be killing 57,500 elk a year. Considering that states like Utah only have 65,000 elk in the entire state and hearing that the Animal Rights groups claim they want five to eight thousand wolves, you don't need a calculator to see that wolves would dramatically reduce game herds and destroy the hunting and ranching industries—and then, after the game herds are gone, the wolves would kill each other and focus on livestock and pets, while considering people. We knew that, even though wolves were years away from Utah, we had better solve this problem before it got to our doorstep. As an engineer and businessman, I clearly knew that an exponential function can get out of hand in a hurry.

Even though the math is pretty straightforward and the models are easy to predict, it still amazed me how many biologists and others told us that wolves would NOT greatly reduce our flourishing game herds or the tens of millions of dollars in economic activity that abundant herds sustain. These biologists had been taught in school that, with good habitat, predators would have minimal impacts on game herds and the rest of the wonder wolf fable.

The Real Fairy Tales

It's easy to see that a huge mistake has been made by the mismanagement of wolves in the Northern Rockies. The original predictions in the Environmental Impact Statement of wolf population goals and what impact the wolves would have on the wildlife and the economics of the states where wolves take up residence were the real fairy tales about wolves. When wildlife biologists in Wyoming looked at the difference between what the original USFWS Environmental Impact Statement for the wolf recovery program said and what actually happened, my predictions were more than confirmed.

In 2005, ten years after the relocation took place, the Wyoming Game and Fish Department did a review of the predictions made by the USFWS in that Environmental Impact Statement (EIS). This is what they found:

- The wolf population in the Greater Yellowstone Area (GYA) in 2005 was at least 3.3 times the original Environmental Impact Statement (EIS) prediction for a recovered population.
- The number of breeding pairs of wolves in the GYA in 2005 was at least twice as high as the original EIS prediction, and the number of breeding pairs in 2004 was at least 3.1 times the original EIS prediction.
- In 2005, the wolf population in Wyoming outside Yellowstone National Park exceeded the recovery criteria for the entire region and continues to increase rapidly.

- The estimated annual predation rate (twenty-three ungulates per wolf) is 1.8 times the annual predation rate (twelve ungulates per wolf) predicted in the EIS.
- The estimated number of ungulates taken by 325 wolves in a year (7,150 is six times higher than the original EIS prediction).
- The percent of the northern Yellowstone elk harvest during the 1980s currently taken by wolves (50 percent) is 6.3 times the original estimate of 8 percent projected in the EIS.
- The actual decline in the northern Yellowstone elk herd (more than 50 percent) is 1.7 times the maximum decline originally forecast in the EIS.
- The actual decline in cow harvest in the northern Yellowstone elk herd (89 percent) is 3.3 times the decline originally forecast in the EIS.
- The actual decline in bull harvest in the northern Yellowstone elk herd is 75 percent, whereas the 1994 EIS predicted bull harvests would be "unaffected."
- Since wolf introduction, average ratios of calf elk to cow elk have been greatly depressed in the northern Yellowstone elk herd and in the Wyoming elk herds impacted by wolves. In the northern Yellowstone elk herd and in the Sunlight unit of the Clarks Fork herd, calf-to-cow ratios have been suppressed to unprecedented levels below fifteen calves per one hundred. The impact of wolves on calf recruitment was not addressed by the 1994 EIS.

Conclusion

To me, knowing that hunters and ranchers have invested hundreds of millions of dollars to restore and sustain abundant game herds and then see a group of people act recklessly—willing to make irresponsible representations in management and models, change agreements to have one hundred wolves in each of three states, then allow for state management, and go to court and have judges use ancillary and irrelevant biological arguments to achieve their agenda of no wolf management—was very disturbing. The people pushing for no wolf management have contributed little, if any, money to growing and sustaining the very food sources of "their" wolves. To sportsmen and women, who know that wildlife belong to the people and that we have fought and worked so hard to have abundant herds, it was very offensive.

I grew up in a single-wide trailer with no indoor plumbing, and I worked hard for fifty years. Many of my sportsmen friends are of the same mold. We have worked hard and sacrificed a lot, and we do not intend to let a handful of activists destroy our abundant game herds and all the intrinsic and economic values that come from having abundant game herds.

I knew the answer to the problem was not biological. It was not legal. I knew the answer was political. Working with Ted Lyon and many others, we have set a precedent that will honor the work of sportsmen for many years, and we have made headway on updating the Endangered Species Act to respect the rights of people as well as wildlife.

Endnote:

1. Sportsmen for Fish and Wildlife (SFW) has played a major role in the establishment and obtaining of the funding for the 750,000-acre Watershed Restoration Act, and it has helped increase funding for wildlife and land conservation by more than two hundred million dollars. SFW gave a voice to sportsmen in Utah, a state that had been dominated by non-wildlife-friendly interests. SFW also worked to restore world-class trophy bull elk on Utah's public lands; has led the aggressive effort to transplant new herds of bison, antelope, wild turkey, mountain goats, and bighorn sheep; and has been instrumental in the Mule Deer Recovery Act.

Photo credit: Debbie Steinhausser/Shutterstock.com

CHAPTER 7

. . .

The Caribou Conservation Conundrum

By Arthur T. Bergerud, PhD

It is assumed that an undisturbed animal community lives in a certain harmony . . . the balance of nature. The picture has the advantage of being an intelligible and apparently logical result of natural selection in producing the best possible world for each species. It has the disadvantage of being untrue.

—Charles Elton

Fifty Years of Research

Caribou or reindeer (*Rangifer tarandus*) are an Arctic and Subarctic species of deer found all around the world's northernmost regions. There are two general groups: the tundra caribou that live in the far north in comparatively barren tundra habitat, and the woodland caribou that reside in northern forests.

Woodland caribou (*Rangifer tarandus caribou*)—a subspecies that is found primarily today in Canada but once was found in the North American boreal forest from Alaska to Newfoundland and Labrador, and as far south as New England, Idaho, and Washington—are now listed as "threatened" with some herds listed as "endangered."[1]

Based on well over fifty years of research on woodland caribou, I can only conclude that the primary reason for their decline is increased predation from expanding wolf numbers. To reduce wolves to lower densities, we need the support of the general public, especially the environmentalists that are most concerned about the caribou. However, there are two huge problems in gaining public support: for decades, biologists and naturalists have argued that habitat determines animal numbers and that caribou are wilderness animals that cannot coexist with logging and industrialization. Secondarily, environmentalists and some biologists and naturalists have accepted the pseudoscientific myth that there is a balance of nature, a stable equilibrium between prey and predators, which prevents extinctions in the absence of anthropogenic factors (of, relating to, or resulting from the influence of human beings on nature). They argue that, if caribou are declining, it must be anthropogenic factors such as logging and human disturbance that are causing these declines.[2]

Predator-Prey Relationship Misconceptions

Fundamental to understanding predator-prey interactions is the debate whether ecological systems are structured from top down (predator-driven) or bottom-up (food-limited). This question has been debated since at least 1960[3] and has been rekindled with the introduction of wolves to Yellowstone National Park and Idaho in 1995–96 and with the rapid decline in recent years of woodland caribou across Canada.

The decline in woodland caribou, however, has been in progress since the early 1900s.[4] Common early explanations were that the decline was the result of the loss of boreal forests from logging and the loss of the lichen's ranges from forest fires and early settlements. That argument continues today.

Based on years of research that began in the 1950s in Newfoundland and Labrador, in 1974, I published that the decline of both woodland and barren-ground caribou resulted from increased predation, primarily from wolves, and not from a decrease in lichen pastures by fire and overgrazing of terrestrial lichen.[5] My 1974 paper is now generally quoted by young caribou biologists as showing that the proximate cause (an act from which an injury results as a natural, direct, uninterrupted consequence and without which the injury would not have occurred) for the decline of caribou is wolf predation with the implication that the real reasons are anthropogenic factors such as logging that have upset the "balance of nature." Many older wolf biologists still believe in the balance of nature concept, have helped indoctrinate the general public to that view, and still do not support management intrusion into natural systems.

Research Findings

To resolve the question of top-down or ground-up, I selected two study areas following all the controversy of the 1974 paper. For the experimental area, we chose an insular caribou herd on the Slate Islands in Lake Superior near Pukaskwa National Park, Ontario. Those islands had no terrestrial ground lichens and also no natural predators of caribou—i.e., no wolves, bears, lynx, or wolverines.

As the control population, we selected the caribou in Pukaskwa National Park (PNP), fifty kilometers distant. In the Park, there was a normal boreal fauna of wolves, bears, lynx, moose, and caribou and little anthropogenic disturbance. We compared the demography to these two populations over thirty years (1974–2004). On the Slates and in the absence of wolves, the population persisted for thirty years with the highest density of caribou in North America, up to ten caribou/km^2 with numbers varying from 150 to six hundred animals. Total numbers were regulated by starvation from a shortage of summer foods, not lichens.[6] The animals went into the winter too weak to survive, regardless of the abundance of winter forage. This density was a hundred times greater than the control herd in PNP where the density was 0.06 caribou per km^2—a hundred times less than the Slate Islands. This original low density was due to the predation of wolves existing at eleven wolves per 1,000 km^2. As the study continued, the PNP herd declined, and by 2009, only four caribou were left. The herd may now be extinct. It was clearly regulated top-down by predation and was favorably critiqued in the prestigious textbook *Ecology* 6th Edition, 2009, by Emeritus Professor Charles Krebs.

Near the end of our Slate Island study, a natural experiment occurred in 1994 when Lake Superior froze and two wolves crossed to the Slate Islands. We lost nearly all the calves born in 1994 and 1995, and in those two years, the wolves caused sufficient mortality of adult females to change the sex ratio from the nominal sex ratio of one male to two females (approximately 35 percent males) to 55 percent males, a decline of a hundred animals, mostly females. This mortality was replicated in 2003 and 2004 when another duo of wolves reached the islands and, again, the wolves killed nearly all the calves on the island.

Caribou Decline

In the 1970s, I researched the demography of mountain caribou in northern British Columbia and published three peer-reviewed papers in 1984, 1986, and 1987.[7] The 1984 paper documented that 90 percent of the calves were killed each spring before the age of six months mostly by wolves and bears. The 1986 paper documented that the decline of caribou in British Columbia occurred when the moose population increased, which brought higher wolf populations to northern British Columbia. Additionally, we compared the recruitment and adult mortality of caribou for all herds in North America where wolf abundance had been measured (a sample size of 750,000 caribou). The wolf density where caribou mortality and recruitment were balanced for caribou [the stabilizing density (r^s)] was 6.5 wolves/1,000 km^2 [finite rate of increase (λ) = (1 percent adult mortality)/ (1 percent recruitment)]. When wolf densities were higher, caribou declined. With caribou, the stabilizing recruitment of calves to balance adult mortality should be 15 percent of the population, or twenty-five calves per one hundred females measured, when calves are ten to twelve months of age. The pregnancy rate in caribou is normally greater than 80 percent (eighty-plus calves per one hundred females).[8] To maintain caribou populations, wolf numbers need to be below the densities of eight or less wolves/1,000 km².[9] In multi-ungulate systems, wolf densities are commonly fifteen to twenty-five wolves/1,000 km².[10]

In the 1980s, I joined Dr. John Elliott, regional fish and wildlife biologist in British Columbia, in studying the demography of caribou, moose, elk, and Stone's sheep in northeastern British Columbia. At that time, all four species were in decline. As in the case of the Ontario studies, we had control and experimental areas replicated in two separate regions: the Kechika region (18,400 km²) and the Muskwa region (19,000 km²). The two control replications were left undisturbed, and in the two experimental areas, we removed wolves—Kechika: 492 wolves in four years, and Muskwa: 505 wolves in three years. The grizzly bear and wolverine populations were left undisturbed. The study was ten years in duration.[11]

In the paper, we showed that in the areas where we reduced wolves, recruitment was greatly increased based on calves per one hundred females at six months to twelve months. For moose, recruitment averaged forty calves per one hundred females; elk, forty-nine per one hundred females; Stone's sheep, forty-one per one hundred females; and caribou, thirty-nine per one hundred females. For those herds with no wolves removed, the recruitment was moose, eleven calves per one hundred females; sheep, twenty per one hundred; elk twenty-six per one hundred; and caribou, seven per one hundred. The elk and moose populations in this study that had wolf management increased from eighteen thousand to thirty-three thousand. In all removals, wolves were harvested in the spring before wolf denning, and in all years, young wolves were already dispersing into the removal area but insufficiently organized for denning so that the four prey species secured positive recruitment. At the end of the ten years, the wolf population was a healthy twenty-plus wolves per one thousand km².

In the other areas in the North where wolves have been managed, the percentage of calves before and after wolf reductions were as follows: Nelchina herd, Alaska, 1960s, 24 percent vs. 40 percent; Forty Mile herd, Alaska, 1970s, 5 percent vs. 31 percent; Delta Herd, Alaska, 1 to 9 percent vs. 25 percent; and Beverly Herd, Northwest Territory, 1960s, 7 to 8 percent vs. 20 to 25 percent.[12]

The Predator Pit

The last study I wish to mention is the George River caribou herd in northern Quebec and Labrador. I counted the herd in 1958 at fifteen thousand animals. In 1958, there were no wolves; hunters that had been going in the country for decades had seen practically none on the land. The wolverine had also gone extinct. Starting in 1958, the herd increased yearly, reaching three hundred thousand by 1980.

In the 1970s, wolves reappeared and increased as the herd grew. By 1980, the wolves finally reached sufficient numbers that their predation halted the growth of the herd (recruitment equaled adult mortality). Then in 1981 and 1982, the wolves got rabies and declined by 80 percent. The caribou herd exploded with numbers reaching 537,000 by 1985 and 650,000 by 1987, the largest herd in the world at that time. The herd started to decline in 1988 when densities reached ten per km^2, similar to the density that resulted in decline on the Slate Islands. However, unlike the Slate Islands, there was no winter starvation. In the George River herd, the females died in the summer from malnourishment with lactation problems and insect attacks with the result that pregnancy rates declined.[13] In addition, wolf predation continued as did hunting, which was unabated. The herd reached seventy-four thousand in 2010, well within the carrying capacity of forage, but it is continuing to decline from heavy predation of calves and adults. The herd is now in what some biologists call "the predator pit," which means that each time the herd starts to recover, the predation intensifies, and the herd remains limited.

The herd could be turned around now and started back up if wolves were managed, but in Canada, there has been little or no wolf management in recent decades. The herd will continue down until the wolves go elsewhere; that was what happened in the decline in the 1890s, from a high of seven hundred thousand to the fifteen thousand I counted in 1958. If the herd was managed now at seventy-four thousand, it could increase rapidly as did the Western Arctic herd in the 1970s. That herd had declined from 242,000 in 1970 down to seventy-five thousand by 1976. Then Alaska Fish and Game took over, and the harvest was drastically curtailed with the natives fully cooperating, and predators were managed. With wolves managed, the herd was back to 113,000 by 1979 and continued higher.[14]

Support of the Population

The only wildlife agency in North America that manages wolf numbers as a standard practice is in Alaska. This management has the support of the majority of the population in a system where the populace commonly depends on moose and caribou as subsistence food.[15] This dependence is protected under both state (state subsistence statue) and federal (Alaska National Interest Lands Conservation Act—ANILCA) and is in agreement with the National Research Council 1997 standards.[16] There has been some sporadic management of wolves in Alberta and of coyotes and bears on the Gaspe Peninsula—150 animals are left at this time. Recently British Columbia had planned to manage wolves for their remaining endangered arboreal caribou (less than fourteen hundred), but environmental groups thought the high abundance of wolves was more valuable than the last remaining arboreal caribou and forced the government to cancel the management.[17]

Going Extinct

Species around the world are going extinct. In Africa, many endemic species are declining from poaching and predation, and population numbers of predators themselves are decreasing, especially the large cats (tigers, cheetahs, and lions).[18] But in North America, wolves are prospering, spreading across the western United States from the Yellowstone and Idaho introductions and increasing in Canada as their prey base of moose and deer expand with climate change.

In 1992 or 1993, I received a questionnaire from a polling company apparently under contract to the US Fish and Wildlife Service (USFWS), asking how wolf introduction to Yellowstone Park would impact the other species and the vegetation in the Park. The participants of this poll became known as the Delphi Committee, which was made up of ungulate and wolf biologists. It was an anonymous committee. I still do not know who else took part, but I believe it was weighted to pro-wolf biologists.

Based on our research in northern British Columbia, I predicted a major decline in elk and moose if wolves were introduced. My major comment was that, if the introduction went ahead, the wolves would have to be managed in Yellowstone Park and prevented from spreading beyond the Park. Well of course, there was no intention to manage wolves; they have now reached California and Colorado, and Utah is trying to hold the line. The Northern Yellowstone elk herd in the Park is mostly gone—from twenty thousand in 1994 prior to the wolf relocation, to 4,635 in 2011—and elk hunters may have seen the end of their hunting in that area. Animal rights activists and anti-hunting organizations have won.

I couldn't believe USFWS would introduce wolves to Idaho where the last caribou herd was classified as endangered (the Selkirk herd), but they did, and the wolves are now hunting these last caribou. An even more unethical act, the USFWS is adding an additional area of 152,000 hectares (61,538 acres) to the preserve surrounding the herd that is off-limits to the local residents. The caribou do not need this additional land grant. They need predator control of the mountain lions and now the introduced wolves, but there will be none by the USFWS. Lion predation has been a problem in the past.

Patrick Valkenburg, a caribou biologist, emailed me stating: "With 300 million people in the United States now, is it realistic to just let 'wolves do their thing'? Wolves belong in the boreal forest, not the Great Plains and the Great Basin where there are no longer any buffalo. Perhaps the USFWS did not know."

Some older wolf biologists—Dr. Victor Van Ballenberghe, Dr. James Peek, Dr. John Theberge, and Dr. Paul Paquet—are emotionally involved with wolves and do not want to see wolves managed. I can relate to an emotional involvement relative to caribou. There is nothing more heart rending than watching a female caribou that has had her calf so completely eaten by a wolf or bear that there is not even any scent left. She does not understand that her calf is dead, and will stand and look and then go back to the last undisturbed location where they were together, and then finally back to the birth site. I have seen a female travel up a long line of cows and calves, scenting each calf, seeking recognition of her lost progeny. Some cows will stay near the remains of their dead calves for many days, waiting for them to get up and guarding the carcass from the lynx that often monitor attack and kill sites.[19] Such cows can be called close by imitating calf bleating.

However, such emotional views will not help us manage the moose-caribou-wolf system. Those wolf biologists that have elevated the wolf to icon status have done a disservice to wildlife management in fostering the balance of nature myth. In contrast, coyote biologists have remained unattached and objective, relating the damage to livestock and the dangers of these animals in cities to small children and pets.

Balance of Nature

Charles Elton—the father of ecology, who discovered the three- to four-year cycle and the ten-year cycle of mammals and who wrote the first ecology book, *Animal Ecology and Evolution*, in 1927—told us at the beginning:

> It is assumed that an undisturbed animal community lives in a certain harmony . . . the balance of nature. The picture has the advantage of being an intelligible and apparently logical result of natural selection in producing the best possible world for each species. It has the disadvantage of being untrue.[20]

In recent years, three widely respected biologists in Canada have concluded that woodland caribou are endangered from the increasing wolf population triggered by the increase in the moose and deer prey base with climate warming: Professor Emeritus Charles J. Krebs, Professor Emeritus A. R. E. Sinclair, and Professor Emeritus Valerius Geist.[21] These men are the leading ecologists in Canada.

The most respected wolf biologist in the world is Dr. L. David Mech. In his 1996 monograph *The Wolves of Isle Royale*, he used the words "stable equilibrium" to describe moose-wolf interactions ("stable equilibrium" is a synonym for the "balance of nature").

In later years, the balance completely disappeared.[22] Mech accepted this disproof, stating in 1998: "The Isle Royale moose and wolf populations have fluctuated greatly over time, and there is little correlation between wolf and moose numbers in any given year."[23]

From 1986 to 1994, Mech and his students studied the Denali caribou herd in Alaska.[24] In *The Wolves of Denali*, Mech states: "The Denali wolf-caribou relationship is a good illustration of the dynamics of populations and why stability or balance at a variety of levels should not be considered inherent in natural systems."[25]

Good scientists try to disprove their own hypotheses, and this is what he has done. This is significant because Mech, with his decades of study and publications about wolves, unintentionally has contributed probably more than any other scientist to furthering the pseudoscientific myth—the balance of nature.

Woodland caribou survived the Pleistocene epoch, more commonly known as the Ice Age, in the Appalachian Mountains, spacing themselves away from the large predators of the megafauna living at lower elevations. These caribou lived with a completely different set of species than they do today; i.e., sloths, peccaries, tapers, and austral faunal elements. In *Science*, Graham and associates stated that the fossil mammal fauna of the Late Quaternary at 2,945 fossil sites supported the Gleasonian community model, which assumes that species respond to environmental changes in accordance to their individual tolerances with varying rates of range shift.[26] They stated, "Modern community patterns emerged only in the last few thousand years, and many late Pleistocene communities do not have modern analogs."[27] Those who say caribou and wolves evolved together for hundreds of eons have not checked the fossil record.

Each species, through individual selection, evolves its own distinct behavior-habitat strategies to persist, but that does not guarantee continual survival. Each species walks its own road down through time—there is no magical balance of nature. An article by Elisa Beninca and associates[28] in the prestigious journal *Nature* stated: "Advanced mathematical techniques proved indisputable presence of chaos in this food web; short-term prediction is possible, but long-term prediction is not."

Critical Habitat

In Canada, when animal species are classified as "endangered," the federal government requests that the provinces involved develop a management plan for the species in danger of extinction, and they require that critical habitat be identified. For barren ground caribou, it is generally accepted that the calving grounds are the critical habitat. However, some still do not recognize the key element of that habitat is reduced predation risk and not forage.[29] The critical habitat for montane and boreal woodland is the areas where the cows calve and where there is reduced risk for newborns.[30] The montane caribou calve on alpine peaks spaced away from moose and wolves at lower elevations, and the boreal caribou calve on the islands, shoreline, and muskegs of the boreal forest.[31] The critical habitat is not the winter ranges where boreal woodland caribou seek lichens. They adapted long ago to rotating their winter ranges in response to lichen destructions by forest fires and over-utilization.[32] However, for the montane caribou, the old-growth high alpine forests with their lichen loads and protective deep snow that serves as a barrier to wolves are the critical habitats. In British Columbia, 2.2 million hectares (890,688 acres) of old-growth forest are protected from industrial forestry activity, sufficient for the present low population of the remaining fourteen hundred animals.

The Canadian Parks and Wilderness Society (CPAWS) in February 2012 delivered a petition of thirty-two thousand signatures to the Environmental Minister to save woodland caribou by protecting their boreal habitat from industrialization. The World Wildlife Fund's solution to the caribou conservation conundrum is to create more parks.[33] None of these efforts will save the caribou. The boreal ecosystem is structured from the top down by predation. The boreal forest can be logged if the calving habitat is left undisturbed and if the wolf population is managed to the same levels as in Alaska, fewer than 6.5 wolves per 1,000 km^2.[34]

Mistaken Beliefs

Some believe that caribou are wilderness animals, cannot tolerate anthropogenic[35] disturbance, and cannot tolerate the presence of man.[36] These are mistaken beliefs. Two biologists who hold this belief—Schaeffer and Vors—are modelers with very little experience studying the behavior of caribou on the ground. I've been on the calving grounds studying calving behavior for thirty-three years in five provinces, as well as the Northwest Territory and Alaska. In addition, I've established thirteen herds and maintained a captive herd for several years. There are numerous examples of caribou adapting to the presence of man. For example, caribou have wintered at the Armstrong Airport in Ontario for the past thirty-seven years so they can avoid the presence of wolves. In Alaska, the caribou have calved for decades in the Prudhoe Bay oil field, the herd increasing from five to thirty thousand animals in the absence of wolves and bears.

Skeptics should Google "Slate Island caribou" and see pictures of the caribou, visiting occupied campgrounds and seeking handouts and the ashes in the fire pits. Caribou adapt well to the presence of benign humans and have been domesticated by the reindeer herders in Eurasia.

In contrast, the caribou are now extinct due to predation in the Canadian National Parks of Banff, Glacier, and Revelstoke, where there is no logging, and anthropogenic disturbances are minimal.[37] The arboreal caribou in British Columbia are facing extinction from predation in the protected habitat set aside for them. Wittmer showed that old-growth forest with arboreal lichens was adequate for the herd.[38] There are possibly only three or four caribou left in Pukaskwa National Park, and wolf predation is the cause that stark decline.[39] The woodland caribou in Newfoundland from 1970 to 1996 increased from 7,000 to 96,000 animals.[40] These were years of intense anthropogenic disturbance: logging near calving grounds, roads built across the caribou habitat, lakes and rivers dammed, and increased mining—yet the caribou were able to increase, despite all this disturbance, because wolves had been extinct for several decades in Newfoundland.[41]

The Caribou-Moose-Wolf System

Woodland caribou spent the Pleistocene[42] south of the Laurentide ice sheet in the Appalachian Mountains, spaced away from the megafauna at lower elevations.[43] They left the mountains 13,000 to 12,000 BCE and reached Ontario at 10,000 BCE, passing through deciduous forests. Fossils have been found near Udora, Ontario, 10,500 BCE,[46] and near Atikokan, Ontario, 9,940 BCE.[45] The caribou were blocked going farther north by the ice sheet and, through time, evolved the anti-predator tactics of using islands and shorelines to calve where they could escape wolves by swimming.

A second segment of the Appalachian caribou gene pool moved though Maine and New Brunswick following the shore around the ice at about 9,000 BCE and turned north in tundra where forests were delayed by the cold temperature of the Labrador current. They reached Indian House Lake at 7,000 BCE, as did the Paleo Indians.[46] This population became the "barren ground" race, the George River Herd. Wolves must have followed these two separate lines, but fossils have not been found (the acid soils of the Canadian Shield erodes fossils).

Moose did not reach the boreal forest until the end of Little Ice Age, 1,850 to 10,000 years later than the caribou. The caribou had persisted in the boreal forest with the single wolf prey system for ten thousand years. However, the arrival of moose signaled the decline of caribou,[47] as moose provided the biomass to sustain much higher population of wolves. Biologists like Fuller calculate wolf numbers based on the total ungulate prey biomass weighted as follows: moose at six, elk at five, caribou at three, and deer at two.[48] Moose provided the biomass for many more wolves than were provided previously with the single-prey system of caribou. Switchover by predators between prey species is well recognized and researched in biology. In the boreal forest with a two-prey system of moose and caribou, wolves commonly switch over to the caribou that are easier to kill than moose, which may stand

and fight, but the large biomass of the moose is what maintains the higher wolf population.[49] This results in inverse density dependence mortality—the Allee Effect—[50]and this generally leads to extinction unless safe refuges exist, like islands.[51]

Now we have a warming climate and more moose and deer than ever that are expanding farther north.[52] The increased predation from wolves is causing the extinction of caribou along the southern edge of their range. Moose have now reached the Arctic Ocean coast.

Caribou can only cope with wolves in densities of six to eight per one thousand km²—this then is a species diversity problem for caribou, which generally do not have a shortage of forage. If wolf populations were managed, we could have caribou densities of greater than two per km², but because of the Allee Effect, caribou can only persist by being rare—densities commonly of only 0.06 per km².[53] The only reason caribou cannot exist in logged areas with their abundance of deciduous forage is the greater abundance of moose supporting more wolves. The sequence is

settlement + climate change
= more deciduous forage = more moose = more wolves = more predation on caribou = more extinctions

In Ontario, the extinction line north for caribou has only halted because of the safety of islands in several large lakes in two provincial parks and Lake Nipigon.

Conclusion

Environmentalists are contributing more to the extinction of caribou than anyone else is. Fish and game departments are not going to manage wolves without the support of the public, especially the vocal environmentalists. Environmentalists, as well as some biologists, are always blaming anthropogenic factors for declines in caribou like roads, seismic lines, pipelines, logging, etc. They say that such disruptions actually assist wolves in locating caribou. To the contrary, caribou have gone extinct in national parks, large parks that have not been logged, and where roads are minimal, due to predation. The supposed anthropogenic disturbances are absent, and the supposed effect persistence is also absent—the cause is not necessary.[54] Caribou have persisted since the Ice Age in a simple one-prey primary predator system and continue to persist with climate change and increased species diversity. The bottom line is that wolves or the alternate prey species (moose and deer) will have to be managed if caribou are to survive.

Endnotes:

1. COSEWIC (Committee on the Status of Endangered Wildlife in Canada) Report, 2002. [COSEWIC is a committee of experts that assesses and designates which wildlife species are in some danger of disappearing from Canada.]

2. Schaefer, J.A., "Long Term Recession and Persistence of Caribou in the Taiga," *Conservation Biology*, 2003, 17: 1435–1439; Vors, L.S., J.A. Schaefer, B.A. Pond, A.R. Rodgers, and B.R. Patterson, "Woodland Caribou Extirpation and Anthropogenic Landscape Disturbance in Ontario," *Journal of Wildlife Management*, 2007, 71:1249–1256.

3. Hairston, N.G., F.E. Smith, and L.B. Slobodkin, "Community Structure, Population Control and Competition," *American Naturalist*, 1960, 194:421–425.

4. Cringan, A.T., "Some Aspects of the Biology of Caribou and a Study of the Woodland Caribou Range on the Slate Islands, Lake Superior, Ontario," Master's Thesis, 1957, University of Toronto, Toronto, Ontario, Canada.

5. Bergerud, A.T, "The Decline of Caribou in North America Following Settlement," *Journal of Wildlife Management*, 1974, 38: 757–770.

6. Bergerud, A.T., W.J. Dalton, H.E. Butler, L. Camps, and R. Ferguson, "Woodland Caribou Persistence and Extirpation in Relic Populations on Lake Superior," *Rangifer*, Special Issue, 2007, 17: 57–78.

7. Bergerud, A.T., H.E. Butler, and D.R. Miller, "Anti-Predator Tactics of Calving Caribou: Dispersion in Mountains," *Canadian Journal of Zoology*, 1984, 62:566–575; Bergerud, A.T. and J.P. Elliott, "Dynamics of Caribou and Wolves in Northern British Columbia," *Canadian Journal of Zoology*, 1986, 64: 1515–1529; Bergerud, A.T. and R.E. Page, "Displacement and Dispersion of Parturient Caribou as Calving Tactics," *Canadian Journal of Zoology*, 1987, 65: 1597–1606.

8. Bergerud, A.T., "A Review of the Population Dynamics of Caribou and Wild Reindeer in North America," 1980, pp. 556–581 (in "Proceedings of 22nd International Reindeer/Caribou Symposium," Edited by E. Reimers, E. Gaare, and S. Skjenneberg, Roros, Norway, 17–21 September, 1980).

9. Bergerud, A.T. and J.P. Elliott, "Dynamics of Caribou and Wolves in Northern British Columbia," *Canadian Journal of Zoology*, 1986, 64: 1515–1529; Thomas, D.C., 1995. "A Review of Wolf-Caribou Relationships and Conservation Implications in Canada," pp. 261–273 (in L.N. Carbyn, S.H. Fritts, and D.R. Seip, editors, Ecology and Conservation of Wolves in a Changing World, Canadian Circumpolar Institute, Edmonton, Alberta, Canada); Lessard, R.B., "Conservation of Woodland Caribou in West-Central Alberta: A Simulation Analysis of Multi-Species Predator-Prey Systems," Ph.D. Thesis, 2005, University of Alberta, Edmonton, Alberta, Canada.

10. Messier, F., "Ungulate Population Models with Predation: A Case Study with the North American Moose," *Ecology*, 1994, 75: 478–488.

11. Bergerud, A.T. and J.P. Elliott, "Wolf Predation in a Multiple-Ungulate System in Northern British Columbia," *Canadian Journal of Zoology*, 1998, 76: 1,51–1,569.

12. Bergerud, A.T. and J.P. Elliott, "Dynamics of Caribou and Wolves in Northern British Columbia," *Canadian Journal of Zoology*, 1986, 64: 1515–1529.

13. Bergerud, A.T., S.N. Luttich, and L. Camps, The Return of Caribou to Ungava, 2008, McGill and Queens University Press, Montreal, Quebec, Canada.
14. Davis, J.L., P. Valenburg, and H.V. Reynolds, "Population Dynamics of Alaska's Western Arctic Caribou Herd," in "Proceedings of 22nd International Reindeer/Caribou Symposium," Edited by E. Reimers, E. Gaare, and S. Skjenneberg, Roros, Norway, 17–21 September, 1980.
15. Titus, K., "Intensive Management of Wolves and Ungulates in Alaska," 72nd North American Wildlife and Natural Resources Conference, 2007, pp. 366–377.
16. Regelin, W.L., P. Valkenburg, and R.D. Boertje, "Management of Large Predators in Alaska," Wildlife Biology in Practice, 2005, 1: 77–85.
17. Wittmer, H.U., A.R.E. Sinclair, and B.N. McLellan, "The Role of Predation in the Decline and Extirpation of Woodland Caribou," Oecologia, 2005, 144: 257–267.
18. Sinclair, A.R., E. Pech, R.P. Dickman, C.R. Hik, S. Hik, P. Mahon, and A.E. Newsome, "Predicting Effects of Predation on Conservation of Endangered Prey," Conservation Biology, 1998, 12: 564–574.
19. Bergerud, A.T., "The Population Dynamics of Newfoundland Caribou," Wildlife Monograph, 1971, 25.
20. Elton, C., Animal Ecology, Macmillan, London, Sidgwick, and Jackson, 1927.
21. Krebs, Charles J., Professor Emeritus, Ecology: 6th edition, 2009, pp. 197–198; Sinclair, A.R.E., Professor Emeritus, co-author (with H.U. Wittmer and B.N. McLellan), "The Role of Predation in the Decline and Extirpation of Woodland Caribou," Oecologia, 2005, 144: 257–267; Geist, Valerius, Professor Emeritus, The Deer of the World: Their Evolution, Behavior and Ecology, Mechanicsburg, Pennsylvania: Stackpole Books, 1998.
22. Peterson, R.O., "Wolf Ecology and Prey Relationship on Isle Royale," Science Monograph Series No. 11, US National Park Service, 1977; McLaren, B.E., and R.O. Peterson, "Wolves, Moose, and Tree Rings on Isle Royale," Science 266, 1994, 1,555–1,558.
23. Mech, L.D., L.C. Adams, T.J. Meier, J.W. Burch, and D. Dale, The Wolves of Denali, Minneapolis, Minnesota: University of Minnesota Press, 1998.
24. Ibid.
25. Pimm, S.L., The Balance of Nature: Ecological Issues in the Conservation of Species and Communities (Chicago: University Chicago Press) 1991.
26. Graham, R.W. et al., "Spatial Response of Mammals to Late Quaternary Environmental Fluctuations." Science 272, 1996, 601–606.
27. Past 0.5 to 1.0 million years.
28. Benincà, Elisa, et al., "Chaos in a Long-Term Experiment with a Plankton Community," Nature, February 14, 2008.
29. Bergerud, A.T., S.N. Luttich, and L. Camps, The Return of Caribou to Ungava, McGill and Queens University Press, Montreal, Quebec, Canada, 2008.
30. Montane means of mountains and other high-elevation regions; Boreal means of or relating to the forest areas of the Northern Temperate Zone.
31. Simkin, D.W., "A Preliminary Report of the Woodland Caribou Study in Ontario," Department of Lands and Forest, Section Report No. 59, 1965; Shoesmith, M.W. and D.R. Storey, "Movements and Associated Behavior of Woodland Caribou in Central Manitoba," Proceedings of International Congress of Game Biologists,

1977, 13: 51–65; Hatler, D.F., "Studies of Radio-Collared Caribou in the Spatsizi Wilderness Park Area," Smithers: British Columbia Spatsizi Association for Biological Research, 1986; Edmonds, E, J., "Population Status, Distribution, and Movements of Woodland Caribou in West-Central Alberta," *Canadian Journal of Zoology*, 1988, 66: 817–826.

32. Bergerud, A.T, "The Decline of Caribou in North America Following Settlement," Journal of Wildlife Management, 1974, 38: 757–770.

33. Petersen, B., A. Iaconbelli, and E.E. Kushny, "The Caribou Conundrum: Conservation of Woodland Caribou and Designing Protected Areas," World Wildlife Fund, Toronto Poster Presentation, 1998, 8th North American Conference, Whitehorse, Yukon.

34. Bergerud, A.T. and J.P. Elliott, "Dynamics of Caribou and Wolves in Northern British Columbia," *Canadian Journal of Zoology*, 1986, 64: 1515–1529; Bergerud, A.T., "The Need for the Management of Wolves," an open letter, *Rangifer*, Special Issue, 2007, 17: 39–50.

35. Of, relating to, or resulting from the influence of human beings on nature.

36. Schaefer, J.A, "Long-Term Recession and Persistence of Caribou in the Taiga," Conservation Biology, 2003, 17: 1435–1439; Vors, L.S., J.A. Schaefer, B.A. Pond, A.R. Rodgers, and B.R. Patterson, "Woodland Caribou Extirpation and Anthropogenic Landscape Disturbance in Ontario." *Journal of Wildlife Management*, 2007, 71: 1249–1256.

37. Serrouya, R. and H.U. Wittmar, "Imminent Local Extinctions of Woodland Caribou from National Parks," *Conservation Biology*, 2010, 24: 363–364; Hebblewhite, M., C. White, and M. Musiani, "Revisiting Extinction in National Parks: Mountain Caribou in Banff," *Conservation Biology*, 2010, 24: 341–344.

38. Wittmer, H.U., A.R.E. Sinclair, and B.N. McLellan, "The Role of Predation in the Decline and Extirpation of Woodland Caribou," Oecologia, 2005, 144: 257–267; Wittmer, H.U., R.N.M. Ahrens, and B.N. McLellan, "Viability of Mountain Caribou in British Columbia, Canada: Effects of Habitat Change and Population Density," *Biological Conservation*, 2010, 143: 86–93.

39. Bergerud, A.T., W.J. Dalton, H.E. Butler, L. Camps, and R. Ferguson, "Woodland Caribou Persistence and Extirpation in Relic Populations on Lake Superior," *Rangifer*, Special Issue, 2007, 17: 57–78.

40. Wildlife Division Newsletter, June 6, 2009.

41. Bergerud, A.T., H.E. Butler, and D.R. Miller, "Anti-Predator Tactics of Calving Caribou: Dispersion in Mountains," *Canadian Journal of Zoology*, 1984, 62: 566–575.

42. Geological epoch that lasted from about 2,588,000 to 11,700 years ago.

43. This ice sheet was the primary feature of the Pleistocene epoch in North America, commonly referred to as the Ice Age. The ice sheet was up to two miles thick in Quebec, Canada, but much thinner at its edges.

44. Storck, P.L. and A.E. Spiess. 1994 "The Significance of a New Faunal Identification Attributed to an Early PaleoIndian Occupation at the Udora Site, Ontario, Canada," *American Antiquity*, 59: 121–142.

45. Jackson, L.J. 1989. "First Ontario C-14 Date for Late Pleistocene Caribou," *Archeology*, 89: 4–5.

46. Bergerud, A.T., S.N. Luttich, and L. Camps, *The Return of Caribou to Ungava* (McGill and Queens University Press, Montreal, Quebec, Canada, 2008).

47. deVos, A., and R.L. Peterson. 1951 "A Review of the Status of Woodland Caribou in Ontario," *Journal of Mammalogy*, 322: 337.

48. Fuller, T.K., "Population Dynamics of Wolves in North-Central Minnesota," Wildlife Monograph 105, 1989.

49. Simkin, D.W., "A Preliminary Report of the Woodland Caribou Study in Ontario," Department of Lands and Forest, Section Report No. 59, 1965.

50. Allee Effect: Mortality accelerates as prey numbers decline.

51. Wittmer, H.U., A.R.E. Sinclair, and B.N. McLellan, "The Role of Predation in the Decline and Extirpation of Woodland Caribou," Oecologia, 2005, 144: 257–267; Bergerud, A.T., W.J. Dalton, H.E. Butler, L. Camps, and R. Ferguson, "Woodland Caribou Persistence and Extirpation in Relic Populations on Lake Superior," Rangifer, Special Issue, 2007, 17: 57–78.

52. Latham, A.D.M., M.C. Latham, N.N.A. McCutchen, and S. Boutin, "Invading White-Tailed Deer Change Wolf-Caribou Dynamics in Northeastern Alberta," Journal of Wildlife Management, 2010, 75: 204–212.

53. Bergerud, A.T., "Rareness as an Anti-Predator Risk for Moose and Caribou," Wildlife 2001 Populations, editors D.R. McCullough and R. H. Barrett, New York: Elsevier Applied Science, 1992, pp. 1008–1021.

54. Hempel, C.G., Philosophy of Natural Science, Englewood Cliffs, N.J., 1966.

Credit: Tarasov/Shutterstock.com

CHAPTER 8

Wolves, a Serious Threat to Livestock Producers

By Heather Smith-Thomas

"Cattle and wolves are never going to be compatible. Some people keep saying that wildlife people and cattlemen are going to have to learn to manage for wolves, but that's impossible. It's hard to 'manage' when you are always on the losing end."

—Len McIrvin

Imaginary Scenarios

Wolves have become the noble, beautiful creatures of myths in the minds of most Americans; it's easy to believe imaginary scenarios when a person has no personal experience with the reality of the beast. The average American has had no reason not to believe the myths about wolves—since most published materials perpetuate the myths. Some Western ranchers, however, remember their parents and grandparents telling about bloody carnage when wolves attacked their herds and flocks, often killing for sport, or leaving half-dead animals in their wake after a night's spree. These stories and accounts of wolf attacks have been discredited by modern "wisdom" about wolves. Now ranchers are paying the price.

Despite predator control campaigns, wolves were never eliminated, and never even endangered; there were, in fact, thriving populations in Canada, Alaska, and Minnesota. Yet most US ranchers from the 1950s through the mid-1990s could sleep easy at nights, not having to worry about wolves wreaking havoc with their livestock. This all changed when government agencies transplanted an "experimental" population of wolves in the northern Rockies in 1995.

Controversy over "Reintroduction" of Wolves

Rural communities in the West fought against wolf "reintroduction," knowing that these animals would have a significantly adverse impact on wildlife and livestock and possibly human safety. Their views and arguments were overshad-owed by the broader voice of Eastern public as well as Western cities (where the average person is completely removed from the realities of where his/her food comes from and wolves live mostly in zoos). Both of these large segments of pop-ulation thought it would be nice to have wolves on the Western landscape again. Reintroducing a native animal, considered to be an important part of this eco-system, sounded like a great idea, but this reasoning was flawed for two reasons.

First, wolves were not historically present fulltime in the Intermountain West, as there was no stable food source. Wolves were primarily plains animals, following bison, elk, and antelope herds. From the records of the Lewis and Clark expedition of 1806, we know there were very few bison, deer, or elk in central Idaho in pre-settlement times. The plains wolves during that era came into central Idaho only when the bison did, on a sporadic basis. The Native Americans of that era nearly starved for lack of meat. They had to travel annu-ally across the mountains, onto the plains of what is now Montana to hunt bison to obtain meat to help them get through winter. Wolves moved into Idaho after the settlers came, and the early livestock industry provided a prey base. The wolves that harassed livestock in the central Rockies during the late 1800s and early 1900s were diligently hunted and eliminated. Big-game herds expanded with less hunting pressure and with the introduction of elk in the 1930s by the Idaho Fish and Game Department.

Second, the wolves that were dumped into the central Rockies were big Canadian wolves that could weigh 130 pounds or more, not the smaller native wolves of the American plains. Wolves are killing machines, and big wolves have big appetites.

Additional Issues

When the federal government made public its plan to introduce wolves, many Westerners feared that wolves were being used as part of a larger agenda designed to wrest control and management of private land away from land-owners. Those who raised livestock were also fearful that the introduced wolves would not stay in their "designated areas," like Yellowstone Park. The elk herds in the Park generally come down out of the high country in winter and often mingle with ranchers' cattle on feeding grounds. Most of the game herds in the West spend a lot of time on private land, sharing pastures and hayfields with domestic animals. Wolves don't just stick to wild prey—livestock are easy tar-gets, and ranchers felt that wolves would simply follow the prey source.

The wolf reintroduction plan not only infringed on ranchers' property rights and ability to protect their livestock (especially if wolves denned on a ranch, that area would be off-limits for traditional uses), but it also boded economic losses due to depredation. Ranchers feared and resented this infringement and were afraid of the personal tragedies that would result. Those fears have become reality.

For people who raise livestock as a way of life, their animals are not just a paycheck. After spending a lifetime delivering baby calves or lambs, doctoring the sick ones, and fine-tuning every ability and instinct to become competent caretakers of livestock, the rancher won't forsake his animals. The rancher is responsible for their well-being, and is morally committed to taking care of them. To find a favorite old cow lying hamstrung and bleeding or young calves totally missing—with their mothers upset and bawling—is unacceptable.

The wolf advocates promised there would be ways to reimburse ranchers for their losses, but such compensation would never cover all the economic losses or any of the emotional losses. If a baby calf is killed, would the rancher be reimbursed for what it would have been worth later as a big steer or herd sire, or as replacement heifer in the herd? How do you put a value on permanent genetic potential lost? And how do you alleviate the personal loss and anguish the rancher suffers when watching his animals being maimed and killed? Or how do you compensate for the fear of wolves attacking ranchers and/or their children?

Ranchers fight bad weather, livestock diseases, and other threats. In good conscience, they cannot stand meekly by and watch wolves slaughter their livestock. To demand that we couldn't shoot a wolf to protect our livestock or wildlife created an inevitable clash that would make lawbreakers out of most ranchers and many sportsmen. Wolf protection according to the Endangered Species Act was an unworkable law. If the majority of people can't abide by or live with a law, then it's a poor law, and this should be reason enough to rethink it. Little is said in popular press about the viewpoint of the people who live with wolves in the northern Rockies. This chapter lets people hear the voices of those of us who live with wolves.

Wolf Depredations

Within days after the first wolves were turned loose in the backcountry of central Idaho, a wolf killed a young calf belonging to Gene Hussey, an eighty-three-year-old rancher on Iron Creek near Salmon, Idaho. The wolf was shot, perhaps by someone driving by the pasture on the road, since in ranch country it is common practice to protect livestock from predators. When the dead wolf and dead calf were discovered side by side, Hussey was afraid of what the feds might think and called a local veterinarian, Dr. Robert Cope, to perform a post-mortem examination on both the calf and the collared wolf.

Later, when the federal experts arrived, they began harassing Hussey, assuming he had shot the wolf. They took both bodies for their own post-mortem examinations and claimed that the calf was stillborn and that the wolf did not kill it but was simply scavenging. What the experts didn't know was that Dr. Cope videotaped the original autopsy as he performed it, which showed that the calf had milk in its stomach and fully inflated lungs. According to Cope, this was indisputable evidence that the calf was born alive. When the feds found out about the tape, they realized they'd gone too far, twisting the facts to fit their own purposes and to keep up the façade about wolves.

The swat-team mentality of the government agents, who threatened to persecute anyone who dared shoot a wolf, was emphatically resented by rural communities. It didn't make sense that the penalty for killing a wolf to protect one's animals or in self-defense was worse than for killing a human for the same reason.

Sheep Depredation

The Soulen family near Weiser, Idaho, owns one of the sheep outfits that early on was significantly affected by wolf depredation. Margaret Soulen Hinson ranches with her father and brother in a family-owned corporation. "My grandfather started our business in the 1920s, and I'm the third generation," she says proudly.

Soulens have about ten thousand ewes and one thousand cows that graze on a mix of public and private land. Margaret explained:

> We own about fifty-thousand acres and also run on BLM and Forest Service, as well as some state grazing leases and private leases. Our range operation covers about 450,000 acres of rangeland located in eight Idaho counties. We had our first loss to wolves in 1996, shortly after they were introduced. Most of our wolf depredations have been on the Payette National Forest near McCall, Idaho. We're only about eighty miles (as the crow flies) from where wolves were turned loose in the Frank Church Wilderness. It didn't take the wolves long to find our sheep. They love lamb!

The first year, Soulens lost thirty head.

> We didn't know what to expect when we started losing sheep. We didn't know if the killing would continue or not. The first night they tore up thirteen head. We then did everything we could to keep them out of our sheep, since it was illegal to kill the wolves. We took extra people and slept on the bed grounds with the sheep, but we still lost thirty head that summer. It was a learning experience, because wolves

have been gone from our area for a long time. Even my grandfather didn't have any experience with wolves.

The next year they lost twenty-eight head. The loss numbers bounced around: thirty-seven head in 1999; five in 2000; none in 2001; two in 2002; fifty-five in 2003; and 332 in 2004.

> In 2004, the Cook pack, about thirteen wolves, hit the sheep before we even made it to the National Forest. They killed seventy-eight head in one night and continued to hit the sheep. This pack was living in an area where all our bands had to pass through, and they hit almost every band. We also had many other wolves in the area that year. Every band had wolves around them.

Finally the Soulens received help from US Fish and Wildlife Services, which came in and removed the Cook pack, but by that time, the wolves had done a lot of damage. "It was almost August by the time they took out those wolves," she says. In 2004, it was still controversial as to whether the government agency would allow removal of any wolves, and many people still didn't think wolves would ever be much problem for livestock.

Soulens also worked with Defenders of Wildlife and other wolf advocate groups to try to do whatever they could to co-exist with wolves.

> We bought extra guard dogs and had extra people sleeping out with the sheep. We tried rag boxes, cracker shells, gun training for the herders, etc. Our herders became more experienced at recognizing when wolves are a danger. When the guys know wolves are around, they stay with the sheep twenty-four/seven. But the reality is that sometimes you have to remove the wolves that are causing too much damage. They become habituated to preventative measures and won't let up. By taking out some of the wolves, you can break that cycle.
>
> In 2005, we lost about 220 head; in 2006, we lost 175; in 2007, we lost 125; and in 2008, we lost twenty-five. We did a graph to look at what the wolf population numbers have been, in comparison with our losses. There was an exponential increase in wolf population—and our losses reflected that increase. One good thing about all this was that the Idaho Conservation League actually sent out a press release in support of removing that pack, which helped people recognize what some of these issues were. I think people are starting to recognize that wolves kill livestock. The stories early on were that they only kill the weak and the lame, or that they only eat deer and elk and leave the livestock alone.

In reality, wolves are predators and opportunists. The idea that they kill only what they will eat is also false. With sheep, especially, they'll just tear them up. I've seen instances where they rip them open and string the guts out. They may eat only a small portion, or they may eat the entire animal to where there's only a skull left that looks like it's been boiled clean. Sometimes you never find anything; the animal is completely gone. Hopefully, the delisting and hunting season will put some fear into the wolves, if they are shot at more often.

It may be difficult for hunters to find them, however, because wolves are often nocturnal in their depredations and are very good at hiding during daylight hours.

Our losses typically occur at night. Where sheep are herded, at least someone is there, and you can get the problem identified right away. It's more difficult for cattle people to protect their animals than it is for sheep producers, because cattle are scattered and not as closely watched like the sheep with their herders. You can't be out with cattle that much; often you never find the animals that are killed and eaten.

In order to have any compensation for losses or to document a wolf problem, livestock owners must prove that it was actually wolves that did the killing.

This can be an issue in our remote range areas. The herders have a challenge to get the word out and to get Wildlife Service people up there to document and verify the kill before the carcass has deteriorated or something else has eaten on it. Wildlife Services and the US Fish and Wildlife have been very responsive when we had problems. They have been very good to work with.

The Soulens were compensated for most of their early losses. State funding—federal money that came through the state—paid for the probable wolf kills.

Defenders of Wildlife paid for every documented wolf kill and were fair on prices. We were one of the first ranchers to experience losses. They've been fair, also, in what they've paid. One issue is the challenge of getting it documented in a timely manner.

There are significant losses when dealing with wolves, however, that are hard to quantify and document and for which the producer will never be compensated.

The compensation doesn't make up for all the depredation losses, because there are some things you never get compensated for—like all the extra labor, the stress on the sheep, and the lower weight gains when wolves are harassing them.

Wolves create a lot of extra work for the herders, the foreman, and others who are dealing with this situation. We've had to herd the bands a lot tighter. They're not as spread out, and it's a little harder on the land. The lambs are not gaining as well, as a result. The herders have to sleep out with the sheep or bring them back closer to camp at night. This is not optimum management. Typically we'd herd them a little looser, and they would graze more.

The wolves have added many additional expenses to their sheep operation.

We've also needed more guard dogs. Our herders know what to expect when they hear warnings from the guard dogs. The dogs make a different sound when it's wolves than when it's bear or coyotes. Our foreman said that, when wolves are around, the dogs cry. The dogs are afraid of them. We've lost guard dogs to wolves. We run four guard dogs per band. If it's just a single wolf, the guard dogs gang up, and the wolf will back down, but if there's more than one wolf, they kill the guard dogs. We used to use Pyrenees guard dogs, but now we use Akbash and Akbash-cross dogs. They are a little more athletic and aggressive. The dogs are expensive; the last ones we bought were $650 apiece.

During that first year, when there was a lot of effort trying to trap the first pair of wolves on Pearl Creek to relocate them, Wildlife Services didn't have manpower to deal with other problems. They did finally relocate the wolves, but in some of the other bands we were experiencing bear and cougar damage, and nobody was available to deal with those issues. The wolves have taken a lot of time and a lot of money.

When wolves were introduced, the USFWS didn't realize how well they would do or how fast their numbers would grow. With total protection for more than a dozen years, wolves became bold.

If you look at it from the wolf advocates' point of view, this is a tremendous success story. Moving forward with delisting and a more proactive management system for wolves must be the next step. With our operation, we were granted "take permits" early on, so we could shoot a wolf if it was in our sheep. But in all those years, we never managed to shoot a wolf. Wolves can be very difficult to actually shoot.

Ranchers like the Soulens hope there won't be too many roadblocks (lawsuits by wolf advocates to stop the wolf-hunting seasons) in progressing toward workable management of these predators.

I was disappointed that Defenders of Wildlife and some of the other groups filed for the injunction to halt hunting. They're not being fair. The wolves' numbers had exceeded their goal, so it's time to move forward with the delisting and start managing them. This blocking effort is not productive. It makes the environmental community look bad.

Ranchers want something logical and workable, and when the environmental and wolf advocate groups refuse to work toward a management solution, it's viewed in a negative light, and cooperation becomes more difficult.

Learning Curve

Nearly twenty years after the wolf project was launched, almost every rancher in Central-Eastern Idaho has seen wolves or their tracks, and livestock losses are all too common. In 1999, Ralph McCrea, a rancher near Leadore, Idaho, lost twenty lambs to a wolf that came into his flock two different nights.

This wolf was collared and had been turned loose in Yellowstone Park at noon on a Wednesday. The next day he went through Dillon, Montana (seventy miles away), and by Friday night, he'd come over the mountains and was at my place killing sheep.

McCrea has also been losing calves; for several years wolves killed some of his big calves on the range, and then wolves began coming into his calving pens at night, killing baby calves right next to his house.

James Whittaker runs a thousand cows near Leadore, Idaho, and experienced his first confirmed losses (three bred heifers) in 2007. In 2008, Whittaker had two confirmed kills within a mile of his house.

In 2009, we had four confirmed kills on our range. It's hard to find them soon enough to get the losses confirmed; our range is thirty-five miles long. All told, we lost forty-five calves that summer. Our annual loss went from 2 percent to nearly 5 percent that year.

When you lose a calf from other causes, you usually find the carcass, but with a wolf kill, you may not find anything.

> It's like the animal evaporated. We get paid for confirmed kills but not for the ones that disappear or that we find too late to determine the cause of death.

Bruce Mulkey, who ranches near Baker, Idaho, recalls wolf advocates on a panel discussion at Idaho State University saying that only 1 percent of livestock are killed by wolves, and that ranchers lose more to disease and other problems. "But if you are part of that 1 percent and happen to lose 15 percent to 20 percent of your calves, that's a huge loss."

Jay Smith, on a ranch near Carmen, Idaho, did some research of his own, looking at ranchers' BLM use reports for several years. He checked as far back as these records go to assess range losses on the Diamond-Moose allotment.

> I documented how much loss ranchers experienced before wolf introduction, compared with losses after the wolves came. Very soon after wolves were brought in, John Aldous and some of the other ranchers on that allotment had huge losses.

Aldous lost thirty-four calves in one small area near Leesburg the first year.

> Then the wolves scattered out and hit other ranchers besides me. They had dens on the border between our range and the Edwards'. Our elk population dropped from three hundred in Moose Creek Meadows fifteen years ago to where now you're lucky to see twenty elk. The USFWS eliminated the Gerano wolf pack three times, and they just kept coming back. I think some of the offspring that go off on their own come back and regroup.

Fay and Eron Coiner have a grazing ranch high in the mountains above Salmon, Idaho, and spend summers there with their cattle. Fay says they had more than one hundred elk on their ranch, but by the third summer of wolves, they were down to thirteen elk.

> Now we are lucky to see any elk. But we lose at least one calf a year to wolves, even with Eron riding out there every day to try to chase them off. He watched one of our calves being killed, and that was gruesome. Eron was given a permit to kill any wolves that were with our cattle, and he's shot a few, but they still came back.

Coiners worked closely with the USFWS and Fish & Game to help them manage wolves rather than have to kill them. Fay explained:

> We did our best to make the wolves understand that this is our territory and that our place should be off limits. When the government agencies got together to do a flagging experiment around our ranch to try to deter the wolves, it didn't work very well because wolves are curious. They came to see what was going on, and that summer we lost three calves. We've used other deterrents, too, such as rubber bullets and firecrackers to scare them away, but the wolves came right back.

Often wolves don't kill an animal outright, but bite at it numerous times (leaving characteristic scratch marks and bruising the underlying tissues), wearing the animal down, and they often start eating on it before it dies. The Coiners have had numerous maimed and injured cattle that don't survive, even with intensive medical care, which costs a lot of money.

Ross Goddard, a rancher near Tendoy, Idaho, says his cattle on summer range are very unsettled—constantly nervous and on the move.

> They don't graze as much or gain as much. And I can't use my range allotments properly. If we put cows in a certain area and they're supposed to be there for a month, two days later, they may all be out of there and in the next unit, and you can't get them to go back. It disrupts your whole season's grazing management.

His grazing use is under strict observation because of fish habitat.

> We have to keep cattle off riparian areas because of endangered fish, so we shove the cattle into higher country right into wolf domain.

Bruce Mulkey says that after wolves harass cattle, you can't gather or move them with dogs anymore. "The cows are all stirred up, and all they want to do is fight the dogs." Even the cattle that have been trained to work with dogs get on the fight and are overly protective of their calves. Coiners saw wolves harassing a cow-calf pair until the cow was utterly exhausted trying to defend her calf. "That cow was bellowing and frothing at the mouth and on her last legs," says Fay.

Whittaker says cattle are much flightier when wolves are around. "One group of cows ran and hit the fence and tore the gate down, getting into the adjacent pasture and mixing with another group of cattle." McRea says that when wolves go through his private pastures, the cows go crazy, bellowing and running around the field trying to find their calves. Many ranchers have lost young calves—trampled and killed by stampeding cows.

> I had a calf run over; he got rolled around and ended up with a crooked neck, but survived. I had a month-old calf chewed up by wolves, with his intestines hanging out. The vet came, and we sewed him back up and gave him antibiotics, but he only lived two days.

Cattle that survive a wolf attack or calves crippled by stampeding cows can't be sold for top dollar with the rest of the group at sale time. The rancher has to butcher them or sell them at a loss. Ranchers are not compensated for the costs of doctoring injured animals or for cows that must be sold because they lost a year of production due to loss of their calves, which eats into the sustainability of the cow herd and makes it necessary to raise or buy more replacement heifers. Ranchers are also not compensated for lower weights on market calves. McCrea said:

> In the winter, wolves have been coming in to kill at night, and my calves lose weight. I have to corral them every night so they're close enough that I can get to them when the wolves come in. By spring the cattle don't want to go back into the corral at night, but I don't dare leave them a mile and a half away from the house with no chance to protect them. When you get up in the night to check cows, it's horrible that part of your calving equipment has to be a rifle on your back. I was within twenty-five yards of one wolf, right in the corral with the cows, with just a flashlight.

Now he takes a gun with him whenever he goes out into the field. McCrea says he realizes that we can never completely control wolves to keep them from killing livestock, but he points out that it would help if ranchers could get full compensation for missing animals. "If the public insists we have wolves, they should pay the price and help us resolve the problems wolves have created." Many ranchers feel that wolf introduction—together with the impositions it has caused—is a taking of private property by the federal government.

Working with the Agencies

Every state's laws are different in how problem wolves can be handled, and ranchers must work with the state as well as federal agencies. When the wolves were still listed as endangered, however, in all instances, a wolf kill of livestock had to be confirmed before any action was taken to remove the wolves. Rick Williamson, Idaho branch of USFWS, said:

When I'm called to a rancher's place to investigate a suspected depredation, we skin out the dead animal(s) and look for evidence such as bite marks and tracks. We have to determine whether the animal was actually killed—not just fed on—by wolves. There are four categories in our evaluation: confirmed kill, probable kill, possible kill, and other. For a while, the Defenders of Wildlife would pay a rancher 100 percent for the loss if it was confirmed and 50 percent for a probable kill.

Once we confirmed a kill, we conferred with state Fish & Game because they have the final authority to tell us whether we can remove a wolf, or if we can trap and collar it. For instance, we collared a wolf on Dale Edwards's place, and he had a telemetry receiver so he could check for presence of that wolf.

We have some other tactics, like radio-activated scare boxes that we can put out near a den site or rendezvous site to deter wolves. But we have to be careful because, in some situations, we may be just moving them from one producer to his neighbor. I think the best thing we can do, if we have depredation, is to remove the problem wolves.

Now things have changed with the wolves delisted and the state involved in managing wolves.

A few years ago, a graduate student, John Oakley, did a study for his master's degree, looking at numbers of livestock killed by wolves. Williamson explained:

Oakley was trying to determine how many kills you don't find. He and his helpers radio-tagged more than 350 calves on the Diamond-Moose allotment. What Oakley discovered was that, for every kill found, there could be up to eight head you don't find.

In our state, most of the wolf problems with livestock are occurring on private property. People are seeing wolves near their homes. A few years ago, the White Hawk pack was causing problems on Carmen Creek, and over a three-year period, there were more than fifty livestock kills—and all but two of those were on private land. When we later removed that pack, it stirred up a hornets' nest of protest around the country. But people need to understand that wolves have no business being on private property.

Wolves: Smart and Cagey

Control of problem wolves has proven difficult and expensive. Problem wolves are generally relocated or removed by USFWS, but this, too, can be labor intensive and frustrating. Whittaker says one year, when the wolves hit him hard, there was a den two miles from his house, and the government trappers worked all summer trying to get the wolves and only killed two.

> They trapped one and shot one from a plane. It's hard to get one from a plane, because if you don't get them the first time, they hide when they hear it coming.

Allen Bodenhamer, a rancher near Baker, Idaho, says wolves were killing calves in his barnyard at night and hiding up on the mountain in the timber during the day. "The government hunters flew over our area six times trying to find the wolves and never got any of them." During calving season, Bodenhamer and his wife were out checking the cows every hour but didn't see the wolves that came in to kill their calves. "For those wolves to come in between checks with our spotlights, they must have figured out our pattern."

Jay Smith runs cattle on seventy-five thousand acres of public land. He says:

> At least one of us (myself, my wife, my mom and dad, or the hired help) is out there horseback almost every day. We've only had close enough encounters for shooting opportunity about ten times since 1995. It's a lucky accident if a person is able to shoot a wolf. The only two wolves I've seen that were close enough to shoot, I wasn't carrying a gun at the time. Also, if they feel pressured by humans, they go nocturnal and kill livestock at night.

Aldous says you have to be lucky to shoot a wolf. "In our area there's so much timber that we may only get a glimpse of one when we're going down a trail with a herd of cows, and then it's gone."

Wolf Hunts Not Enough

Most ranchers and sportsmen feel the wolf hunts won't be enough to control the expanding wolf population. They also feel that wolf numbers are much higher than federal and state game departments estimate. Only a small percent of the population is actually radio-collared, and there's no way to count them all. Officially estimated numbers have increased enough, however, that wolves in the central Rockies were finally delisted and no longer protected by the Endangered Species Act. Some states have been allowed to manage them as a game animal and initiate a hunting season.

In Idaho, an unlimited number of wolf tags can be sold by Fish & Game, but the hunt in each unit closes as soon as the allotted number of wolves is harvested. For the first hunt in 2009, there were 26,384 tags sold in Idaho—for a chance to kill 220 wolves. Due to lack of hunting success, however, several units remained open after the late December deadline and stayed open into spring. Some quotas were still not filled by summer 2010.

Jay Smith says that having a hunting season is a step in the right direction, but, "You could open the season year round, and we wouldn't put a dent in the wolf population."

Whittaker feels the hunting season won't make much difference. "Amateur hunters just make them more elusive. If you shoot at a wolf and miss, he won't give you that opportunity again."

With the increasing impact on elk herds in certain areas of the state, the Idaho Department of Fish and Game (IDFG) now realizes there must be more control of wolves than the hunting seasons are providing. As of January 2013, the IDFG began considering two more ways to work toward its goal of reducing current wolf numbers to a manageable level without risking coming under federal protection again. The first proposal is to take fifty thousand dollars from their coyote control program in eastern Idaho and use it to supplement wolf control efforts in areas where elk herds are in decline. The second idea is to develop relationships with successful wolf trappers and possibly give them financial assistance to offset some of their costs, as long as the trappers are working in those same areas where elk are being significantly impacted.

Wolf-Cattle Interactions Study

The increasing wolf population in the West is having a significant impact on ranchers and livestock, but the extent of this impact had not been well documented. The Oregon Beef Council (with help from USDA) funded a ten-year study project that began in 2008 to look at how wolf activity affects cattle behavior. GPS collars are being used on some of the cattle on ranches and a few wolves in the study area to monitor movements of these animals and when and where they interact. Dr. Pat Clark, range scientist at Northwest Watershed Research Center, USDA-ARS at Boise, Idaho, has been involved with this study from its beginning and was instrumental in helping with the GPS collars. He had used collars on range cattle for an earlier study looking at how cattle changed their use patterns after a fire. Those first commercial GPS collars cost about $5,000 apiece, so Clark decided to build his own. With the help of some Boise State University engineering students, he created a more functional, less expensive GPS collar. By 2008, they had one that could log GPS data every five minutes for more than a year and cost less than $500 each.

Clark began using the collars in a pilot wolf-cattle interaction study in central Idaho in 2005, to determine the effects of wolf presence on cattle range-use patterns. Part of the challenge of GPS collaring a wolf was getting the collar into a wolf-sized package. A cow can carry a fair amount of bulk and weight, but wolf collars had to be smaller and lighter.

The Oregon Beef Council (OBC) heard that Clark was using GPS collars to collect data and to address some research questions about the wolf-livestock issue in central Idaho. The OBC realized wolves would soon be coming into Oregon from Idaho and wanted to evaluate how this would impact rangeland cattle production in their state. The OBC asked Clark to extend the wolf-cattle interactions research he was doing in central Idaho to the border area of northeastern Oregon (where there were no wolves yet) and to western Idaho (where wolves were common). The proposal for the study was developed and submitted to OBC in 2007. The Oregon/Idaho Wolf-Cattle Interactions project began in 2008 with three study areas in western Idaho and three in eastern Oregon. Clark said:

> We selected our Idaho study areas from the western part of the state where range cattle grazing is the dominant land use and where wolves were common. We searched for study areas in Oregon that had similar topography, vegetation types, soil types, cattle breeding, calf age at turn-out, grazing schedules, and so on.
>
> This type of experimental design is called a BACIP (Before-After, Control-Impact, Paired) design. The Idaho study areas served as the Control, since wolves were already present. What would eventually change (and create impact) was wolf presence in Oregon. Wolves were expected to travel from Idaho into Oregon and establish packs in the Oregon study areas.
>
> In Oregon, we wanted to collect at least a couple years of "Before" period data before wolves arrived in enough numbers to have a true impact, and then we could contrast these "Before" data to "After" period data collected after wolves set up shop in Oregon.
>
> We put GPS collars on at least ten cows in each study area. Cow-calf herd sizes in these areas were 300 to 450 cows. The sixty-collar total sample size was limited by available funds from OBC.

The study was later expanded to include two more paired ranches, providing a large-enough sample size for adequate data to detect differences in cattle range use and other behaviors. GPS data tracks the location of collared cows and determines how fast they are traveling, enabling the researchers to classify cattle activity into resting, foraging, or traveling. Clark explains:

> From this, we can calculate how much time a cow spends foraging during the day versus standing looking around or bedding. A

hyper-vigilant animal will spend a lot of time standing and watching, rather than foraging, which could impact nutritional status. With GPS data collected every five minutes, we can detect changes such as increase in vigilant behavior and decreases in foraging time.

One of the Idaho study sites is the OX Ranch near Council, Idaho, managed by Casey Anderson, where data has been collected since 2008. This ranch experienced wolf depredation before the study started, but didn't have documentation.

> In 2009, we were able to confirm more of the wolf kills—eighteen animals—but were also missing five cows, two yearlings, one bull, and about seventy calves that were unaccounted for.

There were not many wolves in Oregon when the study began—just a few passing through and returning to Idaho. Within two years, however, the wolves coming from Idaho established packs in northeastern Oregon, and the study moved into the "After" design—comparing high wolf pressure with low wolf pressure (rather than no wolf pressure). "Having wolves and cows collared enables us to match up their location and time and to determine where and when wolf-cattle interactions are occurring," says Clark. Data on wolf presence and activities is augmented by wolf scat sampling along forest roads and use of trail cameras.

Interesting Study Results

Even though only one wolf was collared in 2009, this gave some interesting results, said Clark:

> We found interactions between all ten of the collared cows and that single GPS-collared wolf within a very extensive study area (over fifty thousand acres). The ten collared cows were dispersed throughout the landscape, part of the larger herd of 450 cow-calf pairs, yet the data showed this wolf was within five hundred yards of every one of the ten collared cows many times throughout the summer (for a total of 783 contacts).

The collared wolf's area was 210 square miles with a fifty-five-mile perimeter. The least distance he traveled was about six miles per day, and the most, twenty-nine miles per day. "On a day of confirmed depredation, we could look back to the data and see that the wolf was in the immediate area at that same time," he says. This wolf was a member of an eleven-wolf pack. There were three different packs, totaling thirty-four wolves, roaming that study area during 2009.

The rancher on that study area suffered more than forty confirmed or probable wolf depredations.

The collared cow with the least number of interactions with the collared wolf had twenty-three interactions within five hundred yards during that grazing season. The cow with the most encounters with that wolf interacted 140 times over the course of 137 days (mid-June through early November 2009—when the collar was removed from the wolf to retrieve the data. All ten collared cows had interactions within 250 yards, and only one of the ten cows didn't have interaction at a hundred yards or less. Envision 450 cows with more than thirty wolves moving among them. If one of those wolves had 783 encounters with just ten of those cows, think how many encounters there must have been by thirty wolves among the 450 cows; in other words, the cattle were constantly impacted by the presence of wolves. Two of the collared cows' calves disappeared that summer. Clark said,

> If the behavior of this wolf was representative of the larger local wolf population, it is easy to imagine that this cattle herd, and likely other neighboring herds, was exposed to an intensive wolf predation threat.
>
> We wondered how many of these GPS-detected wolf-cattle interactions resulted in depredations, but we will never know. We did go back and look at the wolf GPS data and compare the tracking logs with locations of known 2009 depredation sites. We could see tight, spiral patterns in the wolf movements that occurred at those sites. These circling activities may illustrate prey-appraisal or pursuit events.
>
> The following spring of 2010, we hiked into the sites where spiral patterns had occurred in the 2009 wolf log and inspected the sites, looking for signs of depredation, and found fresh cattle bones at one of those sites.

Rendezvous Sites

Certain areas are used by wolves to park their pups with a babysitter wolf while other members of the pack go back and forth on hunting excursions. Dave Ausband, a researcher with the Montana Cooperative Wildlife Research Unit at the University of Montana, developed a wolf rendezvous prediction site model and, from that, created a predicted wolf rendezvous site map for Idaho. Clark said:

> These rendezvous sites anchor the distribution patterns of the entire wolf pack to specific points. Out of curiosity, we plotted both our wolf GPS data and our cattle GPS data from one of our Idaho study

areas onto Ausband's map and found the clusters of concentrated wolf location data lined up with areas predicted on the map as being high quality wolf rendezvous site habitat. We wondered if the map and model could be used to predict where wolf-use patterns and cattle-use patterns frequently overlap—increasing the potential for wolf depredation.

Wolf rendezvous sites tend to be in grassy areas close to water—not necessarily riparian areas, but grassy meadows without a lot of overhead cover. These areas are also good foraging areas for cattle. Rendezvous site information could enable ranchers to be aware of certain danger areas and to check for signs of depredation in those areas. They might choose to not put cattle into those areas until later in the season when the pups are grown and the pack is not tied so closely to those sites.

Indirect Effects

Clark explained about the indirect effects of wolves upon cattle. "There was some earlier research done on death and injury losses due to wolf predation on cattle, indicating that these impacts are underestimated," says Clark. Some cattle just disappear; a pack of wolves can consume an entire carcass overnight.

> However, no one has researched the indirect effects of wolf presence on rangeland livestock production. These indirect effects may have much larger impacts on livestock producers than livestock death and injury losses to wolves. Indirect effects of wolves on livestock range-use patterns include impact on foraging efficiency, disposition, and stress levels.
>
> These effects could cascade down to affect cattle diet quality, nutritional status, and disease susceptibility. Cascading further down, wolf presence may indirectly affect (and reduce) calf weaning weights and cow body condition in the fall and result in increased winter feeding costs, along with conception and pregnancy failures, increased veterinary care and supply costs, and death loss to disease. This complex cascade of effects could substantially impact a ranch's bottom line.

Casey Anderson at the OX Ranch was weighing and assessing his cows when they were collared in the spring before turnout and again when they came off the range to be hauled down to winter pasture and the collars removed. "What we were seeing across the board in our herd was that the cows were coming home a full body score less than they had been in the past. With a mature cow, each score is about a hundred pounds," he explains. This means extra feed costs to get them through the winter. "We were also seeing the conception rate plummet. With our herd health program and mineral and protein supplement, there was no reason for this other than the wolves. What normally would be 90 to 95 percent conception rate in our herd went down to as low as 82 percent," says Anderson.

No Single Effect, But Significant Losses

Some of the direct impact went beyond actual kills; additional animals were severely maimed. "Although we spend a lot of time doctoring these and trying to save them, we end up with a bunch we can't sell because they are crippled," Anderson says. During the summer of 2012, he had nine calves with wolf bites in just one week. "The alpha female in the pack is using the calves to teach pups how to hunt. These injured calves may live, but you have to doctor them." In addition, four young cows in the first calf-heifer group had big abscesses behind the shoulder and/or above the flank area, due to infected wolf bites.

Wolf depredation is generally not a single effect; it usually stimulates a change in cattle use patterns and different areas for grazing. "We are also seeing changes in the way cattle use the range. They are bunching up more against fences, and wolves are using the fences to corner them," says Anderson.

This may be a shift from a high-quality foraging area to a low-quality foraging area just because cattle are worried about wolves as Clark explains:

> Cattle may get up on an open hillside where they can see farther to detect if wolves are approaching, but the time of year they are doing this might be when the grass is so dried out it doesn't have much protein—so cattle diet quality declines. Cattle under threat bunch up, standing in open areas where they feel safer, but these areas soon become dry and dusty from concentrated trampling, and the rising dust can lead to respiratory disorders—especially in stressed cattle that have become more vulnerable to disease.

If cattle are being bred on the range, as is the case for the majority of outfits raising commercial cattle, all these factors work toward decreasing the number of cows that get bred on time and increasing the number that end up breeding late or coming up open in the fall, according to Clark. Late calves change the uniformity of the calf crop, which adversely affects the price received for those calves, which is another financial impact. This may also affect the number and quality of replacement heifers a rancher is able to keep, and the future productivity of the herd is adversely affected.

Cattle Temperament Changes

There are also changes in cattle temperament. Clark explains:

> In one of the central Idaho herds, a wolf followed cattle down to the calving areas, and the cows were very aggressive in trying to protect their newborn calves. The rancher couldn't tag a calf without being attacked by the mother cow; he had to get the calf into the back of a pickup to safely do anything with it.

Anderson says that before wolves came, their cattle were all dog-broke and easy to work with dogs.

> Now our cattle chase dogs; we can't use dogs anymore to herd them. In this steep, rugged country, dogs have always been a good tool. Now it is extremely difficult to move cattle.

The OX Ranch calves in late May through June. The calves are branded when the cattle are gathered in the fall. "By that time, the calves are 350 to four hundred pounds, and when they come into the corral, they size you up and take you," he says. They are totally focused on defending themselves, attacking a dog or a person. "This is not normal behavior. These cattle were all nice to handle before the wolves came. Now they attack anything that comes close to them," says Anderson. He had a horse that had to be sold after it was run through a fence by wolves. Even though the horse's injuries healed, it was no longer safe to have dogs near the horse as it would strike and kick at the dogs. Similarly, threatened, stressed cattle are unpredictable and often become more aggressive, especially when people are working on foot, but also in rangeland situations where ranchers are trying to herd them on horseback. If cows are prone to fight or take off and go through fences, this affects ranchers' costs, and injury risk for ranch personnel also becomes an issue, as Clark explains:

> There are more handling-related injuries to cattle that have been exposed to depredation threat, along with increased frequency of bunching and flight events. Handling flighty, nervous cattle may lead to more trampling deaths of small calves and broken legs in larger calves. Nervous cattle are more prone to chute-crashing, ramming fences, crawling up corral walls, and other flight-type behavior, which can result in spinal injuries, rendering the injured animals unmarketable.

Dr. John Williams (OSU Extension) says one of the purposes of this ten-year study is to try to understand how cattle react differently due to presence of wolves. "We are finding that cattle temperament changes drastically when they have to live with wolves," says Williams.

> This winter (2013) we hope to put collars on cattle, do blood tests on the ones that have been living among wolves, and compare their stress level (measuring cortisol levels in the blood) with cattle that have not been living with wolves. We know that cows are individuals, and some will be stressed more than others.
>
> We want to find out how long this effect lasts. If we were able to measure the stress on a cow that is attacked, her cortisol level the next day would be very high. Our ranchers have been telling us that many of these cows' changed behavior lasts for a while. In some cows, it becomes less pronounced, in some cows it disappears, and in others it does not.

Neil Rimbey, a range economist with University of Idaho, has been evaluating economic impact on ranches in the study, using economic models to assess the economic impact of management alternatives. The depredation loss (killed animals) is what gets attention, but the first years of this study have shown that indirect losses have more impact on ranch profitability and sustainability. Rimbey says:

> A recent study in Wyoming indicated that ranchers are able to find only about one of every seven animals killed by wolves, so in Wyoming they are compensating ranchers on a one to seven basis. This research in Wyoming also suggests that some of the indirect losses—more open cows, reduction in weight gain on calves, more veterinary treatments for injured calves or for stressed calves that got pneumonia—should raise the compensation rate (from one to seven) up to one to thirteen or fourteen.

The change in cattle behavior in the presence of wolves impacts management. Cattle may not stay in the areas they are put—for instance, coming right back down off the mountains to try to get away from wolves. They may use some areas more heavily while avoiding others. Cattle also crash through fences and are harder to handle. Rimbey says:

> The Oregon and Idaho ranchers have mentioned increase in time, labor, and costs associated with managing cattle. There's more travel involved; they have to go more frequently to try to check cattle. There is also more time spent meeting with Fish and Wildlife Service to try to get confirmation on death loss, etc.

The study has confirmed things ranchers suspected regarding behavioral changes in cattle, and it will be interesting to see how the economic impacts calculate out.

> For instance, each 1 percent change in the percentage of calves weaned in Idaho (using conservative prices) is amounting to about a $1,750 change in gross revenue, or nearly six dollars per cow . . . And this is just a starting point of this component of the project.

Wolf Threats to Humans

Casey Anderson at the OX Ranch says that when wolf numbers are controlled, they have a tendency to stay farther back in the high country. Anderson's wife has found wolf tracks in fresh snow less than fifty feet from the house. Anderson says:

> We've had wolves lie on the county road in the snow at night, watching cattle under a light at the end of the barn. You could see where the wolves had come down next to the corrals and were lying in the snow watching the cattle.

Data from the collared wolf showed how close these wolves are coming to homes and human activity. "We have several houses here on the ranch. The collared wolf came within five hundred yards of one house 307 times that summer," says Anderson. The pack of twelve came within three hundred yards of the ranch lodge and spent all day there.

> The people who take care of the lodge have three little boys. The wolves were there all day, right above the county road in a little clump of timber, and watched the lodge. We had proof because the collared wolf was there with them in what we call a rendezvous site.

Many ranchers are reluctant to be out doing routine tasks, such as hiking around their pastures fixing fences, without a gun. Says Jay Smith, a rancher near Carmen, Idaho:

> This maneuver by our federal government is stealing from me, my children, and our future. It may destroy our livelihood, and our entire lifestyle is also in jeopardy. I have small children, and I like to take them with me to the hills when I check cattle or fences, and now it's a risk.

Ralph McCrea near Leadore, Idaho, tells of one incident a few years ago when his grandchildren were sledding on the hill behind his house.

> A couple hours later, I went out to check the cows and found a wolf standing in the herd, just forty yards from where the kids had been sledding. They had our old dog with them, and if the wolf had arrived a little sooner, it would have been attracted to that dog and the kids. We're being pulled right back into the 1890s when everyone had to be armed to defend against predators.

Wolf advocates say that wolves are shy and stay away from people, but this is proving to be untrue. There have been increasing numbers of incidents the past dozen years in which wolves have threatened, attacked, or killed people. Wolf advocates try to discredit many of these stories, however, as they did an earlier attack of a vacationing family on a pack trip into the Middle Fork of the Salmon River when the wolves were still under full protection.

From Fishing Vacation to Nightmare

Tim and Diana Sundles, who lived on Carmen Creek near Salmon, Idaho, had taken their teenage sons into a remote area where they had been camping for many years. They had ridden the sixteen miles from the trailhead with their horses and mules and set up camp. Sundles recalled:

> We woke up the next morning before daylight, hearing wolves howling and horses screaming. I ran out of the tent with a flashlight and small revolver, hollering and shooting into the air to run the wolves off. I assumed it was over, that the wolves were afraid and left. We cooked breakfast, sat around camp an hour or so, and then got our fishing gear ready to hike to one of those remote lakes to spend the day, leaving the horses and mules at our camp, with a couple of them turned loose to graze. I picked up my rifle as we started to walk out of camp and commented to my wife that I had an uneasy feeling.
>
> We got about a hundred yards where two pack animals were grazing, and a big grey wolf was sneaking up on them. In the back of my mind, I was wondering where the rest of the pack was. I fired a shot over the wolf to scare it, and it turned on me. I didn't know at the time it was the alpha male and radio-collared. When I fired the shot over it, the wolf came straight at me, full speed. I fired two more rounds, trying to hit it, but a wolf running at you through the trees is hard to hit!

The wolf came within ten feet of Sundles, then veered off, and circled around the family, possibly because he'd been shot in the foot, as Sundles discovered later, looking at the body.

> But he didn't act lame or even slow down. He turned toward my wife on the trail behind me, and I finally had a good broadside shot and killed the wolf. It dropped about ten feet from my wife. Then I saw the radio collar and ear tags and realized our vacation was over.

Sundles went home and tried to find an attorney familiar with the ESA but had no luck.

> The law said you've got to report the shooting of a wolf within twenty-four hours, but it took us forty-eight hours to get out of the wilderness so I was already in violation of the law by the time I found out about it.

Sundles was going to keep quiet about shooting the wolf, because of the heavy penalty involved, but then a few weeks later decided to tell people—at a public meeting in Salmon that August when US Senator Mike Crapo was holding a hearing to get local input about delisting the wolves. Sundles said:

> If the wolf attack happened to us, it could happen to anyone, and I felt the public needed to know that wolves can be dangerous. One of the big lies the US Fish and Wildlife Service and the environmentalists are telling us is that wolves are not dangerous to humans. If I kept quiet, the next time it might be a little kid, and then I would have felt partly responsible. Here we were, on vacation, suddenly thrust into this situation, and now I'm feeling guilty! A lot of folks who are out camping or fishing don't have a gun. If I had not picked up my rifle when we headed out from camp, we would have had a serious problem!

Protected

Wolf packs in Washington are on the rise, migrating in from Idaho, Oregon, and Canada, preying on livestock and game herds. Ranchers in northeastern Washington have been especially hard hit. Len McIrvin, his son Bill, and grandson Justin, who run Hereford cattle on their Diamond M Ranch, have suffered alarming losses, as Len explains:

> We had wolf damage starting a few years ago, but not this severe. In 2011, we had sixteen head killed; in 2012, it was forty head. One of

our herds on Forest Service land and another herd of 350 on our private land took the brunt of it.

Wolves are efficient killing machines. It's easy for them to take down mature cattle.

> In 2011, we had five bulls lost to wolves on our range. In the fall after breeding season is over, they tend to go off by themselves and are easy targets for a pack of wolves. When wolves find one bull by himself, he doesn't have a chance. Though most of the time they rip the animal open, we've had many calves killed that they never took a bite of. They sometimes just chew him up to kill him, and once their plaything is dead or dying, they go on to the next, like a cat playing with a mouse. For a long time, our Fish and Wildlife people refused to believe that wolves were killing those calves. You don't see any blood when a cat kills a mouse, and it's the same with these calves; they are mauled to death.

There may be a few scratch marks on the hide from their teeth and sometimes no marks at all, but often there's no open wound—just a lot of bruising and internal bleeding. The McIrvins found one five-hundred-pound calf dead and couldn't find a mark on it.

> We hesitated to call Fish & Wildlife, but we did, and when they opened that calf up to do a necropsy, they saw the muscles under the skin were pulverized, like hamburger. The calf died of internal injuries.

NAR-Wolves

Then there's the problem of relocated wolves, wolves already in the habit of killing livestock, as Len explains:

> Wolves are showing up in every little hamlet and community, but it's scientifically impossible for this many breeding pairs of wolves to crop up simultaneously in an area two hundred miles by two hundred miles. These wolves are suddenly everywhere. The Fish & Game call them NAR-wolves—non-authorized release. One group of wolves in our area has been particularly aggressive on cattle, and I have a feeling they are problem wolves that were gathered up from somewhere else and dumped out here.

Even though wolves have been delisted in the wolf management areas in Idaho, Montana, and eastern Washington, the state of Washington is still protecting them; it is illegal for a Washington rancher to shoot one. Len said:

> The eastern third of Washington is part of the Rocky Mountain Management Area where the Feds delisted wolves, but Washington State made them stay endangered, with all the ramifications of federal protection.

And in the western two-thirds of the state, the wolves are still on the federal endangered list, which causes its own set of problems, as explained by Len:

> We had help from the State Senate and House of Representatives as we were dealing with our depredation problem, and the county sheriff was there for every kill we found, documenting it. The county commissioners were there, and we got a lot of press from our local newspaper and radio station. The US Fish and Wildlife finally had to recognize the problem and made the decision to eliminate this pack of eight wolves. They killed some, but didn't get them all, and there are plenty of other wolves.
>
> We have very little game left. The only deer and elk we see now are in our back yards. Mule deer that are usually in the high country spent this past summer in our hayfields and pastures by the house. The same thing has been happening in Idaho the past dozen years; elk and deer have been coming down out of the mountains to take refuge on private land, amongst the cattle, and congregating in populated areas.
>
> We try to keep our cattle up high in the good feed on summer range, and the wolves run them right back down to the bottom fences. Even if we make an eight-mile drive one day to take cattle back up, the next day they are down again.

This makes it hard to manage a range. The Forest Service people are not happy if cattle are spending the summer down in the riparian areas, but when wolves keep harassing them, the cattle won't stay in the high country. Len stated:

> Ranchers in Idaho, Montana, and Wyoming have been having wolf problems for nearly twenty years, but it's fairly new here—within the last five years. One of our neighbors runs about eighty cattle. The first seventeen cows he brought home from the range last fall had only two calves with them. The other fifteen calves were almost certainly killed

by wolves during the summer. His losses are horrific and will probably put him out of business.

The winter before, wolves harassed that rancher's cows all winter; he finally fenced them into a two-acre pasture next to his house and barn. At that point the wolves hadn't killed any yet, but the cows had trampled several newborn calves when they were upset by wolves. "I've estimated our loss last year at about a hundred thousand dollars but it's hard to know exactly what it is," says Len. Losses go far beyond actual kills, when injuries, weight loss, and lower conception rates in the herd are taken into consideration.

Cattlemen in eastern Washington have tried using non-lethal means to scare off wolves, but these methods haven't worked. Ranchers in Washington are hoping for delisting of wolves as an endangered species before another problem pack has to be removed at taxpayer expense. There are at least seven or eight other packs in this three-county area. If ranchers could be allowed to address wolf depredation as they address other predator issues, major problems could be minimized.

Political Pawns?

Len McIrvin said:

> My son Bill went to the state capitol in Olympia and testified on the wolf issue. He said the room was full—more than five hundred people, predominantly women—and he could feel their hatred. They were screaming at him and crying.

They really believe that wolves are harmless and that ranchers are evil in trying to control them. Len recently saw an elementary school assignment, given by a teacher in the Colville School District north of Spokane, Washington.

> This was a class assignment paper for third graders to fill out. It was titled "The Three Little Wolves and the Big Bad Pig: Building Vocabulary." The students were to fill in the blanks in these sentences: "The three little wolves look soft and _____. In order to sneak up on the wolves, the Big Bad Pig came prowling through the trees. The Big Bad Pig grunted because he was big and bad. The wolves were scared and began trembling. Each time one of their houses crumbled, the wolves were determined to build a better one.
>
> We grew up with the Big Bad Wolf and the Three Little Pigs, and this example is a total turn-around. It's a warped mentality that would come up with this for a third grade worksheet in public school.

He feels the wolf introduction into the western United States and their protection is NOT about the wolves. It's about depopulating the West.

There are many people—including some high up the government—who want a ninety-mile corridor from Yellowstone to the Yukon (Y to Y) for wildlife. They have an agenda, and wolves are part of the plan. Ranchers don't have much say in these issues; all of agriculture is less than 2 percent of the population in this country. All we do is feed the world, but most people don't care about that.

Before the wolf plan was finalized in Washington State, I was in a meeting and asked the Fish and Wildlife person, 'If I go home tonight and see a wolf pulling down a cow, can I shoot the wolf?'

They said, 'Absolutely not, unless you want to pay a fifty thousand dollar fine, spend ten years in jail, be a felon the rest of your life, and never be allowed to own a gun.' By contrast, if I was sitting at home and someone knocked a window out with a baseball bat and came in after me, can I shoot him? They said, 'Absolutely! That's your Constitutional right! You can shoot a human in self-defense, but you can't shoot a wolf that's killing your animals.'

They told us that, if they find a dead wolf in our area, they would have a complete investigation. Every firearm the rancher and crew own would be confiscated until ballistic tests had been run. By contrast, right across the state line in Idaho, it's legal to shoot wolves that are harassing or killing cattle, and anyone can buy a wolf tag for the wolf hunting season.

We've had a lot of damage this year, yet we've only seen the wolves a couple times and never had a chance to get a shot. It was usually a brief glimpse in the headlights of a vehicle in the middle of the night or when we didn't have a gun.

The wolf problem will make ranching very tough, especially in areas like Washington and northern Idaho where there's a lot of timber for them to hide in. It's a little easier to hunt them in open areas like eastern Montana. However, it's impossible to keep wolves under control just by hunting them. Idaho's first wolf hunt was a perfect example. Everyone thought it would be great, and the Fish & Game Department sold thousands of tags but it was unable to fill the quotas. Lots of people were eager to shoot a wolf. The hunters bought tags thinking they could shoot a wolf when they were out hunting deer and elk. But they didn't get much opportunity because wolves are elusive. Now the novelty of hunting them has passed, and the wolves are still expanding. In Idaho, the Fish &

Game Department can't begin to keep up with the natural population increase.

Cattle and wolves are never going to be compatible. Some people keep saying that wildlife people and cattlemen are going to have to learn to manage for wolves, but that's impossible. It's hard to "manage" when you are always on the losing end.

Ranchers have always had challenges, explained Len:

We've fought cougars, bears, and coyotes killing baby calves, but those problems were nothing compared to wolf depredation. And this has become a political thing. Our state Wildlife Commission says they've had ten thousand e-mails from people who want wolves and only a few from people who don't. They forget that we have a Constitutional right to defend our property. But the Wildlife Commission is swayed by the mob, and there goes our rights.

There were always plenty of wolves in Minnesota, Canada, and Alaska (at least fifty thousand to sixty thousand in Canada). They were never endangered. Now Alaska is in trouble because wolves have decimated wildlife in many areas. The state was going to start eradicating wolves with planes and helicopters. Then environmental groups and wolf advocates started a boycott on tourism, and Alaska backed down. The wolf is now the spotted owl for the ranching industry.

Conclusion

Even though the wolf was finally delisted, there are still political agendas at hand, and certain elements are fighting for continued protection. Some people believe the wolf issue is a ploy to eventually do away with hunting (since, theoretically, wolves can keep game numbers in check without human management) and then our guns.

Laws that supersede our other laws and rights, such as property rights and the right to protect one's family and animals, are dangerous. There is growing concern among many Westerners, especially ranchers, that the wolf, like the spotted owl, is being used to put unacceptable restrictions on land use, both public and private. The Endangered Species Act (ESA) is a dangerous law. Few laws have been so divisive or driven by such emotional forces. The ESA is being used by preservationists and pseudo-environmentalists to lock up natural resources and halt logging, mining, grazing, and traditional use of private property.

Hero worship puts movie stars, sports, and music superstars in a league of their own, untarnished by their actual failings and foibles. Wolves have come to hold that same god-like status, despite the realities of what living with them is really like.

> "Historical and current evidence indicates that one can co-exist with wolves where such are severely limited in numbers on an ongoing basis so that there is continually a buffer of wild prey and livestock between wolves and humans, with an ongoing removal of all wolves habituating to people and domestic animals. The current notion that wolves can be made to co-exist with people in settled landscapes is not tenable."
>
> —Dr. Valerius Geist

CHAPTER 9

• • •

The Unspoken Costs of a Growing Wolf Population

By Rob Arnaud and Ted B. Lyon

"It is dangerously thin ice on which environmentalists expose the economic costs of public lands ranching while at the same time selling wolves as a potential economic boon."

—Martin Nye, PhD[1]

Wolf recovery in the Northern Rockies from 1973 through 2005 has cost about $21 million. Wolf recovery now costs about $2.7 million per year in federal tax money. The government has now spent about $100 million over the last several years propping this population of wolves up.

—Fox TV News, March 19, 2012[2]

"We don't want them (wolves) here."

—Idaho Governor "Butch" Otto, Fox TV News, March 19, 2012[3]

When developing a natural resources management program plan, it's customary to consider economic benefits and costs. By now you've learned a great deal about some of the consequences of living with wolves—ecological, social, health, and economic. This chapter will summarize the economic costs, which are considerable and are given precious little attention in the media.

The economic benefits of wolves are mostly limited to the expenditures associated with wildlife watching. According to a 2001 US Fish and Wildlife Survey, there are about sixty-six million "wildlife watchers" in the United States, 85 percent of whom pursue birds. In contrast, only about 56 percent of these "wildlife watchers" pursue large land mammals, and their favorite target is deer. According to the survey, they spend about $38 billion a year in pursuit of wildlife.[4]

The epicenter of the 1995 Northern Rockies gray wolf transplant was Yellowstone National Park, although wolves were also released in central Idaho. Early economic forecasting by the US Fish and Wildlife Service predicted that increased tourism from people hoping to see a wolf would account for approximately twenty-three million dollars a year in additional income for the Yellowstone economy.[5]

This forecast claimed to take into account decreases in income from reduced hunting opportunities due to wolf predation and promised livestock replacement costs. However, this study was based on populations of only one hundred wolves in each of Idaho, Montana, and Wyoming. As of December 31, 2011, the USFWS estimates that there are at least 1,774 wolves in Idaho, Montana, Wyoming, eastern Washington, and Oregon.[6] Many others believe the wolf population of the Northern Rockies is actually closer to three thousand, perhaps even higher.[7]

Do more wolves translate into more or less economic value?

Most definitively less.

The original US Fish and Wildlife Service's economic study did not contemplate the costs associated with wolf populations six to ten times larger than the anticipated three hundred wolves, nor did they consider all costs of wolves to people living near them, such as earlier chapters have described. The primary focus on this chapter is to provide an overview of the economics of wolf populations in Montana, Idaho, and Wyoming—the transplanted "experimental, non-essential" population. Additionally, we will also include some economic data from other areas with wolves: the Northern Rockies population is spreading to Washington, Oregon, Utah, and Colorado; the Mexican wolf restoration program underway in Arizona and New Mexico; there is a growing population of wolves in New England; and the at least three thousand wolves that reside in Minnesota, Michigan, and Wisconsin.

Overall, there may be as many as one hundred thousand wolves in North America today, and the population is growing. For the most part, aside from Yellowstone National Park, wolf watching will never be a significant economic draw because habituated wolves living near people are the ones that are removed one way or another, and the survivors in the wild quickly learn to be invisible to survive.

Projected Economic Benefits of Wolves

An often-quoted 2005 study of the value of wolf eco-tourism in the Yellowstone area by University of Montana economist John Duffield estimated that at least $35 million per year economic benefits accrue from wolf ecotourism in

Yellowstone National Park. Duffield and his co-researchers further claim that the ultimate total benefits could be doubled as a "multiplier effect," as wolf eco-tourism money circulates in the local economy. The study was funded by the Yellowstone Park Foundation, which "is the official fund-raising partner of Yellowstone National Park," and a pro-wolf advocate.[8] The study did not seek to establish if people would come to Yellowstone regardless of if wolves were there or not.

Over two million visitors per year have come to Yellowstone since the 1960s. Actually, from 1995 to 2006, annual visitation to Yellowstone declined from 3,125,285 to 2,870,295. Visitation rose in 2009 and climbed to a high of 3,640,185 in 2010; however, this reflects a national trend in increased domestic outdoor recreation during tough economic times. In 2011, annual visitation declined 6.75 percent overall and in every month of the year to a total of 3,394,322. In 2015 it rose to just over four million.[9]

The largest visitation by far is in the summer months of June, July, and August, when seeing wolves is least likely to occur. Additionally, many wildlife watchers who do see a wolf are satisfied with a one-time sighting of a species for their bucket list, and then they move on to their next quest.

Duffield's data were gathered from a random survey of 2,992 visitors to Yellowstone National Park between December 2004 and February 2006 selected from two locations: park visitors (sampled at park entrances) and Lamar Valley visitors (sampled randomly at parking locations throughout the valley where wolves are most likely to be seen). Nearly three thousand potential respondents were contacted for Duffield's survey, but only 1,943 questionnaires were completed and returned, an overall response rate of 64 percent from a tiny fraction of all visitors—.05 percent. The 46 percent rejection rate also suggests a bias in respondents toward people who wanted to make a pro-wolf statement. Additionally, there is no way to check the validity of their responses.

The 2005 study took place during a time when the wolf population in Yellowstone was rapidly approaching its maximum number, close to at least two hundred, and many pro-wolf articles and TV programs were flooding the media. Relatively few of these reported how local ranchers and hunters were concerned about the impact of the rapidly growing wolf population on livestock and big game, especially elk. Survey respondents who were in support of the wolf restoration program and were aware of the growing controversy would have had a compelling reason to participate and even enhance their reports in favor of wolf-watching benefits.

Duffield also asked people what animals they would like to see while in Yellowstone. While 44 percent said that they wanted to see wolves, an even more popular sighting goal was seeing grizzly bears.[10]

When Duffield estimated economic losses to hunters, ranchers, and farmers from wolf populations in the Greater Yellowstone Area, he concluded that the benefits outweighed the costs by between six million dollars to eight million dollars. As we shall soon see, his calculation of costs leaves much to be desired. In 1992, Duffield co-authored a study predicting a 4 to 5 percent increase in tourism due to wolves. He claims that his 2005 study "proves" the accuracy of his 1992 predictions.[11]

A 2005 study of wolf-watching tours in Yellowstone led by private eco-tourism operators was conducted by outdoor educator Jim Halfpenny, PhD (author of two books on wolf watching, and who leads wolf-watching tours). He identified thirty-four outfitters who offered tours to see wolves. Twenty-seven of the outfitters responded to this survey. The twenty-seven tour guides offered 569 departure dates for 6,165 participants, at an average cost of $761 per person. The costs varied from $45 for one day to $3,300 for seven days; a total of $4,690,134 for 2005. Other programs where wolves were part of wild-life-related experiences generated another $234,348 (494 people). Halfpenny concluded these programs generated $4,924,482 in 2005.[12] It is not known if tours continue to be as popular, and again, many wildlife watchers only want to see one of a species to add it to their Life List.

Estimates from Duffield's Yellowstone USPS visitor survey indicate that conservatively 325,000 park visitors saw wolves in 2005, which is about 1 percent. Since 2005, Yellowstone National Park wolf populations have declined almost 50 percent in part due to a combination of factors: a dramatic decline in the local elk and moose populations (the favorite food of wolves in that area), deadly battles between wolf packs for dominance, possibly drought, and wolves contracting distemper and mange, and suffering from malnutrition. Thus, the visitor to Yellowstone today is less likely to see wolves than when the Duffield and Halfpenny studies were conducted, and those wolves that are seen often are not as healthy appearing. In addition, elk and moose numbers in Yellowstone, which were a major tourist attraction, have plummeted.

Some outfitters have developed wolf watching tours as a way to compensate for losses in elk hunts, but as rancher-outfitter John Robbinette, who has tried offering tours to offset lost hunting revenues, reports, by 2005 tourists were often more interested in seeing grizzly bears and geysers than wolves and interest in wolf watching has ebbed.[13]

When the Wyoming Sierra Club surveyed visitors to Yellowstone in 2005, asking them why they came, the number one reason given for visiting the park was the scenery. The number two reason was watching wildlife. While 44 percent said they would like to see wolves, seeing grizzlies was more important to most. Only 3.5 percent said that they would not come if there were no opportunity to see wolves.[14]

It's likely that the Yellowstone wolf population may never again rise to the 2005 level. This in itself will result in declining economic benefits of wolf watching. The 1994 federal EIS predicted that presence of wolves would result in a 5 to 10 percent increase in annual visitation to Yellowstone National Park, which would generate twenty million dollars in revenue to the states of Idaho, Montana, and Wyoming.[15] The drop in visitation from 1995 through 2009 suggests that prediction was inaccurate.[16]

In addition, since 1995, the elk herd in Yellowstone has dropped 70 percent—from 16,791 elk to 4,635 in December of 2010, and moose have virtually disappeared.[17]

As the wolves spread out from Yellowstone, they are more difficult to see, because access will not be as easy, and because outside of the park, wolves are fair game, which means they will be much more wary of people.

In the upper Midwest, there are even larger wolf populations; however watching these wild wolves is more difficult due to dense forest cover and different predation patterns. Wolf watching in this region is popular but is focused at educational centers like the International Wolf Center in Ely, Minnesota, and Wolf Park in Battle Ground, Indiana, where there are captive wolves. The International Wolf Center draws fifty thousand visitors a year that generate three million dollars to the local economy and support sixty-six jobs.[18]

If people really want to see wild wolves, it seems very likely that the chances of doing so in Yellowstone will diminish unless the park becomes like one of the wolf educational centers, or promotes one nearby. Wolf biologist Dr. David Mech has said that opportunities to see wolves without professional assistance are rare and limited to areas of open terrain.[19] In keeping with this prediction, it should be noted that several studies have found that the presence of wolves in Yellowstone has driven elk away from frequenting open fields and into thicker cover to have a better chance of evading wolves. Thus, the presence of wolves is actually decreasing the overall wildlife-watching experience, which in turn will result in tourism declines.

Reviewing wolf eco-tourism, University of Wisconsin-Madison sociologist Matthew Wilson agrees:

Wolves are elusive, difficult to see, and occur in low population densities—all characteristics that inevitably confound the eco-tourist's desire to get close to a wild wolf . . . only with the aid of science, technology, and unique adaptations in recreational strategies have current opportunities for viewing and hearing wolves in the wild been created. In sum, despite the growing desire by Americans to see, hear and photograph wolves, it is likely that wolf-oriented eco-tourism will be limited to relatively uncommon environmental, social, and economic contexts for the foreseeable future.[20]

There is one other significant source of revenue arising from wolves—sales of wolf-hunting licenses. In Montana in 2009, when a wolf-hunting season was permitted, license sales brought in $325,916. Unfortunately, this is only about a third of the annual costs of the state's wolf management program, and a tiny contribution to the annual budget of Montana Fish, Wildlife & Parks, which is fifty-seven million dollars. Sixty-six percent of that budget comes from hunting and fishing license sales. Wolves are essentially crippling overall fish and game management in Montana, as well as Idaho, Washington, Oregon, and elsewhere, by taking away money that could have been used in other programs.

Economic Costs of a Growing Wolf Population

The economic costs of growing wolf populations are diverse and include many costs that are not easily calculated or quantified. It would take several books to identify and analyze every cost imposed on citizens by the introduction of wolves into just the Northern Rockies, and each region seems to have some costs that are both similar and unique. This chapter simply provides an overview of some of the general economic problems caused by wolves and identifies how certain groups of citizens are forced to pay a disproportionate amount of the costs.

Federal Agencies Costs

The principal federal agencies involved in wolf management are the US Fish and Wildlife Service, the US National Park Service, the US Geological Survey, and USDA Wildlife Services. The following is a brief overview of some of the major expenses involved with wolf management.

1. Environmental Impact Statement

Before any resource management program is undertaken, an Environmental Impact Statement must be conducted. The Environmental Impact Statement for wolf relocation to the Northern Rockies took ten years and cost approximately six million dollars.[21]

2. Annual Management of Wolves

In the original EIS for the Northern Rockies wolf, relocation program costs for the Greater Yellowstone area were estimated to be $3,077,500 for a five-year reintroduction effort (1994–98) and about $1.3 million for monitoring and wolf control (1999–02), or about $320,000 per year after that. Reintroduction in central Idaho was expected to cost about another two million dollars for the same period. In reality, reintroduction took only two years, not five, and cost only about $870,000 for both Yellowstone and Idaho combined. However, the combined annual continuing costs of wolf monitoring and management the US Fish and Wildlife Service now estimates has been about $1.5 million per year and it has continued to rise.[22] Between 1974 and 2000, the total spent for wolf recovery and management totaled fifteen million dollars.[23]

Early estimates of costs have proven to be far too conservative. Annual combined costs to USFWS and NPS on wolf management in the Northern Rockies in 2004, for example, were $1.2 million.[24] More wolves, more costs. In 2009, federal agencies in the Northern Rockies spent $3,763,000 for wolf management. Moreover, in 2010 (Oct 1, 2009–Sept 30, 2010), Federal agencies spent $4,556,000, including $1,103,000 spent by USDA Wildlife Services, to investigate reports of suspected wolf damage and to control problem wolves.[25]

The Wildlife Services division of the US Department of Agriculture Animal and Plant Health Inspection Services seeks to manage damage to crops and livestock due to predation by wild animals. Each state comes out with an annual report. In fiscal year 2009, USDA-APHIS Idaho Wildlife Service reports approximately $517,000 of appropriated and cooperative federal funds responding to complaints of reported wolf predation, conducting control and management actions, and for other related costs, such as equipment and supply purchases, meeting attendance, etc.

Of the estimated ninety-seven wolf packs in Idaho in FY 2009, USDA APHIS WS was able to verify that at least forty of them were involved in livestock depredations. Based on Idaho APHIS WS investigations, the minimum number of confirmed livestock depredations due to wolves in FY 2009 was seventy-six calves (killed), seven calves (injured); fourteen cows (killed); 344 sheep (killed), twenty sheep (injured); sixteen dogs (killed), eight dogs (injured); one foal (killed); and one goat (killed).

When livestock and ranchers' animals are killed, USDA Wildlife Service investigates. As Catron County Wolf Inspector Jess Carey showed us, sometimes multiple predator species tracks are found at a kill site, making it difficult to determine what species killed the animals, versus came to feed on carrion after the kill was made. Probable kills for 2009, which should be added to the previous list, were twenty-six calves (killed), three calves (injured); one cow (killed); 156 sheep (killed); four dogs (killed), two dogs (injured); and one goat (killed).[26]

The number of both cattle and sheep killed and injured by wolves in FY 2009 in the Northern Rockies was the highest ever recorded.[27] In 2010, it climbed to 199 cows and 249 sheep killed by wolves.[28] And since it's been estimated that only one kill can be confirmed for every 6.3 kills that are probable, the actual predation by wolves on livestock is much, much higher. Add to that as wild prey populations, especially elk, plummet in regions throughout the Northern Rockies, as the experience of Russia shows, wolves will increasingly turn to livestock for food.

When a "wolf" is killed by a private individual outside of a sport hunting season, that animal must be determined to be a 100 percent wolf for protections under the Endangered Species Act to be enforceable. The only sure way to do this is to submit the wolf for DNA testing. The USFWS Forensic lab is swamped with requests for DNA testing wild canids that are often hybrids with dogs and/or coyotes. Not only is this a cost of wolf management but it cuts into time and money available for work on other species.

A seldom recognized cost to the federal government from wolf management occurs when non-governmental organizations or individuals sue the US government concerning an environmental issue. If the citizen or group prevails, the 1980 Equal Access to Justice Act (EAJA) (5 USC. § 504; 28 USC. § 2412 provides for the award of attorney fees (up to $125 per hour) and other expenses to eligible individuals and organizations that are parties to litigation against the government when they prevail over the government, unless the government's position was "substantially justified" or special circumstances make an award unjust.

The EAJA was originally intended for use by seniors, veterans, and small businesses, but environmental groups have also been able to capitalize on it, although just how much money is actually being paid out to the groups is not certain, according to a study by the Notre Dame Law School and the Government Accountability Office. The GAO has identified $4.4 million per year of EAJA payments from 2000–10 for just ten units of the USDA and Department of Interior. When they looked at additional federal court records and public tax returns from twenty environmental organizations, they found that $9.1 million had been reimbursed during 2010 alone.[29]

According to Wyoming attorney Karen Budd-Falen (see later chapter), through spring 2010, wolf litigation has cost US taxpayers at least $436,762 in attorney fees paid to pro-listing animal rights and environmental groups. This is in addition to US government attorneys for these cases, whose fees likely cost at least an equal amount.[30]

State Agency Costs

The Natural Resources and Agricultural agencies of each state where wolves are found are responsible for an annual management plan, conducting research, responding to depredation reports, removing individual wolves that have been deemed dangerous, and compensating ranchers, farmers, and pet owners for livestock and pets killed or injured by wolves. The natural resources agencies also feel the brunt of loss of revenues from hunting license sales due to declining big-game herds, for such license revenues are a major source of income for state fish and game agencies. The following are highlights from states with significant wolf populations.

Montana

Montana Fish, Wildlife and Parks estimates that it costs from 913,000 to one million dollars a year to manage wolves in the state.[31] This amounts to currently spending approximately two thousand dollars per year per wolf for wolf management.[32] The federally funded budget for wolf monitoring and management has increased by 8 percent since 2005, while the Montana Fish, Wildlife and Parks budget for all big-game monitoring has declined by 15 percent since 2006. Currently the wolf program budget has become two-thirds of the size of the entire state big-game program.[33] The state says that it can't afford this, especially if they cannot sell wolf-hunting licenses (66 percent of the Montana Fish, Wildlife and Parks budget presently comes from hunting and fishing license sales). And even if they can continue to sell hunting licenses, that revenue will not pay for the wolf management program. So other programs and species will suffer. In 2010, in an attempt to bolster sagging hunting license sales revenues, Montana chose to increase the cost of a non-resident big-game combination license from $628 to $897, while a non-resident deer combination license increased from $328 to $527. When Idaho tried raising license fees to increase its cash flow from big-game licenses as deer and elk numbers plummeted, they ended up with a million dollar shortfall.[34]

Wyoming

In 2004, Wyoming Game and Fish Department spent just under $119,000 to manage the gray wolf in Wyoming, even though the state does not have jurisdiction to manage wolves. Game and Fish directors originally estimated that wolf management costs for Wyoming would approach one million dollars annually after delisting.[35] In 2007, Wyoming state officials revised their estimate, stating that it will cost about $2.4 million to manage the state's wolf and grizzly programs during the year after delisting, and two million dollars for each year after that.[36]

Idaho

Idaho's Wolf Conservation and Management Program is expected to cost between $900,000 and $1.2 million per year. No single source of funding to cover these costs has yet been identified.[37]

Former Idaho Senator Gary Schroeder requested Idaho Department of Fish and Game to conduct a study of costs to hunting from wolves in Idaho. The report finds that:

a) Wolves in Idaho kill about 9,517 elk per year. The economic value of each elk taken legally by hunters is about eight thousand dollars including direct and indirect benefits. Consequently, if those 9,517 elk had been available to hunters, typically about 20 percent of them, or 1,903 animals, would have been harvested. So, wolves killing the elk represent an economic loss of about $15.2 million.[38]

A 2009 economic analysis of wolf management indicates that Idaho could be losing from $7 million to $24 million per year in hunting revenue due to the introduction of wolves.[39]

In Idaho, resident hunting license sales declined 3 percent in 2009 but big-game tags have dropped dramatically, especially for non-residents. Non-resident hunters in Idaho are 30 percent of the hunting population but bring in 70 percent of the license revenues. The declining economy is one explanation, but other states have seen increases in hunting license sales in the last few years. The decline in Idaho, therefore, is more likely associated with declining elk herds in some districts. When Idaho Department of Fish and Game tried to increase resident and non-resident fees to make up for the decline, non-resident license sales dropped even more. In 2009, with higher fees, deer license sales fell 36 percent and elk license sales fell 26 percent resulting in a loss of income approaching two million dollars in revenues for Idaho Fish and Game.[40]

In February of 2010, Idaho Fish and Game Commissioner Randy Budge reported on a study of why out-of-state hunter numbers are declining. Budge stated, "Our out-of-state hunting numbers were down 25 percent in 2008, 31 percent in 2009." Fish and Game polled previous visitors to the state to find out if the economy was the culprit or if it was some other reason. "The number one reason listed for not coming to Idaho was, 'You haven't taken care of your wolves and your wild animal populations are down,'" Budge recounted. "The number two reason was, 'Your license fees are unfair.]"[41]

Oregon

Oregon has a growing wolf population in the northeastern part of the state. The state wolf management plan projects they ultimately will have fourteen pairs of wolves and a total population of 207 wolves in the state. The 2009 Oregon Wolf Conservation and Management Plan estimates that those 207 wolves will translate into 4,844 deer and 1,615 elk killed annually by wolves. The total loss in hunting benefits (license fees and all other expenditures by hunters) to the state from 207 wolves is estimated at $2,168,500 per year.[42]

Wisconsin

The Wisconsin Department of Natural Resources has paid more than one million dollars in reimbursements to those who have had livestock and pets killed by wolves since 1985.[43]

Costs to Businesses

Ranchers

Coyotes, wolves, bears, and cougars are responsible for over sixty million dollars a year in livestock losses nationwide.[44] One-tenth of a percent of all livestock in the United States are killed by confirmed wolf predation, but this is not spread equally as wolves are found in only a few states. In states where wolves are abundant, livestock predation by wolves may be as much as almost half of all predation. According to the USDA 2010 Cattle Death Loss report, losses due to wolves are found in five states. In those states, the following predation losses due to wolves are Idaho at 47.4 percent, Montana at 20.3 percent, Wisconsin at 47.5 percent, Wyoming at 14.6 percent, and Minnesota at 37.7 percent. Considering the relatively small number of wolves, as compared to ten to one hundred million coyotes and a national population of well over a hundred thousand black bears in the Lower 48, these losses are very significant.[45]

Losses related to wolves may be small when considering the entire industry, but they can be very significant in certain regions and for individual ranchers. And, the magnitude of losses due to wolves is further amplified by a study by Collinge that found individual wolves in Idaho are approximately 170 times more likely to kill cattle than are individual coyotes or black bears. He also found that individual wolves were about twenty-one times more likely to kill cattle than individual mountain lions.[46]

As one indication of what a difference in depredation a thriving wolf population makes, between 1987 and 2009, ranchers in the Northern Rockies received more than $1.3 million in compensation for almost four thousand animals killed by wolves, mostly sheep and cattle.[47]

From 1995 through 2009, in Northwest Montana, the Greater Yellowstone, and the Central Idaho Recovery Areas, confirmed wolf kills of livestock included: 1,278 cattle, 2,842 sheep, and 141 guard dogs. From 1987 to 2009, wolves were also responsible for killing twenty-five llamas, thirty-one goats, and ten horses.[48] These are confirmed kills. Sommers and associates found that 6.3 calves were likely killed by wolves for every one calf that was confirmed.[49]

And wolf predation keeps rising. According to the United States Department of Agriculture (APHIS) Wildlife Services, during 2010 wolves killed 4,440 cattle and calves in Idaho, Montana, and Wyoming.[50]

Wisconsin has a growing wolf population. Verified wolf depredations in Wisconsin from 1976 through 2005 included five horses killed, one horse injured, fifty sheep killed, 184 cattle killed, seven cattle injured, thirty-eight deer killed, 264 poultry killed, ninety-nine dogs killed, and thirty dogs injured. Over half a million dollars has been paid to livestock, hunters, and pet owners in Wisconsin since 1985 as compensation for losses due to wolf attacks.[51]

In Wyoming, wolf depredations reached a peak in 2009, when 222 domestic pets and livestock, mostly sheep, were killed.[52]

Officials with the Montana Department of Livestock add that for every confirmed wolf kill, they believe another seven cattle, sheep, or horses are killed by wolves that are not found immediately and are fed on by other wild animals, so those deaths are known as unconfirmed kills.[53]

As opposed to other predators, on occasion wolves engage in mass spree killings of animals, such as in the summer of 2009, when 120 prize rams were killed by wolves in one night on a farm near Dillon, Montana.[54]

Spree killings are found among wolves worldwide. In July 2011, in France, one wolf attacking one herd of sheep in one night killed ten sheep, sent sixty-two panicked sheep plunging over a cliff to their death, and thirty more were missing in the woods. In the first six months of 2011, France has had to compensate ranchers and shepherds $530,000 for sheep killed by wolves.[55]

Wolf advocates are quick to assert that these are isolated incidents that are blown out of proportion, resulting in needless fears. Actually, mass killings of livestock by wolves are well-documented and unpredictable. Therefore the possibility that wolves might attack and engage in mass killing is especially likely to ignite fear among farmers and ranchers, especially if a documented case of this has happened recently and in that vicinity. Managing livestock, one seeks to prevent probable losses. Preventative measures to counter wolf predation include buying more guard dogs, hiring more security guards to protect the herds, moving livestock to pastures closer to dwellings, and increased feeding due to restricted grazing areas.

Additional negative effects of wolves residing in areas where there are ranches extends far beyond killing livestock.[56] One of these is that wolves are a vector for *Neospora caninum*, a parasite that causes cattle to abort. Over thirty-five million dollars per year in damages to livestock are associated with this parasite. When wolves move into an area, they typically chase out coyotes, however, as wolves also carry this parasite, there is no reduction in damage, just a new vector. (See chapter on zoonotic diseases.)[57]

The mere presence of wolves affects behavior of cattle. A study in Oregon finds that livestock herds in areas where wolves are found are at least 10 percent lower in weight when rounded up in fall. The weight loss is associated with stress caused by wolves howling and simply being near cattle, as well as making attacks.[58]

A quick list of some other costs:

- Public land ranchers in New Mexico and Arizona have been forced out of business because of wolf predation and increased costs associated with trying to protect livestock from wolves. In areas where wolves are found, ranches decline in value and/or cannot be sold, as they have been blacklisted as bad investments.
- Guard dogs for livestock typically cost $600 to $1,500 each. Coyotes tend to leave guard dogs alone, while wolves attack and kill them. Llamas and donkeys that guard sheep from coyotes are also killed by wolves.
- The 2002 Oregon Cattleman's Association Annual Survey found that labor costs to hire extra guards for livestock herds ranged from $1,800 to $2,500 per month per guard.
- The 2002 Oregon Cattleman's Association Annual Survey also found that 50 percent said that their cattle were pastured on rangeland not normally closely attended in the summer months. When they had to bring cattle closer to ranches for protection, the costs for feed, labor, and equipment ran as high as $1.80 per day per head of cattle.[59]

Farmers

Farmers grow livestock that may not be free roaming, such as dairy cattle, chickens, pigs, turkeys, and ducks, and they all are prey for wolves. In Wisconsin, where farmers are more likely to raise dairy cattle that are milked daily and kept close to barns, at least seventy-five farms have lost at least one cow or sheep to wolf depredation from 2001 to 2006. In 2010, according to the Wisconsin Department of Natural Resources, wolf depredation to livestock occurred on forty-seven farms, compared to twenty-eight farms in 2009. This exceeds the previous record of thirty-two farms in 2008. Total livestock depredation in 2010 included: sixty-three cattle killed (forty-seven calves), five cattle injured, six sheep killed (four lambs), one goat injured, and six farm deer killed. Compensation paid to farmers was $113,586.50.[60]

Dr. David Mech has plotted out two scenarios for maintaining wolf populations in Minnesota at a minimum of twenty-five hundred wolves: first, a population of about fourteen hundred primarily in the wilderness and semi-wilderness, and second, allowing wolves to move into agricultural areas for five years after their removal from the endangered species list. Under the first option, Mech estimates that each year twenty-seven farms would suffer livestock losses and wolves would be expected to kill about three dogs. In response, thirty-six wolves would be killed and the cost per wolf in the total population would be $86 to $215,000 for twenty-five hundred wolves.

Under the second strategy, there would be an estimated thirty-five hundred wolves. In this scenario each year ninety-four to 171 farms would suffer damage; and wolves would kill eight to fifty-two dogs. In response, 109 to 438 wolves would be killed for depredation control. In short, the annual cost averaged over the total population would be eighty-six dollars for each of the 1,438 wolves living primarily in the wilderness and an additional $197 for each wolf outside the wilderness, for a total cost of about $530,000 per year.[61]

Businesses

The Duffield study speaks of money coming into the local economy to restaurants, gas stations, sporting goods stores, grocery stores, and motels as a result of wolf watchers. The same model can also be reversed in terms of money lost to the local economy from hunters choosing to hunt elsewhere. A 1986 US Forest Service publication indicated that one day of elk hunting generated $39.10 in economic activity. The 1994 USFWS environmental impact statement used that figure to estimate that one hundred wolves would cost the state between $572,000 and $857,000, based on 14,619 to 21,928 days in lost hunting opportunities. After adjusting the 1986 figure to a 2008 exchange rate, and assuming a linear relationship between reduced hunting opportunities and mushrooming wolf population, a contemporary estimate of the cost of lost hunting days due to wolf predation on big game in the North Rockies is between seven million and eleven million dollars.

The Oregon Wolf Conservation and Management Plan uses a figure for the net economic value of one hunter hunting elk for one day as ninety-two dollars, expressed in 2009 dollars. With this standard, they calculate that with 207 wolves in the state, the economic loss from lost hunting revenues per year would be $1,114,440.[62]

Hunting outfitters also incur costs associated with wolves reducing herds of prized big game such as elk, moose, caribou, wild sheep, and deer. Those clients in turn would have frequented local businesses—motels, restaurants, bars, grocery stores, sporting goods stores, and gas stations. Fewer clients also mean less revenue from license sales, which is definitely a problem in Northern Rockies states.

Private Individuals

Hunters

Hunters spend a considerable amount of money to pursue wild game. The US Fish and Wildlife Service says that the annual value of hunting to just Montana is worth $237 million. Studies in Alaska have clearly shown that predator control of wolves and bears results in increased herd size of ungulates—caribou and moose—that translate into more hunters and hunting dollars.[63]

Hunters traveling farther to hunt are a cost of wolf depredation, and there are other costs to hunters. Sixty-one hunting dogs were killed by wolves in Idaho between 2003 and 2009, with the number of dogs killed increasing every year. Trained hunting dogs may be worth a thousand dollars or more each. Bear hunters in Michigan, Wisconsin, and Minnesota have lost dogs to wolves. In Wisconsin in 2010, twenty-four dogs were killed and fourteen were injured by wolves. Compensation payments were $61,193.[64]

We have already mentioned pets lost due to wolf attacks. Sometimes there is compensation paid, other times the family pet is gone and it cannot be proven conclusively that wolves are responsible.

Attacks on People
As we have already covered, there is a long history of wolf attacks on people in Eurasia, especially in Russia, Pakistan, India, and Kazakhstan, including thousands of fatal ones. There is also a significant number of attacks North America. Putting a price on a human life is not easy, but clearly there are many human and economic costs, and the danger of people being attacked and killed by wolves, which is unpredictable, generates anxiety and many precautionary measures that all come with costs.

The fear of being attacked, especially for children, surely is a cost to the human soul that can be translated into dollars spent. In Catron County, New Mexico, after children waiting for school buses were stalked by wolves, parents built special wolf-proof shelters for their children to wait for the school bus, or simply chose to take the time to drive them to school. Those shelters easily cost $1,000 or more.[65] In that same area, there are documented cases of children suffering from Post Traumatic Stress Disorder (PTSD) caused by wolf encounters who require psychological treatment.

Public Health
Anytime there is a potential public health problem associated with wildlife, local, state, and federal public health agencies become involved and conduct ongoing monitoring to strive for prevention of the disease before humans contract it. If people do contract the disease, then there are medical costs to be paid, as well as preventive measures undertaken to curb the spread of the disease.

To all these costs, we should include diseases carried by wolves that can be spread to humans, pets, and livestock, including *Echinococcus granulosus*, *Neosporum caninum*, and rabies.

Conclusion

The bottom line is that the Duffield study makes for good press, but the methodology and results are so questionable that it would not be admitted as evidence in any court in the United States. In reality, the asserted economic benefits of wolves have not been heavily documented or reliably calculated. And they tend to be focused in relatively small areas, such as in and around Yellowstone National Park.

However, the substantial costs of wolves can be found virtually wherever wolves are found, affecting the lives of many primarily rural Americans, and those costs cannot be disputed. It should also be obvious from this chapter that when wolves move into modern society, there are many economic costs that collectively far outweigh any economic benefits from wolf watching.

How do we begin to calculate *all* the economic costs of wolves?

Former Montana Fish, Wildlife and Parks biologist Allen Schallenberger states that wolves kill twenty-three elk per wolf during the six winter months. Assuming that wolves will kill at about the same rate throughout the year, wolves kill an average of forty-six elk per wolf during the entire year. If there are six hundred wolves in Montana this will result in 27,600 elk being killed by wolves each year in Montana.[66]

Based on Schallenberger's study, former Montana State Senator Joe Balyeat, who is also a CPA, has come up with a very simple method of calculating the total cost of lost wildlife due to six hundred wolves in Montana. In a 2010 presentation to the Montana Senate, former Senator Balyeat calculated the cost of lost wildlife devoured by six hundred wolves in Montana to be $124,200,000 per year. The value of each elk that Balyeat uses is $4,500, which is based upon the values established in Montana statutes for bull and cow elk restitution for illegal takes—$8,000 for bulls, $1,000 for cows—or, an average $4,500 per elk. The six hundred-wolf estimate is very low . . . with perhaps as many as fifteen hundred wolves in Montana alone, but using six hundred wolves X forty-six elk per year X $4,500, one comes up with the total cost of elk killed as $124,200,000 per year. If you take the Idaho-Montana-Wyoming area low-end estimate of fifteen hundred wolves, the lost wildlife cost per year would be fifteen hundred X forty-six X $4,500 = $310,500,000 per year. If you take a higher estimate—say twenty-five hundred wolves in the tristate area, the cost would be $517,500,000 per year . . . over half a billion in lost wildlife per year. This does not include any wolves outside the tristate area. Nor does it include any of the economic losses of outfitters, hunting communities, agricultural communities, area restaurants, hotels, etc. Balyeat says, "This model simplifies things as it is based on the assumption that each wolf only eats elk. If wolves consume Bighorns or moose, the value of the lost wildlife would greatly increase, if they consumed only deer the value would be less . . . I used elk as a good average-value species; and also felt it was the species best documented with respect to numbers killed per wolf."[67]

If wolves consumed bighorn sheep or moose instead of elk, the value of the lost wildlife would increase since Montana law sets the value of an illegally taken bighorn sheep at thirty thousand dollars and the value of an illegally taken bull moose at six thousand dollars. Other values set by Montana law are: six thousand dollars for a mountain goat; eight thousand dollars for a trophy antlered deer; two thousand dollars for an antelope; one thousand dollars for a cow moose; five hundred dollars for a buffalo or antlered deer; and three hundred dollars for deer, antelope, or any other game animal not specifically given a statutory value.

Balyeat only looks at the Northern Rockies. But now wolves are now found in the upper Midwest, New England, the Carolinas, and in Arizona and New Mexico, where a program to restore Mexican wolves has been underway for thirty years. As of 2014, it is estimated that more than twenty million dollars has been spent on this program, which has resulted in over one hundred wolves in the wild.[68] The twenty-million-dollar figure doesn't include costs for losses of livestock or pets from wolf predation, costs for parents to build wolf-proof school bus shelters for their children, costs to treat children suffering from PTSD associated with encounters with wolves, costs associated with declining values of ranches where wolves are found, or costs associated with extra protective measures to attempt to prevent wolves from preying on livestock. Even fifty wolves can quickly result in considerable economic costs, and with so few animals in such a large area, the potential benefits of wildlife watching for Mexican wolves are insignificant.

Wildlife management should be rooted in science, but the reality is that wildlife management today is a social and political decision. Wolves have received enormous positive publicity in the last three decades, which has improved public opinion of wolves in mainstream America, and even around the world. However, surveys of people who actually live near the wolves consistently report negative opinions about wolves.[69]

The listing of costs associated with living with wolves should make it abundantly clear why negative attitudes about wolves are far more prevalent in areas where people must try to co-exist with wolves.

This dissonance between the opinion of locals who are bearing the cost of wolves, and trying to adapt to living with them and the opinion of mainstream America formed by media portrayal of a romanticized notion of wolves, powerfully demonstrates why wolves need to be managed by the states where wolves occur.

State management of wolves is the only way to ensure that the voices of the local residents paying these costs are not overwhelmed by the media campaigns and lobbying efforts of well-funded, powerful national pro-wolf organizations that thrive on creating crises that often do not even exist, or flood the media with romantic fables about wolves, rather than presenting the stark reality of the costs, as well as the benefits, of living with wolves.

Endnotes:

1. Nye, Martin. *Beyond Wolves: The Politics of Wolf Recovery and Management.* University of Minnesota Press, 2008.
2. http://www.pinedaleonline.com/wolf/wolfimpacts.htm
3. http://video.foxnews.com/v/1518892269001/idaho-gov-wants-more-federal-money-to-hunt-wolves
4. http://www.census.gov/prod/2002pubs/QFBRO.pdf
5. www.pinedaleonline.com/wolf/pdf/WG&FWolfAnalysis.pdf
6. http://www.fws.gov/mountain-prairie/species/mammals/wolf/annualrpt11/index.html
7. www.bozemandailychronicle.com/users/profile/noeconuts/
8. Duffield, Nehr and Patternson. *Wolves and People In Yellowstone: Impacts on the Regional Economy."* 2006. http://wyoming.sierraclub.org/WOLVES%20AND%20ECONOMICS.pdf
9. http://www.yellowstone.co/stats.htm
10. http://www.nature.nps.gov/stats/viewReport.cfm
11. Duffield, Nehr and Patternson, 2006, Op. Cit.
12. http://trib.com/news/state-and-regional/article_408de8b4-a1dd-5be3-a55d-79c853d9122e.html
13. http://www.forwolves.org/ralph/wolf-economic-impact.htm
14. http://www.yellowstonepark.com/MoreToKnow/ShowNewsDetails.aspx?newsid=182
15. http:forwolves.org/ralph/wolf-economic-impact.htm
16. http://trib.com/news/state-and-regional/article_408de8b4-a1dd-5be3-a55d-79c853d9122e.html
17. http://wyoming.sierraclub.org/WOLVES%20AND%20ECONOMICS.pdf13
18. http://www.pinedaleonline.com/wolf/wolfimpacts.htm
19. "Winter Count Shows Continued Decline in Yellowstone Elk," *New West News.* January 12, 2011.
20. David T Schaller, "The Ecocenter as a Tourist Attraction: Ely and the International Wolf Center," Ely, MN: IWC, 1999.
21. Mech, L. D. 1995, "How Can I See A Wolf?" In *Wolf.* 5(1)8–11.
22. http://www.wolf.org/wolves/learn/intermed/inter_human/wilson_ecotour.asp
23. Nye, M. OP. Cit, p.161.
24. King, Nelson. "Wolves in Yellowstone: A short history," *Yellowstone Insider,* 200: http://www.yellowstoneinsider.com/issues/wolves/wolves-in-yellowstone-a-short-history/all-pages.php
25. Daerr, Elizabeth. "A Howling Success." *National Parks* November 2000.
26. http://www.fws.gov/mountain-prairie/species/mammals/wolf/annualreports.htm

27. http://westinstenv.org/wp-content/ID%20WS%20FY%202009%20Wolf%20Report .pdf
28. Northern Rocky Mountain Wolf Recovery Program 2011 Inter-Agency Annual Report.
29. http://archive.sba.gov/advo/laws/sum_eaja.html and http://www.boone-c rockett. org/images/editor/ND_EAJA.pdf
30. Budd-Falen, Karen, "It's not About Saving Species—It's about Spending Taxpayer Money and Making Some Groups Wealthy." Budd Falen Law Offices: Cheyenne, WY. May 26, 2010.
31. http://www.newwest.net/topic/article/wyoming_lawyer_environmental_groups _using_taxpay ermoney_for_legal_fees/C37/L37/
32. http://fwp.mt.gov/wildthings/wolf/wolfQandA.html
33. http://westinstenv.org/wildpeop/2009/02/09/the-high-costs-of-wolves/
34. Hamlin and Cunningham. "Monitoring and Assessment of Wolf-Ungulate Interactions and Population Trends within the Greater Yellowstone Area, SW Montana, Southwestern Montana and Montana Statewide." *Montana Fish, Wildlife and Parks.*
35. http://www.nrahunterrights.org/blog/Default.aspx?id=482
36. http://trib.com/news/state-and-regional/article_408de8b4-a1dd-5be3-a55d-79c853d9122e.html
37. http://www.klamathbasincrisis.org/wolves/fedswontkill012207.htm
38. Idaho Fish and Game News. June 2011.
39. http://wolves.files.wordpress.com/2009/02/wlfecon-impct.pdf
40. http://media.spokesman.com/documents/2009/02/wolves.pdf
41. http://www.idahoreporter.com/2010/hunting-tourism-dives-with-rise-in-license-fees-economy-wolves-also-blamed/
42. http://www.journalnet.com/news/local/article_639aacda-1232-11df-87ef-001cc4c03286.html
43. http://www.dfw.state.or.us/Wolves/management_plan.asp
44. http://www.jsonline.com/news/wisconsin/126334093.html
45. Gese, E., Sean P. Keenan, and Ann M. Kitchen. "Lines of Defense: Coping With Predators in the Rocky Mountain Region." USDA, 2005. http://www.aphis.usda. gov/wildlife_damage/protecting_livestock/downloads/predators_b ooklet7.pdf
46. USDA Cattle Death Loss, May 2010, USDA National Agricultural Statistics Service.
47. Collinge. "Relative risks of predation on livestock posed by individual wolves, black bears, mountain lions and coyotes in Idaho." Proceedings of the Vertebrate Pest Conference. 23: 129–133.
48. http://www.bozemandailychronicle.com/news/article_480432c6-3ae0-11df-bcd8- 001cc4c03286.html
49. http://fwpiis.mt.gov/content/getItem.aspx?id=26932
50. Sommers, A.P, C. Price, C.D. Urbigitkit, E. Peterson. "Quantifying Economic Impacts of Large Carnivore Depredation on Bovine Calves." *Journal of Wildlife Management.* 2010.

51. United States Dept of Agriculture, Cattle Death Loss in 2010. Released May 12, 2011, by the National Agricultural Statistics Service (NASS), Agricultural Statistics Board, United States Department of Agriculture (USDA).
52. http://dnr.wi.gov/org/land/er/publications/pdfs/wolf_impact.pdf
53. http://billingsgazette.com/news/state-and-regional/wyoming/article_258dda80-fb15-11df-9f08-001cc4c002e0.html
54. http://helenair.com/news/local/article_654089dc-f5d7-11de-aa95-001cc4c03286.html
55. http://news.yahoo.com/ravenous-wolves-colonise-france-terrorise-shepherds-064719014.html
56. http://digitalcommons.unl.edu/cgi/viewcontent.cgi?article=1015&context=wolfrecovery
57. Lehmkuhler, J., G. Palmquist, D. Rud, B. Willgrey, A. Wyderem. "Effects of Wolves on Farms in Wisconsin Beyond Verified Depredation." Wisconsin Wolf Science Committee, WL Department of Natural Resources, Madison, WI.
58. http://www.capitalpress.com/orewash/ml-wolf-plan-sidebar-2-100810
59. Oregon Wolf Conservation and Management Plan, Op. Cit.
60. Wis. Dept of Nat Resources, Wisconsin Endangered Resources Report #140, Year-end Summary of Wolf Population Monitoring in Wisconsin in 2010 (Feb. 2011).
61. http://www.mnforsustain.org/wolf_mech_wolf_recovery_costs.htm
62. http://www.dfw.state.or.us/Wolves/management_plan.asp page 101.
63. http://www.nap.edu/openbook.php?record_id=5791&page=1
64. http://dnr.wi.gov/org/land/er/mammals/wolf/pdfs/wolf_damage_payments_2010.pdf
65. See chapter by Laura Schneberger.
66. http://westinstenv.org/wildpeop/2008/11/17/montana-fwpd-wolf-management-fiasco/
67. Personal communication from Senator Balyeat, 2/13/2011.
68. www.livescience.com/49842-mexican-wolf-population-grows.html.

CHAPTER 10

. . .

Reality Bites: Mexican Wolf Impacts on Rural Citizens

By Laura Schneberger

"It's disturbing to see that there is a consistent trend from government agency personnel working within the program to promote the extremist notion that wolves on the ground are genetically special and cannot be removed or controlled. This is contrary to all scientific and policy documents available."

—Laura Schneberger

Mexican Wolf Program Background

The Mexican wolf recovery project did not begin with the passage of the Endangered Species Act in 1973. Nor did it happen in 1976 when the gray wolf was listed as an endangered species. It didn't take hold until 1979 when the US Fish and Wildlife Service (USFWS) approved a recovery team to assist the agency in mapping out a potential recovery strategy for the Mexican wolf, the smallest and rarest subspecies of gray wolf in North America. The USFWS approved the subsequent Mexican Gray Wolf Recovery Plan in 1982. This plan called for captive breeding and establishment through reintroduction of two viable wild populations of Mexican wolf in the Blue Range Wolf Recovery Area—a 6,800-square-mile area that encompasses the Apache National Forest in Arizona, the Gila National Forest in New Mexico, and 2,400 square miles managed by the White Mountain Apache tribe.

Unlike other predator recovery plans, this one failed to contain numerical criteria for recovery and delisting of the Mexican wolf, and when the actual reintroduction project was on the ground, that aspect was put on the back burner. The majority of biologists believed that success would not be feasible with the genetic limitations that existed due to the small number of Mexican wolves, none of which were in the wild. A determination was made that this wolf would never be recovered enough for delisting.

Lack of action on the Mexican Wolf Recovery Plan by USFWS provoked litigation by an environmental organization then called the Southwest Center for Biological Diversity, to force immediate implementation of the recovery plan. This suit resulted in a settlement between the plaintiffs and the USFWS, with undisclosed conditions and parameters. Although the USFWS agreed to the settlement in federal court, neither the public nor the state agencies have ever seen the terms of that settlement. As of 2014, USFWS still claimed there was a court order to proceed with the program, though clearly there was not. Somewhere under those hidden terms, "no legal win, no loss," were the words, "agreed to attempt reintroduction of the Mexican wolf, *Canis lupus baileyi.*"

The current Mexican wolf program is based on a 1982 agreement and is woefully out of date. It is not scientifically current and lacks even the basic backbone of most recovery plans—delisting criteria that would eventually remove the Mexican wolf from Endangered Species Act protection. By 1996, a proposed experimental rule was written and by 1998, the Final Environmental Impact Statement (FEIS) was published directing the program's beginning. In 1998, designation of a "Nonessential Experimental Population Rule" with the required section 10(j) special rule on managing the reintroduced population of large predators allowing lethal control and authorized take was written. March 29, 1998, the first eleven wolves were released into the Blue Range Wolf Recovery Area.

Wolf Lineage

At the time of the 1982 agreement, the captive breeding program consisted of three different lineages—McBride, Aragon, and Ghost Ranch. The only certi-fied animals were the five Mexican wolves captured in Mexico in the late 1970s and early 1980s by Roy T. McBride. McBride, a trapper who often contracted for the federal government, spent years in Mexico hunting out the last of the wild packs of Mexican wolves. He came back with a single female and four males all from the same pack. McBride was in a unique position for this project with both expertise in trapping wolves and a master's in biology.

During the initial planning for captive breeding of the lineages, it was determined that only the lineage captured by Roy McBride in Mexico was cer-tifiable as pure Mexican wolf.[1] In 1997, controversy arose when a captive pack at Carlsbad Caverns National Park designated for release was found by Roy McBride to be largely composed of wolf-dog hybrids.

Zoo employees initially argued that the animals' odd appearance was due to several generations of captivity and diet, but diet and captivity alone could not account for blue eyes (that wolves do not have) or curly tails. Scientists involved in the captive breeding program and Recovery Team decided to eutha-nize the line. Unfortunately, there remained a few animals from the Ghost Ranch lineage in private collections around the Southwest.

With only one purebred female, the USFWS began seeing inbreeding depression in the McBride lineage. Some of her male offspring showed signs of cryptorchidism (undescended testicles) and were not fit for breeding, although crossings with her sons and brother did supply more females for further breeding. The program seemed dead in the water, but in the mid-1990s David Parsons, the Mexican Wolf Recovery Coordinator, made the startling decision to include both the Ghost Ranch private collection and the Aragon lineages in the Captive Breeding Program in an attempt to improve the genetic diversity of the animals. McBride, who was livid at the idea of including what were consid-ered by all experts involved to be wolf-dog hybrids, challenged Parsons on this decision. On June 2, 1997, just six months before the first captive wolf release, McBride sent a letter outlining his concerns about the decision.

> Dear Mr. Parsons: In reading the recent status report of April 1997, I was shocked to see that the wolves from the Ghost Ranch Lineage were being included in the captive breeding program. In the early days of Mexican Wolf Recovery, the origin and genetics of the Ghost Ranch animals were discussed and investigated ad nauseam. In fact, the con-clusion by all members of the early recovery team was that the animals were wolf-dog hybrids. This was the primary factor behind the deci-sion to seek and capture the remaining wild population, because it was the only pure genetic stock available.[2]

McBride's letter goes on to discuss the genetic and legal ramifications of the inclusion at length, citing the flawed science behind the decision as "right out of the *Twilight Zone*." He questioned whether the ESA would even protect the animals that would be created by USFWS if the Ghost Ranch lineage were included in the breeding program. He was adamant that the Ghost Ranch and Aragon lineages were cosmetically, genetically, and completely different from the original McBride wolves captured in Mexico.

Twilight Zone indeed! When geneticists at the University of California, Davis, confirmed that the so-called "Chupacabra dog" coyotes found in Texas were related to the Mexican wolf everyone laughed. But neither McBride nor Parsons could have known of their existence when the original argument erupted over Ghost Ranch. In 2009, DNA tests on a startlingly ugly blue-eyed coyote cross-dog jokingly known as the Chupacabra in south Texas proved conclusively that this anomaly was genetically related to DNA on file in the Mexican wolf program.[3] Several specimens of this hairless blue dog with blue eyes are available and although there is yet no proof, they are likely related to the dog that is one of the unknown founders of the Ghost Ranch lineage. The odd-looking coyote hybrids show marked physical similarities to the Mexican hairless dog, or Xolo, an ancient breed known to have occupied South America for more than three thousand years. Ghost Ranch "wolves" also had some specimens with blue eyes. However, there has been no investigation into the relationship between either the Texas blue hairless (Chupacabra), or the Xolo, or the different lineages of what are now called *Canis lupus baileyi* (Mexican wolf) as of yet.

McBride's concerns were ignored and in 1997, he ceased working with USFWS on wolf-related issues. Unfortunately, none of the original specimens of either Ghost Ranch or Aragon exists to use for modern DNA testing and the original records of the recovery team that McBride was a part of have disappeared. Instead, USFWS set up a committee to review the Ghost Ranch and Aragon lineages and verify their approval, although even that team also appeared to have serious misgivings. McBride writes:

> As of December 7, 1995, Mr. Parsons had yet to view any living descendant of Ghost Ranch lineage firsthand. Nevertheless, and without examining any of these animals firsthand, Mr. Parsons pronounced them fit, to be purely "Mexican wolf," and to be appropriate for inclusion with the certified lineage in July of 1995. Mr. Parsons' determination was based entirely on the results of Hedrick's and Wayne's yet unpublished 1995 studies which remain not generally available for public review as of January, 1996.[4]

One of the first lawsuits against the implementation of the Mexican wolf program was based on the inclusion of hybrid wolf-dogs into the gene pool. This argument was swiftly lost in federal court and the historical samples that may have proved Ghost Ranch and Aragon founders were dogs or wolf-dog crosses were never found again.

The DNA baseline for the Mexican wolf is now from the three combined certifiable lineages. It must be explained that the only way to determine wolf DNA as different from dog DNA is to have family markers. Mexican wolves are purebred because they are all related to wolves in the captive breeding pool, not because they have distinctly different genes than dogs. In other words, it did not matter that they were hybrids, as long as the USFWS claimed they were not and historic records were missing, then the program could go forward as if they were pure wolves.

The unpublished study that Parson used to include non-wolf lineages was eventually published, but the committee still had misgivings about the methodology. The study never used coyotes from Mexico or the Southwest for their comparisons with the questionable lineages but it does admit that:

> Wayne et al. (1995, p.6), cannot eliminate the possibility that the Ghost Ranch lineage originated from other North American grey wolves or a dog whose offspring had backcrossed to wild wolves for several generations. Wayne et al. (1995, p.6), cannot eliminate the possibility that the Aragon lineage originated from other North American grey wolves or a dog whose offspring had backcrossed to wild wolves for several generations.[5]

At the time and probably still today there is no other person alive who knows more about wild Mexican wolves than Roy McBride, yet the Service ignored his plea not to combine these wolves with the McBride lineage because the genetics have been fouled and the likelihood of problems such as a propensity for livestock predation would likely be substantial.

These facts may very well explain why USFWS has documented higher than expected livestock predation, human habituation issues, and the presence of hybrid pups on at least three different occasions in the Mexican wolf reintroduction area.

The entire issue of the existence of the Mexican wolf needs to be reviewed by independent scientists and geneticists. Many of the un-wolflike behaviors that have been exhibited by this population is in part due to the fact that this court-approved population is not pure Mexican wolf and their dog genes have likely impaired their ability to function, as did historic Mexican wolf populations. This leads to the question as to whether these wolves should even be considered an endangered species.

The Modern Mexican Wolf versus the Modern Rancher

The Mexican wolf reintroduction program launch was also the beginning of almost two decades worth of struggle for equality and justice for ranching operations and small rural communities in the recovery area. The goal of one hundred wolves originally listed as the initial goal for the Blue Range Wolf Recovery area is clearly an over-estimate. In over a decade, there are fifty-eight in the wild in spring 2012 according to USFWS.[6]

Regardless of the number of the Mexican wolves, a substantial number now on the ground have left the wilderness recovery area, preferring to eat livestock, pets, and garbage, and frequent homes and campgrounds. This may seem to lead anyone not connected to the relocation program hierarchy to believe that wild prey populations are not sustaining the food needs of the current population, particularly in the wilderness areas.

Logically, this leads to the belief that at any one time prior to the turn of the century, there were never one hundred Mexican gray wolves in southwest New Mexico and Arizona combined. Historical accounts by early explorers such as Emory, Carson, and various Spanish expeditions rarely mention encounters or even sightings of wolves. Further, the same early expeditions by Europeans found the area so lacking in game species, that the threat of starvation loomed almost constantly. Kit Carson, who led several expeditions through the area, survived by eating horses he traded out of the local Native American populations. The lack of game and the few documented sightings of wolves by these folks clearly show that the Mexican wolf was, at best, rare in the area and probably rare throughout its entire range.

Where did the USFWS get their population goal estimates? By the turn of the twentieth century Mexican wolves were relatively common, likely because a new prey animal had been introduced to the area—livestock in the form of both sheep and cattle had been steadily increasing throughout country set aside for ranching that is now considered the Blue Range Wolf Recovery area. The ranching businesses contributed meat needed by the mining towns springing up all over the region but that same meat led to an increase in breeding capability and larger litters from the few wolves that were historically documented on that area. The same situation occurred in California at the turn of the century when livestock became prevalent in the territory and the grizzly population exploded with the expanded food supply.

With increased nourishment, the Mexican wolf populations did increase significantly as did increased wolf/livestock conflict. Early governments authorized the PARC service (early USFWS) to begin the removal of the Mexican wolf. Although several dozen wolves were indeed killed from 1900 to the 1970s, this number was the culmination of seventy years and should not have been used as a population goal, as it in no way represented a natural ecological number to estimate historic population levels.

Note: the gray wolf recovery goal that was set for Montana, Wyoming, and Idaho was one hundred wolves and ten breeding pairs in each state. Wolves prey on large ungulates, hoofed mammals, and snowshoe hares. When they are available, elk are the preferred food for wolves. According to the Rocky Mountain Elk Foundation, New Mexico has a total elk population of eighty thousand (over half of which are found several hundred miles of desert north from the recovery area) and Arizona had 17,500 elk in 2009. Wild elk populations in western New Mexico and Arizona are not large enough to sustain one hundred wolves, if the wolves would only prey on elk.[7]

Ranches and Rural Home Impacts

Excerpt from the case study "Wolf Habituation as a Conservation Conundrum" by Diane K. Boyd:

The expanding wolf distribution has caused an increase in wolf-human encounters and generated concerns among wolf managers and conservationists . . . . This may sound like a tabloid headline, but the attacks were well documented by wolf authorities. Several factors may have led to the attacks including a lack of available wild prey, domestic livestock that were well protected, and many small children playing in the vicinity of the wolves. The common factor among nearly all reported wolf attacks was that wolves had become increasingly bold around humans (perhaps because of food scarcity, or possibly as a new strategy to exploit resources brought by humans into wilderness areas). North American wolves involved in recent attacks were repeatedly seen stealing articles of clothing, gear, exploring campsites, and sometimes obtaining food items—behaviors nearly identical to those reported by early frontiersmen.[8]

The current DNA mix of what are now known as "Mexican wolves" supports all of the above, so much so that many adults and children residing in the BRWRA have been seriously impacted by wolf encounters.

Often the general public, particularly those that do not live in the Southwest, are unaware that while the Blue Range Wolf Recovery Area (BRWRA) is a large area, there are hundreds of private land in-holdings in it where people live, work, and recreate in what the FWS shows as a vacant, empty green spot on a map. Sixty-six percent of the BRWRA is open to raising cattle, mining, and forestry. All of these private in-holdings are over a hundred years old and a large percentage of them are working family livestock operations, hardly good habitat to be releasing and building up a large population of predators. I now share some stories of what it has been like to live with Mexican wolves to illustrate the realities non-residents seldom ever experience.

- In 2007 Mary Miller and her husband Mark were forced to witness their eight-year-old running from a wolf attack on a family dog, an attack that occurred next to the child and likely was an attempt by the dog to protect the child. The dog recovered with extensive veterinary treatment. Two months later the same wolf killed eight-year-old Stacy's pretty black horse, named Six, in his corral while the family was out on a grocery trip. Six's remains were almost all gone after four days and five wolves.
- In 2005, Carlie Gatlin suffered a concussion in an auto wreck and was forced to walk home from a wrecked vehicle with two small children. Her son was bleeding from a head injury and her daughter was small enough that she had to be carried. When Carlie went to the

hospital for treatment the next morning, the tracks of the Luna pack overlapped hers and her son's in the snow beside the road. The Gatlin children have since suffered through several maimings and killings of their family dogs by wolves.

- Ivy Schneberger, thirteen, was riding her mare a half mile from home when the Sycamore Pack of two animals attempted to corner her. She fired off a round from her single-shot .22 and they finally meandered away, choosing instead to kill the family's calves for several weeks before the first legally killed Mexican wolf, AF 1155, was shot on that ranch. This is one incident where the wolf was considered a threat to human safety. One story circulated that a few days prior to the Schneberger incident wolves had threatened one of the program volunteers near their wilderness release site some forty-five miles to the southwest. But when the annual report came out both incidents were downplayed and the wolf was categorized as having been shot for livestock kills, not human safety and removal problems.

- AF 1155 was the same wolf that the Schneberger family had dealt with three years previously during the winter of 2000 when her mate left her to starve at their home and agency personnel chose to allow her to stay there from December until April. While living on the Schneberger ranch, the wolf subsisted on dog feces from their kennels and occasionally scavenged off coyote kills in the area. After she had been at the ranch headquarters off and on for three months, the family's milk cow died on a tributary leading into the home, which provided an excuse for the agency to justify to the public about the behavior of the obviously habituated wolf that fed on the dead cow.

- J. C. Nelson was fourteen years old when the Luna pack of five wolves crowded him against a tree and circled him for fifteen minutes. J. C. made no move to run, instead choosing to find as much cover as possible in an area damaged by a wildfire. He was only able to back against a charred ponderosa pine. The pack apparently smelled his rifle and, having better things to do, it eventually meandered away. This incident was reported to USFWS by Catron county law enforcement. USFWS investigated by sending volunteers to try and entice the pack to copy the behavior. This did not occur and the USFWS chose to deem the incident unimportant. J. C. said he was afraid to shoot the wolves because he thought his father would lose his grazing allotment if he did.

Proximity

Proximity is a problem when dealing with Mexican wolf presence and your family. The onset of the 2012 breeding season and the rise in the wolves'

population numbers, sightings, close encounters, and home encounters have once again created a difficult situation for managers of the Mexican wolf program and they are not getting much slack from local governments and citizens.

In December 2011, the program issued its first lethal control order after a female wolf with a long track record of livestock depredations and human habitation was found circling a private home at regular intervals where small children were exposed to her close presence. The same wolf had birthed a litter of hybrid Labrador pups the prior spring and USFWS was still on the lookout for the one Mexican wolf-dog hybrid that got away. They have not found it. Presumably, it will add to the genetic mix that is the "rare Mexican wolf." The remarkable thing about this control action is the fact that despite dozens of human safety encounters since the beginning of the program, many of which involved Mexican wolves' attraction to children, this was the first time the agency admitted lethal control was warranted for human safety reasons.

The encounters with wolves did not end after AF 1105 was removed from the picture, nor did the controversy of the necessity to remove her. Despite the claims from radical environmental organizations that the wolf was merely lonely and only needed to find a male, that there just were not enough male wolves in the wild for her to mate with, the next three encounters at homes and highways in the same area were with male wolves looking for a mate.

Numerous photos show that these animals are clearly very close to people. Also, they are in the area where pairing with AF 1105 was possible. Instead, the wolves appear more interested in easy prey or a handout at a home than pairing with a female that was making itself readily available. It makes a reasonable person wonder what these big males are breeding with since they showed little interest in AF 1105. Perhaps coyotes? It is possible, as the canine DNA is nearly identical for all species with the exception of family marker identification.

The photos taken on the highway of the large male wolf occurred when a young mother stopped alongside the road to allow her four-year-old daughter some fresh air after she was carsick. The other child, a two-year-old still strapped in the car seat, became impatient and began crying. At that moment, this wolf stepped out of the woods and approached the woman and her children. She was able to put the sick child back in the car and get herself in the car with about twenty feet between herself and this enormous animal. However, this wolf was intent on stalking her, and her children were only goading it more with their distressed behavior.

Since that time, the wolf was killed, and the mother has suffered an unbelievable amount of slander in the local news media. This is simply because she lives in proximity of the expanding Mexican wolf population. The activists who have repeatedly attempted to destroy her credibility and reputation have avoided the factual reports of the situation, of was which was made available to them. Instead, they chose to blame and attack a mom over the death of a habituated aggressive wolf, even though she certainly was not responsible for the decision. The message these people promote is that this wolf was special, presumably more special than the woman's small children and their safety and more important than their freedom to exercise their rights on their own land at their own home.

This wolf is not unique. Hybrid controversy aside, genetically this wolf was redundant to the population of Mexican wolves, which include over four hundred in captivity. Only genetically redundant wolves are legally allowed to be used in the releases on federal lands in Arizona and New Mexico. Dozens exactly like her exist in captivity, ready to enhance the wild breeding pool.

It's disturbing to see that there is a consistent trend from government agency personnel working within the program to promote the extremist notion that wolves on the ground are genetically unique, and so cannot be removed or controlled. This is contrary to many scientific and policy documents available. This biased advocacy indicates that not only are there close ties between the program and the most extreme environmental advocates for wolves in the Southwest, but also that members of the USFWS appear to be coordinating media and press release strategies with those same organizations.

The behavior of the wolf population that the federal agencies are in charge of (as well as the radical wolf advocates) is such that local governments in New

Mexico and Arizona counties that contain wolves are examining their options to protect human safety. Human health and safety is an area that the agencies themselves are supposed to maintain over any policy that they have in place, including promoting increases in the wolf population. With the large number of incidents involving children in the BRWRA, the counties feel they must be ready to step in and do the job the federal government is somewhat lackadaisical about doing. We do not know what will happen if Catron County, New Mexico, kills a wolf in a constituent's yard, but the majority of the small human population in the county are supportive of the idea.

Megan Richardson witnessed the habituated-livestock-killing Middle Fork pack coming to her home at regular intervals, and began to wonder if the noises her small baby son made drew the wolves even closer. The wolves can be seen coming into her driveway in photos from game cameras she has installed. She put it bluntly: "Is it going to take someone getting seriously hurt or killed before something is done?" Megan deals with the situation by keeping game cameras out in the yard to document the animals' frequent presence on her land.

The Middle Fork wolf pack, implicated in dozens of unconfirmed livestock deaths, has frustrated ranchers with uncontrolled killing. The Middle Fork pack was the first, and perhaps only, group of wild-born Mexican wolves to pair and raise successful litters on their own. For the first four years of their existence, there is no record of them killing cattle and they stayed in the wilderness.

In 2007, USFWS discovered the female had a broken leg that had healed poorly and was cosmetically unattractive, although not a deterrent to her hunting ability. They trapped her, took her into captivity, and amputated the leg. She was in captivity for over a month. Her pack, sensing that she was alive, made wider and wider circles out of their normal territory looking for her. By the time the female was placed back into the wild, the wolves had discovered a livestock operation miles to the north. Later that year they began killing calves. Sometime later, the agency also chose to amputate the leg of the alpha male that had been damaged in a trap likely belonging to the same personnel that had trapped the female. The Middle Fork pack was never well-behaved after that, and the USFWS once again contributed habituation behavior of wild-born Mexican wolves that had previously displayed normal behavior patterns.

In spite of a massive hazing effort that took place on the Adobe and Slash Ranches in 2009, in southwest New Mexico, the Middle Fork pack was allowed to kill and maim livestock undeterred. Early that spring the pack was suspected to have killed several yearlings, but none were found early enough to get a confirmation on the kills. Ranch manager Gene Whetten grew more frustrated by the day. He says:

We work our tails off managing this place for elk and livestock and we do a good job here. However, when you read the papers you see the implication that something is managed badly on the ranch and that is why there are wolves here killing our stock. We have even read where someone makes the claim that there were sixteen livestock carcasses that we supposedly left lying around the den. I guess they meant deliberately. We did not even know where the den was until the first confirmed kill happened. If there were yearling kills at the den it was wolves that dragged them there after they ran the elk out of the country.

Here, Whetten is referring to the tendency of the USFWS to release information to environmental organizations on wolf incidents first, thereby allowing these organizations to filter news releases regarding the program, particularly livestock conflicts. The agencies hang back on releasing their own press statements. When caught between the environmental extremists' false accusations against the rancher and the facts, program spokespeople tend to refuse to correct deliberate misinformation that they know is false.

Adobe Ranch Case Study—Performance-Related Losses

From 2000 to 2003, the Adobe Ranch knew of only two wolves on the ranch. In 2004, the number of wolves increased to nine, until 2006 when the total dropped to six. By the fall of 2007, fourteen wolves (three packs) were known to be on the ranch (as per the Adobe Ranch management, personal communication, 2008). Wolves were also in close proximity to the ranch headquarters and branding pasture beginning in February.

This level of wolf activity led to eight confirmed and one probable depredation of livestock. Total depredations for 2007 included confirmed (thirteen animals), probable (one animal), and possible (four animals) on the Adobe Ranch. The Adobe Ranch alone accounted for 46 percent of the total confirmed depredations reported to the Bailey Wildlife Foundation Wolf Compensation Trust in New Mexico for 2007. In addition, 50 percent of the possible depredations and 100 percent of the probable depredations for 2007 occurred on this ranch. Although depredations were confirmed, no compensations were paid by the BWFWC in 2007 and 2008.[9]

The effect of wolves on livestock, however, goes much further than kills. Stress placed on cattle and sheep by the presence of wolves results in weight loss, which translates into income lost. Weaning weights, shipping weights, and site-specific precipitation were available for the Adobe Ranch from 2002 to 2007. Growing-season precipitation was correlated to steer performance, as forage production is closely related to growing-season precipitation. Cumulative precipitation from April through October is considered growing-season precipitation. There was little variation in steer ADG from 2002 through 2006, with an average of 0.08 pounds per day, which is considered normal performance in the region when fall-weaned calves are retained (as per C. Mathis, personal communication, 2008). However, ADG in 2007 was much lower than in previous years although calves were managed similarly between weaning and shipping; ADG fell well below the lower limit of a 99 percent confidence interval of -0.75 pounds per day. Using actual market values from the Clovis Livestock Auction in Clovis, New Mexico, the cost of the estimated impact of weight loss in 2007 was -$108.83 per steer weaned.[10]

Overall, impacts of wolves on ranches is not only significant, it is destroying ranch viability (see chapter 11, "Collateral Damage" by Jess Carey). Actual data on current losses is virtually ignored by USFWS.

Prey Preference

A 2006 study using GPS collars to analyze the prey preference of Mexican wolves showed the agencies' bias and flawed scientific techniques. In the study, the Nantac pack male was implicated in thirteen livestock kills in 2006 over several weeks. The data gathered during the livestock kills was not used in the study. Instead, the study concluded a majority of Mexican wolves killed elk. This occurred simply because the Nantac male with a GPS collar was shot for livestock depredation because the agency could not stop him any other way. The pack's kills were not included in the study simply because the agency did not want it to reflect a livestock surplus killing spree, which they consider abnormal wolf behavior. In one twenty-four-hour period, the two Nantac wolves killed two cows and their calves, while eating nothing.

More importantly, the USFWS needed the study to reflect elk as the major food source in order to ask for continued funding for the program, and to continue federal planning efforts in the scoping process for an eventual rule change. Proving that the wolf surplus killed privately owned livestock was simply not on their agenda. Had the Nantac pack's livestock kills been left in the study, USFWS would have had science that is more accurate.

Carcass Removal

The Mexican wolf management program, from the beginning, has been complicated by environmentalists. The Center for Biological Diversity, who originally sued to push recovery forward after biologists had voted against it, demanded that ranchers remove dead livestock or have no wolf damage control occur on their allotments or land. Ranchers balked at this demand in part due to costs, logistics, and the fact that most allotments have areas that are completely inaccessible for such maneuverings. The ground is always frozen in the winter leaving them with no access to pastures or no way to bury carcasses. This demand is simply unattainable.

Ranchers also own the waters and rights to access that land through beneficial use of the water. Removal or manipulation of livestock access to those rights is a federal take of their property and often leads to serious management interference with their grazing rotation plans. Neither the experts nor the science agrees with the CBD. For example, The University of Minnesota conducted a study in early 1999 to determine if any livestock management practices could prevent wolf depredation. The study could find no management practices certain to prevent wolf depredation. The only method proven to prevent wolf depredation was removing the depredating wolves from the farm.[11]

Wildlife biologist Ed Bangs, who until 2012 was the Wolf Recovery Coordinator for USFWS, and the principle author of the 1993 Environmental Impact Statement for the wolf recovery program in the North Rockies, wrote in a 2006 email that he did not believe that removal of carcasses would make a difference in wolf behavior:

> I thought the idea that wolves eat a dead cow, think beef tastes great, and then start attacking cattle is mythology—as eating carrion and killing prey are two totally different wolf behaviors. Wolves often scavenge all they can. However, I do think that having bone pile next to calving pasture can increase potential for conflict by attracting wolf activity in the vicinity of livestock [research shows cattle near wolf dens are more likely to be the ones attacked simply because level of wolf activity and interacting with livestock is highest there] and that anything that helps wolves become more familiar with livestock can increase the chance they might test them as prey. But normal range practice out here makes it nearly impossible to find and bury [or blow up for human safety concerns as they do for G. bears issues and livestock carcasses along trails] every carcass so if livestock carcass disposal is within 'normal' and traditional livestock husbandry practices we don't consider them an attractant that we would withhold wolf removal. We do advise ranchers not to have a bone yard next to livestock and we have removed carcasses in pastures to prevent wolves from coming back into concentrated livestock to feed on their kills. But feeding on livestock carcasses is a very different thing than attacking livestock— one doesn't necessarily lead to the other.[12]

In spite of the lack of scientific evidence or publications that confirm livestock carcass removal will benefit the program, it is still being pursued by Region 2 USFWS as an option. In addition, there is the notion that the mere presence of livestock on ranches more than a century old is impacting the success of the wolf recovery program.

Recent changes in the program seem to have all been aimed at the cessation of mitigation for problem animals. The only exception in the past five years has been the removal of AF 1105 for human safety reasons. Defenders of Wildlife has stopped reimbursing the immediate cost of ranchers confirmed livestock kills as they did in Wyoming when bears became established and the organization felt ranching should become a thing of the past in those areas. New Mexico is far too poor to appropriate state funds to reimburse ranchers for the take of their livestock by wolves. It is hard not to conclude that the promises outlined in the 1996 EIS and final rule have simply been discarded when it became apparent that wolves or wolf hybrids will wipe out ranchers and ranches if left unmanaged long enough. Let us look at some specifics.

Conflict of Interest, Collusion, and Corruption

Wolf team full cooperators do not include local organizations or local people and do not efficiently mitigate problems arising from the program. Instead, non-government organizations such as the Turner Endangered Species Fund (TESF) and Defenders of Wildlife (DOW) enjoy full cooperator status. This status is unofficial, but the results of their activities in the program speak for themselves. It has also become apparent that other environmental groups funded by TESF have internal knowledge of the day-to-day operations and internal planning of the reintroduction and recovery. The draft of the program's five-year review was released to environmental groups before being made available to the public or local government.

Dave Parsons, former program coordinator and author of the original EIS, and the person behind the introduction of the Ghost Ranch hybrid genetics, is the coordinator of the Southern Rockies Wolf Restoration Project, another TESF grantee. Involvement and support of the ideology of these organizations is in part why the original EIS projections have never been met. It was apparent from the beginning of the program that Parsons had exerted entirely too much influence and was promoting a plan that would eventually be detrimental to the livestock industry as whole. It was also apparent that his team was determined to ignore socio-economic comments, barely referring to those issues or arbitrarily discarding the notion that the program would harm people. In fact, USFWS claimed in the EIS that there was not going to be significant harm done to the region although they had the data from historical Mexican wolf damage records, such as the following:

The estimate of economic damage in New Mexico caused by forty to fifty wolves in 1918 was $60,000—equivalent to about $960,000 in 2007 dollars. From 1915 to 1920, wolf-induced economic losses were estimated at half a million dollars—comparable to $9.4 million in 2007 dollars. In a 1921 US Department of Agriculture news release, the Bureau of Biological Survey estimated annual economic losses in livestock of $20 to $30 million ($205 to $308 million in 2007 dollars) to all predators throughout the West. According to Brown (1992), average destruction by predatory animals during this same period was estimated to be $1,000 worth of livestock annually ($10,000 in 2007 dollars) for each wolf and mountain lion, $500 ($5,000 in 2007 dollars) for each stock-killing bear, and $50 ($500 in 2007 dollars) for each coyote and bobcat. He also illustrated cases where substantial damage was caused by just a few predators. For example, one wolf in Colorado killed nearly $3,000 worth of cattle ($30,000 in 2007 dollars) in one year, two wolves in Texas killed seventy-two sheep in two weeks, one wolf in New Mexico killed twenty-five head of cattle in two months, and another wolf killed one hundred fifty cattle valued at $5,000 ($51,000 in 2007 dollars) during a six month period.[13]

The 2010 Mexican Wolf Recovery Team has room for only three to four members that represent economic, livestock, or hunting interests. Those members are the only members of the recovery team that are true volunteers and not reimbursed or paid for attendance. USFWS, TESF, DOW, and other NGOs, as well as the state wildlife departments, pay other team members. The 2010 Recovery Team is not much different from the 2003 team. There are so many divisions to the team that only the top tier will have significant input into the program. The following is from the official recovery team page in the USFWS website for the program:

The Southwest Region has initiated the revision of the 1982 Mexican Wolf Recovery Plan. In December 2010, we charged a new recovery team with the development of a revised recovery plan for the Mexican wolf. The team includes a Tribal Liaisons Subgroup, Stakeholder Liaisons Subgroup, Agency Liaisons Subgroup, and a Science and Planning Subgroup. When completed and approved by the Service, the plan will include objective and measurable recovery criteria for delisting the Mexican wolf from the List of Threatened and Endangered Wildlife and Plants, management actions that will achieve the criteria, and time and cost estimates for these actions.[14]

Is this collusion, corruption, or simply cooperation?

Currently there are no actual stakeholders involved in the recovery planning, only the Stakeholder Liaisons Subgroup; this is a transparent method of keeping impacted parties from putting information into the recovery plan that will protect and assist actual stakeholders.

It appears that far less influence by non-affected wolf advocacy parties is necessary to further the goals of the ESA in reference to that specific and emphatic statement in the law that the ESA "will not be used to engineer social change." This is recognized in various federal documents, for example the 2011 USFWS publication, "ESA Basics: More Than 30 Years of Conserving Endangered Species" states:

> Two-thirds of federally listed species have at least some habitat on private land, and some species have most of their remaining habitat on private land. The FWS has developed an array of tools and incentives to protect the interests of private landowners while encouraging management activities that benefit listed and other at-risk species.[15]

Wolf activists, however, seem to believe that it is necessary to specifically foster changes in the economic and social structure of the region and to keep required and necessary wolf control and management at a dangerous minimum. The NGO involvement promotes keeping problem wolves on the ground, and further harms the local communities that are relatively defenseless. This is a clear violation of the ESA. USFWS should be developing protocol to enhance affected party participation in decision-making and management and to limit non-affected NGO interests whose goals are contrary to the ESA's requirements for avoiding social change.

Collusion with Non-Government Organizations (NGOs) Over Future Planning and Implementation of Current Procedures

The technical team of Paul Paquet and Mike Phillips, both advisors to the NGO Southern Rockies Wolf Restoration Project, made three-year review recommendations. There was no data collected the first three years for the tech team to go on. The scientists admitted as much and stated as much publicly. They made socio/political recommendations instead: 1. Force ranchers to remove livestock carcasses, holding US Forest Service permit removal over their heads; 2. Boundary removal allowing wolves to spread beyond the original Blue Range Wolf Recovery Area (BRWRA); and 3. Manipulation of problem wolves and control of depredating wolves was claimed in their report to be counterproductive to reintroduction efforts.

These recommendations were admittedly socio/political with no scientific basis.

Due to the non-scientific nature of the technical recommendations, then–wolf reintroduction coordinator Brian Kelly put together a rather large team of stakeholders to make recommendations for changes to the program in a way that would enhance the program, yet eliminate or mitigate the problems and make things work for the majority of the stakeholders. It was a good idea but the NGO participants were not happy about it. The individuals invited to this large working group comprised many different facets of the local communities—economic, human dimension, environmental, and scientific—as well as many wolf experts and biologists. The document developed from that group represents the last time a concrete attempt was made to cooperate with local communities and governments. However, it never went into effect.

The conclusion to be drawn here is simply that those recommendations were not politically expedient to meet the unstated goals of the agencies, and more importantly the cooperating NGOs, because it cooperated with both sides of the wolf issue. Because of this, those recommendations were shunted aside when lead agency status changed in the spring of 2003. Brian Kelly, former USFWS Mexican wolf team leader, left the program and with him went the strategic removal of public input into the program.

This removal of public input allowed the agencies and NGOs to ignore their obligations to stakeholders as set forth in the final Environmental Impact Statement and Final Rule. There was no one involved at upper levels to hold their feet to the fire and be fair to the local stakeholders.

What happened next was entirely predictable. Suddenly, those three-year review recommendations by Paquet and Phillips were massaged into the basis for the new five-year review by new wolf managers. Ranching and community sustainability swiftly took a back seat to wolf recovery.

For the past several years, there has been very little effort to follow the final rule as it relates to complying with wolf management habituation control or livestock protection. Instead funding is focused on planning total recovery in the whole recovery area, planning more releases, and on promotional programs marketing wolf reintroduction to the un-informed and unaffected public.

Allowing the NGOs to serve as experts in the wolf reintroduction and recovery has given them unique power over the landowners in the areas affected by the Endangered Species Act. The NGO scientists have the full force of the federal government behind their plans. When rural people are forced to sue the USFWS, judges always defer to the expertise of the USFWS, however, in reality they are deferring to the expertise of the NGOs who are so heavily involved in structuring the day-to-day management and planning of the program. There is no input or redress for affected people either at the agency level or in the courtroom.

Management and Personnel Bias against Ranching

H. Dale Hall, USFWS regional director until 2006, admitted that when he was first placed in position, members of his wolf staff brought him an anti-grazing NGO published book and told him that grazing should be removed to make way for the wolf recovery. There has been no exposure of this action, nor were there any personnel changes. These persons were not removed from the team and placed elsewhere in the agency. Clearly, there are still many wolf program employees that are openly promoting an anti-grazing agenda.

Reintroduction of Mexican wolves into their historic habitat or historic DPS area would have required reintroduction into northern Mexico, extreme southern Arizona and New Mexico, and southwest Texas. Currently, Mexican wolves have been released into Arizona and New Mexico. Recently, five wolves were released in Mexico in the state of Chihuahua and four of them were either killed by poison or died within the first six months. Suspicious death and disappearance is also a factor in the US side of the program. Cornered people simply do not respond well to threat.

Damage Control: Habituation Management or Management for Maximum Habituation?

Mexican wolves have been shown through DNA analysis to include DNA from three different wolf species from North America as well as coyote and domestic dog genes, again showing that Ghost Ranch and Aragon were completely inappropriate additions. Hybrid wolves are known to be far more aggressive and to kill more livestock and domestic animals than non-hybrid wolves, yet the Mexican wolves we now know were created in a kennel and deemed a pure subspecies in a courtroom.

USFWS numbers indicate there are more than three hundred Mexican wolves with the same DNA in the captive breeding program available for release. Naïve animals without a history of livestock kills exist; still any discussion of removing depredating packs has hit a brick wall within the program, mainly due to the Center for Biological Diversity's insistence through lawsuit threats that each individual wolf is important to the gene pool. This is not a scientifically valid position, as the rule states clearly that only genetically redundant wolves are used in the release program. Yet that false premise has recently been taken up by the USFWS as if that claim had legitimacy, when in truth it is merely another politically motivated decision to kowtow to extremists in order to keep to their own agenda.

The seriousness of the habituation of Mexican wolves can be seen in the documented incidents involving children, many backed up with photographs, including:

Wolf-proof school bus shelter.

Young boy in Catron County waiting for the school
bus in a wolf-proof shelter.

- Child encounters wolves while on horseback.
- Child finds pet dog slaughtered by wolves on private land.
- Child surrounded by wolf pack on hunting trip.
- Children followed by wolves from bus stop.
- Dog defends child in yard from wolf.
- Child's horse slaughtered by Aspen pack in corral.
- Child watched wolf kill kitten in yard.

- Children kept captive in rural communities due to wolf presence.
- Children subjected to wolves coming in their yard while mother unloads groceries from her truck.
- Children calling in a wolf by crying when stopped on the side of a rural highway due to motion sickness.
- Wolves laying in yards and defecating in places where children play, possibly as a method of marking territory where there are dogs present. (Also introducing possible exposure to hydatid disease found in wolf feces.)

USFWS wolf managers still insist that nothing is amiss in this program. Some of them have even called these accounts hyperbole and blame parents for instilling fear into their children. Fear is not always a bad thing. J. C. Nelson responded to that fear instilled in him and reacted calmly. Had he reacted to his natural fear and run from the Luna pack, he likely would have been killed.

Little to nothing is done to mitigate these types of behaviors. Does the name-calling of a family matter if their children are deliberately being placed in harm's way by a mismanaged, government-sponsored program if the community was adamantly against it to begin with? Not really, not on a local level, since the majority of rural residents impacted understand the circumstances that led to the incident. However, marginalizing the family is a common method of downplaying the incident that actually occurred. The Mexican wolf reintroduction area residents share an attitude that is not all that unusual in other cultures where wolves are actually present. Proximity to wolves affects your mindset. As chapters in the book on *Wolf Attacks and Wolves in Russia* clearly show, in Norway, Finland, Sweden, France, Spain, Russian, Karelia, and throughout Asia, historically documented wolf attacks on people were common enough that on occasion they became epidemic. In fact, in a 2003 study in Scandinavia, 85 percent of the wolf-related deaths that were verified were children with no adult present.[16]

These incidents lead researchers to another question: Why did wolves become child snatchers? The researchers came to believe that wolves understood that children were vulnerable prey and made efforts to avoid adults who would hunt them. In fact, most of those killed were doing some very familiar things such as herding livestock—near a house or farmyard, in a forested area, or by outlying barns or fields.

This sounds a lot like the community and culture of the Mexican Wolf Blue Range Reintroduction area. It really doesn't matter that for generations there has not been a wolf attack in the lower forty-eight states. None of the talking points about wolf behavior or the demonization of parenting abilities fazes rural people actually living among and dealing with Mexican wolves. Rural people instinctively know there is danger, and distrust those who downplay their experiences. Parents keep their kids confined due to danger from wolves, so we are actually seeing rural residents changing their behavior due to this program.

Perhaps the question shown by the Scandinavian study should not be why people are afraid of wolves, but rather in the face of such overwhelming evidence, why so many people appear to be unafraid of wolves.

What the Future Holds

The current situation in the United States in regards to the recovery of wolf populations in historic habitat is very bleak for rural Americans. States such as Montana, Idaho, Wyoming, Oregon, and Washington are being overrun with an unmanaged and possibly overprotected population of gray wolves. Wolves are spreading into Utah, Illinois, Indiana, and Colorado. While the Southwest program has been floundering along for many years without significant successes, such as those seen in the Northern Rocky mountains, the very fact that the Mexican wolves are not listed as a subspecies indicates that Congress never intended to allow them significant status for full recovery instead preferring to allow them a footnote in history as a small population of gray wolves with distinct features adapted to the arid regions of the Southwest and Mexico.

The agenda of the environmental movement and federal bureaucrats however is significantly different. The delisting of the gray wolf by Congress was the beginning of the end of federal funding for wolves and these entities are losing their cash cow. It makes the Mexican wolf that much more important as a tool for federal programs, as well as manipulating landowners and federal lands ranchers, and a replacement poster animal for the previously lucrative gray wolf of the Northern Rockies and the Midwest.

Editor's Note

As this book goes to its second printing, the office of inspector general for the US Department of Interior has released a scathing report on the Mexican wolf program. All of the information regarding Mexican wolves in the first edition of *The Real Wolf* and in Laura Schneberger's chapter is supported by the report titled, "Investigative Report of the US Fish and Wildlife Service's Mexican Wolf Program, July 11, 2016."

The investigation began in July 2013 when the Catron County Board of Commissioners submitted a complaint to then-Congressman Doc Hastings requesting the US Department of Interior's Office of Inspector General investigate alleged misconduct by US Fish and Wildlife Service staff overseeing the Mexican Wolf Recovery Program (MGWRP). Congressman Hastings was replaced by Steven Pearce upon the former's retirement in 2015. Pearce submitted the complaint to the Department of Interior in February 2014.

The Board of Commissioners complaint requested the Office of Inspector General investigate violations of the Federal Misconduct Policy, which included that the coordinator of the Interagency Field Team (IFT) for the MGWRP had failed to properly document nuisance complaints about wolves, had not communicated effectively with county residents to address public safety concerns involving wolves, had mismanaged livestock depredation investigations and compensation, had destroyed a wolf DNA sample, had mishandled a wolf bite incident with a MGWRP volunteer, and had refused to assess ranchers' property losses associated with wolf kills. The board also asserted that the Mexican gray wolf is actually extinct, its DNA contaminated with that of coyotes and/or domestic dogs.

The Office of Inspector General narrowed their investigations to: "mismanagement of nuisance wolves; failure to communicate effectively with the public and Catron County; depredation-related impacts; and administrative management issues."

Excerpt from the OIG Report:

In February 2014, the US Department of the Interior's Office of Inspector General received a request from US Congressman Steven Pearce (R-NM) to investigate alleged misconduct by US Fish and Wildlife Service (FWS) staff overseeing the Mexican Gray Wolf Recovery Program (MGWRP). The Congressman referred us to the Catron County, NM, Board of Commissioners, which had submitted a complaint in July 2013 to then-Congressman Doc Hastings. The county's complaint made numerous allegations against MGWRP, particularly involving a former coordinator of the Interagency Field Team (IFT) charged with implementing the program. The county alleged that the former IFT coordinator and MGWRP had failed to properly document nuisance complaints about wolves, had not communicated effectively with county residents to address public safety concerns involving the wolves, had mismanaged livestock depredation investigations and compensation, and had destroyed a wolf DNA sample. The county also alleged that FWS mishandled a wolf bite incident involving a MGWRP volunteer and refused to assess ranchers' property losses associated with wolf kills. In addition, the complaint asserted that the Mexican gray wolf is actually extinct; its DNA is contaminated with that of coyotes or domestic dogs.

The investigation substantiated many of the allegations against the former IFT coordinator, but learned that USFWS had been aware of these issues and had already reassigned her to another position by the time they received the complaint. Since then, MGWRP employees informed OIG that USFWS has been documenting nuisance complaints and has attempted to improve communication with county residents; however, many of the county residents we spoke to said they were still concerned about poor communication with MGWRP and a perceived lack of concern for public safety. The OIG also found that local ranchers have not consistently received full compensation for their livestock losses, but we did not substantiate the remaining allegations.

There were several allegations that were substantiated by the IG report, including a finding that ranchers have not been adequately reimbursed for livestock losses. Among the more serious allegations, the IG's report confirmed an agency employee falsified the location of multiple wolf kills, allowing wolves to kill further livestock and remain in close proximity to humans.

The former IFT coordinator, a wildlife biologist for the program, "Acknowledged that she gave genetically valuable wolves more care, allowed their nuisance behavior to continue, and provided them more opportunities to breed," reads the report. This happened in violation of the agency rule, which required removal of repeat-offending wolves that kill livestock.

The former coordinator was simply transferred to work with rule changes and recovery planning for Mexican wolves. Dr. Benjamin Toggle considers this transfer a full repair for the damage done by the program, although the former coordinator's actions are certainly not the only problems the program has burdened the public with over the nearly two decades it has been in operation.

The OIG report also found that in general there is mismanagement of nuisance wolves, which continues to be a public safety issue of concern; failure to communicate effectively with the public and county government; depredation-related impacts and uncompensated losses; and administrative management issues, including that the Fish and Wildlife Service in general protected "genetically valuable" wolves in the wild.

According to Fish and Wildlife spokesman John Bradley, "We have been working to improve all aspects of our work with the county and the people who live there" and added that the current field office coordinator "considers this matter closed and resolved."

Congressman Pearce called the OIG findings "incendiary" and said it pointed to problems at the highest management levels. "They claim they have fixed the problem by reassigning one person," said Pearce. "Their problems are much bigger than one employee and extend to the highest levels of the agency. Those at the top levels at FWS tolerated a culture of lies, falsification, mismanagement, and manipulation of scientific data, ultimately at the cost of livelihoods and the public trust. This mismanagement has caused economic harm to the state, county, and individuals."

Examples of failure to document problem or nuisance wolves were cited throughout the complaint document and have been verified by the OIG report. In one instance during an interview with the former Interagency Field Team coordinator, the former IFT coordinator acknowledged that under her supervision data occasionally went uncollected, either because staff was "busy" or because the information "slipped through the cracks." Although she knew that MGWRP used the data to make program decisions, she still did not fill out a form for every call about a wolf, even if the call resulted in relevant scientific information. She blamed this on "workload demands" and "human error." She also explained that if the IFT member collecting the data for the report decided that the animal being described was actually a coyote or a dog, not a wolf, the information did not have to be entered. In later interviews this coordinator was said to have not believed that nuisance reports were necessary and did not require her staff to fill them out because the incidents were inherently ambiguous and based on perceptions of the witness, that one person's "attack" may be another person's "spiritual experience."

This type of obfuscation is apparent throughout the report and while many of the allegations remain unproven, those that were indicate the continuing trend that the IFT is willing to spin each situation in a manner that benefits the program and themselves has not ended.

On September 15, 2016, downtown Eagar, Arizona, had such a "spiritual experience" with what, biologically, can only be a pack of wolves. Downtown Eagar consists of a main street and about two miles of small-town stores including fast food restaurants, schools, and grocery stores all bordered within a block by open fields and farms. Incredibly, even after the agency fix of the OIG verified reports, Fish and Wildlife Service claimed a dead cow also found in the same area that day, one block from the main street, was a coyote kill. The cow had 40 mm bite marks; the average coyote bite is 33 mm. Mexican wolf bite spread averages 39 to 44 mm.

This author received the following email from a person who lived in the Eagar area:

> Wolf incident and a dead cow, along with a very scared young lady who was feeding the calves, right in the middle of our town. The details—4 wolves behaving aggressively towards a young woman feeding livestock early in the morning. The horned young bred cow was found dead near the Safeway, although the bite marks were 40 mm wide, FWS claimed the grown cow was killed by coyotes. Something that no

rancher in the Arizona-New Mexico region has ever seen occur with coyotes.

Doyel Shamley, Apache County, Arizona, coordinator says,

> Here is the long and short of the whole incident, whether it was a wolf (nobody in any "team" or "agency" will provide a report yet) or a coyote, the agencies are mishandling all aspects of public health and safety, a constitutionally delegated power of the 10th Amendment to the states and their political subdivisions. The county's "wolf team" didn't even notify their own agency regional chief and former IFT-L, Chris Bagnoli. I had to tell him there was a depredation to begin with, and the location, etc. The "wolf team" never notified the local city government (remember, it all occurred in downtown Eagar), or Apache County, or the Apache County Sheriff, all requirements of their system and business as usual for the program. I had to do that function. The "wolf team" never notified our AZ Department of Agriculture or our AZ Department of Agriculture Livestock Inspector, a requirement of their system and business as usual for their program. I had to do that function as well. The "wolf team" never notified any of us of the outcome of the Wildlife Services investigation with reports, data, etc.—instead it was a one-sentence phone-to-email message from Bagnoli, stating the WS determined it was a coyote with a canine spread the size of a wolf. No reports have since been delivered despite requests. THAT, is exactly how they are running this fraudulent system.

The report claims all problems were resolved with the transfer of the former IFT coordinator; however in one single incident, two, possibly three, separate issues are shown to still be standard operating procedure: blaming coyotes for wolf kills; allowing wolves in areas where the public can be harmed and threatened without thorough notification investigation or mitigation; and possibly allowing coyote/wolf hybrids to exist and breed in the Mexican Gray Wolf Recovery Area, which is a violation of the Endangered Species Act and all applicable regulations associated with the program.

Taking Action

The new IFT coordinator says that he now requires every MGWRP staff member to complete a form when a complaint is received. According to Office of Inspector General report, on July 11, 2016, the Office of Inspector General reported that it had "substantiated many of the allegations against the former IFT Coordinator." By that time, the USFWS had reassigned the former IFT Coordinator to another position. Since that occurred USFWS has been documenting nuisance complaints and attempted to improve communications with county residents. "However, many of the county residents we spoke to said they were still concerned about poor communication with the MGWRP and a perceived lack of concern for public safety. We also found that local ranchers have not consistently received full compensation for their livestock losses . . . we could not substantiate the remaining allegations."

One of those allegations was in regards to the genetic purity of the Mexican wolf. The USFWS senior wolf biologists said that extensive research has been done on the Mexican gray wolf, and the Mexican wolf is the most genetically distinct gray wolf subspecies. The wolf biologist said that on rare occasions when a Mexican wolf had bred with a dog, the pups were euthanized at the den. The genetics of the Mexican wolves will, however, continue to be monitored.

The Investigative Report concludes with this summary of changes that have taken place since the July 2013 letter from the Catron County Commissioners:

> The MGWRP coordinator gave us a list of changes made by the MGWRP since Catron's County's original complaint. Perhaps most significantly, she noted that the former IFT Coordinator had been reassigned in August 2013 to another position in the USFWS. She was still involved with MGWRP, but at an administrative level. Also in August of 2013, the current IFT Coordinator assumed that role, which he fulfills from Albuquerque, NM. MGWRP also added new biology staff and an outreach specialist to work with partner agencies and the media. The reorganization, according to the MGWRP Coordinator, resulted in improvements to the MGWRP, included shorter response times to nuisances, better interactions with livestock owners, improved documentation of nuisances and depredations, and improved communication and working relationships with owners and partner agencies.

The Investigative Report goes on to say that concerning IFT's communication with the public:

- More than fifty signs listing IFT's toll-free number have been posted in and around the Gila and Apache National Forests for years, predating the current coordinator's own employment at FWS. He noted that the number connects directly to the IFT office.
- Two public websites, the MGWRP site and the Arizona Game and Fish Department site, provide contact information.

- MGWRP holds approximately six public meetings annually to discuss wolf recovery issues. Meetings are advertised, held in local communities, and open to the public.
- The current IFT coordinator regularly meets with constituents at their homes to discuss wolf issues.
- Every week, IFT notifies ranchers if they are near current wolf locations.
- IFT's contact information is included in pamphlets provided to hunters.
- IFT members regularly provide wolf education at schools, zoos, and other venues.

In short, the July 2013 complaint has proven to be largely substantiated by the Investigative Report of the USFWS Mexican Gray Wolf Program. But, there is much more at stake for the people of Catron County.

The negative effects to livestock producers caused by Mexican wolves are a wide spectrum not addressed and/or ignored by the US Fish and Wildlife Service. Prior negative data and documentation of wolf recovery from other states were not utilized to mitigate the same negative effects of Mexican wolf recovery in New Mexico and Arizona.

Wolves are continually killing, and prey testing in a herd produces chronic wolf stress in cattle. Chronic wolf-caused stress in cattle leads to loss of body condition, cows birthing weak calves, pre-mature birth of calves, abortion of calves, immune suppression, decreased pregnancy rates, increased susceptibility to disease, weight loss, and altered demeanor from docile to aggressive.

1. True livestock losses are not reflected in confirmed and probable investigative findings;
2. Few livestock depredations are actually compensated;
3. Cumulative effects of wolf predation makes livestock production untenable;
4. Impact on individual family ranchers is devastating, even though the impact of the entire livestock industry of the state may be small;
5. Wolf depredation disrupts grazing management plans;
6. Increased uncompensated hours tending injured cow/calves;
7. Increased uncompensated hours checking livestock;
8. Increased uncompensated hours mending fences when wolves attack/ run livestock through them;

9. Increased uncompensated hours gathering livestock and returning to proper pasture;
10. Loss of market value for maimed and disfigured calves;
11. Loss of replacement heifers/production;
12. Loss of revenue while new herd takes several years to acclimate;
13. Loss of revenue while replacement heifers take three years to acclimate into an existing herd.

Several ranches have been closed as a result of the effects of the wolves on Catron County and elk and deer hunting in that area have suffered. The New Mexico Game and Fish Department has opposed the Mexican Wolf Program and supported Catron County.

The full Investigative Report can be found online at: https://www.doioig. gov/sites/doioig.gov/files/MexicanGrayWolfProgram_Public.pdf.

Endnotes:
1. Hedrick, Philip W, Philip S. Miller, Eli Geffen, and Robert Wayne. "Genetic Evaluation of the Three Captive Mexican Wolf Lineages." *Zoo Biology* 16 (1997), 47–69.
2. http://graywolfnews.com/pdf/the-courts-were-wrong-these-wolves-are-hybrids.pdf
3. http://www.cuerochupacabra.com/id25.html
4. http://graywolfnews.com/pdf/the-courts-were-wrong-these-wolves-are-hybrids.pdf
5. http://graywolfnews.com/pdf/the-courts-were-wrong-these-wolves-are-hybrids.pdf
6. http://www.fws.gov/southwest/es/mexicanwolf/BRWRP_notes.cfm
7. http://www.rmef.org/NewsandMedia/NewsReleases/2009/ElkPopulations.htm
8. Diane K. Boyd, (Case Study) "Wolf Habituation as a Conservation Conundrum" http://www.sinauer.com/groom/article.php?id=24
9. http://www.defenders.org/resources/publications/programs_and_policy/wildlife_conservation/solutions/full_list_of_payments_in_the_northern_rockies_and_southwest.pdf
10. http://aces.nmsu.edu/pubs/_ritf/RITF80.pdf
11. Mech, L. David, Elizabeth K. Harper, Thomas J. Meier, William J. Paul. "Wolf Depredations on Cattle." *Wildlife Society Bulletin* 28:3 (Autumn, 2000), 623–629.
12. Ed Bangs 09/01/2006 (email excerpt from correspondence with other wolf managers in Mexican Wolf Program referring to call from reporter concerned about livestock carcasses attracting wolves).
13. http://aces.nmsu.edu/pubs/_ritf/RITF80.pdf
14. http://www.fws.gov/southwest/es/mexicanwolf/
15. http://www.fws.gov/endangered/esa-library/pdf/ESA_basics.pdf
16. https://www.researchgate.net/publication/228805074_Is_the_Fear_of_Wolves_Justified_A_Fennoscandian_Perspective

Photo credit: FloridaStock/Shutterstock.com

CHAPTER 11

. . .

Collateral Damage Identification: Mexican Wolves in Catron County, New Mexico

By Jess Carey

Introduction

The 558,065-acre Gila Wilderness in western New Mexico, which is part of the Gila National Forest, was the first designated wilderness area in the world in 1924. The Gila Wilderness is also located in Catron County, the largest county in New Mexico. There is little human activity and little livestock grazing in the Gila Wilderness compared to the human use, activity, and livestock grazing in the surrounding Gila National Forest, which is fragmented by homesteaded family ranches, subdivisions, and isolated homes.

Since 1998, the US Fish and Wildlife Service (USFWS) has released numerous Mexican wolf (*Canis lupus baileyi*) packs into the interior of the Gila Wilderness, as this is prime wolf habitat, rife with wildlife for food and timber for cover. Not one of the released wolf packs, however, has stayed there. The packs leave the Gila Wilderness within a short time and travel towards human activity—ranches, homes, and communities where they interact with people and their livelihoods in a number of mostly negative ways.

I was hired to investigate wolf depredation for Catron County, New Mexico, in April 2006, as there were major complaints from resource owners that the results of depredation investigations by federal agencies were lacking. Without such confirmation, no reimbursement fee for depredation losses can be paid. Additionally, no investigations of wolves were being conducted around children, yards, and homes, except by the wolf recovery team. They omitted reports and did no documentation. In the past it often took days before an agency representative came on site to assess the kill to see if it can be

192

unequivocally confirmed a wolf depredation. By this time, it is often difficult or impossible to tell for certain whether wolves or another predator is responsible for the cause of death due to lost evidence by scavenging canines, birds, insects, and advanced decomposition. This is why a study by the US Fish and Wildlife Service concludes that only one in seven wolf depredation livestock incidents by wolves can be proven.

Since I started as the county wolf investigator, the depredation confirmations of wolf-caused deaths of livestock and pets have doubled. This is in part because I could respond to the reports immediately and this caused other agencies to respond in like time.

To put it bluntly, despite what you might hear, Mexican wolves are destroying family ranchers' ability to survive, resulting in their selling off their ranches for a fraction of what they once were worth—if they can sell them at all.

The conflict between wolves and humans began during the colonial period when wolves killed settlers' livestock and dogs. Nothing has changed since that time and wolves will continue to kill livestock and pets no matter what non-lethal scheme is used by the USFWS wolf recovery team. This is a scientific fact.

When I looked for a title for this chapter, I had to look at the folks most impacted by Mexican Wolf Recovery: the many rural family ranchers who have lost their peace of mind, lost their dreams, lost their pursuit of happiness, lost their livestock, and lost their ranches. "Collateral Damage Identification" seemed appropriate. Damage in question is due to non-compensated wolf-caused livestock kills, attacks, and harassment, with little or no compensation.

Some Background

The Mexican wolf (*Canis lupus baileyi*), a subspecies of the gray wolf (the smallest, rarest, and most genetically distinct wolf subspecies), once ranged the Sonora and Chihuahua Deserts from central Mexico to western Texas, southern New Mexico, and central Arizona. They may have once ranged as far north as Colorado. Most wolves are bigger than a German shepherd and weigh seventy to ninety pounds, but some are smaller and weigh forty-five to fifty pounds. The head of the wolf is blockier than a coyote, they have a broader nose than a coyote, and the ears are more rounded. The front feet are larger than the rear feet. Color ranges from a grizzled gray, reddish-brown, whitish mixture to reddish-brown. Their behavior is highly unpredictable.

The Mexican wolf was driven to extinction in the wild by the 1950s as wild herds of deer and elk declined due to increased settling in their range, and so the wolves began preying upon livestock and pets. In 1976 the Mexican wolf was listed on the Endangered Species Act.

In March of 1998 the US Fish and Wildlife Service began releasing the Mexican wolf in the Blue Range area of Arizona. Eleven wolves were initially released. The restoration program was based on raising wolves in captivity in forty-nine different breeding facilities in the United States and Mexico, and then releasing them into the wild in hopes that they will establish wild breeding. There are currently about three hundred wolves in the forty-nine breeding facilities. Prior to releasing them, the wolves are sent to release sites such as Sevilleta National Wildlife Refuge in La Joya, New Mexico, where contact with humans is minimized to attempt to discourage habituation. The goal of the program was to establish a wild population of a hundred Mexican wolves in the Apache and Gila National Forests of Arizona and New Mexico by 2008. As of 2014, the "wild" population is one hundred nine animals, with six breeding pairs.[1] Since the outset this program has been fraught with problems as the wolves arrive habituated to human feeding when in captivity. When they are released into the wild, agencies continue to feed them with horse meat and roadkill drops beside roads—"supplemental feeding"—because these wolves do not have the hunting skills of wild wolves. The result has been wolves standing beside roads, expecting handouts as bears once did in Yellowstone, and wolves preying on livestock, farm animals, and people. Some habituated wolves will stand and look at you even after you fire a firearm into the air.

In addition, Mexican wolves have taken to breeding with house dogs and stray dogs, resulting in three confirmed litters of wolf-dog hybrids that are even more prone to prey on livestock and approach people. Wolf-dog hybrids are not protected by the Endangered Species Act, however telling the difference between a wolf and wolf-dog on sight is hard, if not impossible depending on the breed of dog. Wolves can breed with any other canis species, and wolf-coyote hybrids are also hard to distinguish. Only DNA analysis can give conclusive proof.

The Endangered Species Act only protects pure-bred species. If tracks are all you have to go on, it is near impossible to tell if wolves, coy-wolves, wolf-dog hybrids, or even some animals with wolf, coyote, and dog DNA are responsible for a livestock depredation or a dead pet dog.

One case of a suspected wolf-dog hybrid occurred north of Luna, New Mexico, on December 27, 2006. I was notified that feral dogs had just killed an elk calf. The elk calf had been killed by bites in the throat, just under the jaw. The calf had been caught against the barbwire fence and killed. A small hole was opened in the throat area where the dogs started feeding on the calf. On December 27, 2006, just after daylight, three dogs were feeding on the elk calf carcass. When the dogs detected the presence of humans they ran. Two of the dogs were killed and the third got away.

On December 29, 2006, J. Brad Miller of Wildlife Services came to my residence to inspect Canine #1. He stated that Canine #1 had unusual characteristics and should be DNA tested as he felt it could be a wolf-dog hybrid. I gave Miller DNA material from samples taken on December 27, 2006: Canine #1C, Canine #2C, and a CD with the photographs that I had taken at the scene. I understood that Miller would turn over these items to Dan Stark at the Alpine office, and Stark would send out these samples for DNA tests.

On December 31, 2006, Dan Stark came by my home and I gave him a blood sample of Canine #1 to send in for DNA analysis. Also Mr. Stark reaffirmed the results of the DNA report would be given to Catron County. I requested the DNA test results from USFWS Wolf Field Coordinator John Oakleaf several times in 2007, 2008, and 2009. Oakleaf stated the technician at the DNA lab could not find the results.

Finally, in September 2009, Oakleaf stated that Dan Stark had thrown the DNA samples away and he never had sent them in to the DNA lab. I told Oakleaf that this was not satisfactory behavior of a USFWS biologist, and the importance to know if Mexican wolves are breeding with domestic dogs producing hybrids is paramount.

As a result of experiences and conditions such as those I've just described, on June 20, 2011, the New Mexico Game and Fish Commission voted unanimously to discontinue collaborating with the Mexican wolf recovery program.[2] They also have approved trapping in the area in New Mexico where wolves now reside. So now it's solely up to the US Fish and Wildlife Service in New Mexico to manage wolves.

On January 13, 2012, during the Arizona Game and Fish Department's nongame activities briefing to the Arizona Game and Fish Commission, the Commission voted unanimously to amend its policy on the release of Mexican wolves in eastern Arizona. If the US Fish and Wildlife Service wants to release more wolves in Arizona, Game and Fish's director will now have the authority to approve a wolf release in cases where an animal is lost from the population due to an unlawful act. When a wolf is lost to any other cause, the commission must approve a release.[3]

Nonetheless, the wolves are present in and around the Gila. So, it is necessary to investigate all wolf-human interactions, especially suspected depredations, as there currently is no way to control Mexican wolves unless it can be proven they are killing livestock and/or menacing or attacking people.

Confirming Wolf Depredation

Livestock killed by predators usually can be distinguished from those dying from other causes by several factors: the presence of external hemorrhaging; subcutaneous hemorrhaging and tooth punctures; damage to the skin, other soft tissues, and skull; blood on the soil and vegetation; and carnivore tracks, scats, or territorial marks near dead animals.

Wolves primarily attack cattle on the hindquarters targeting tail, vulva, lower thighs, hocks, and hamstrings. Occasionally they will attack on the neck, face, and jaw, behind the front legs, in front of the rear legs, and on the belly. Wolves will chase cows, calves, and yearlings, stressing the animal until it can no longer stand; normally there will be capture bite and rake marks on the skin with corresponding hemorrhage.

Newborn livestock killed by predators and only partially consumed can be distinguished from stillborn livestock by characteristics not found in stillborn animals, which include: a blood clot present at the closed end of the navel, pink lungs that float in water, fat around the heart and kidneys, milk in the stomach and intestines, milk fat and lymph in the lymphatic vessels that drain the intestinal tract, a worn soft membrane on the bottom of the hooves, and possibly soil on the bottom of the hooves.

Normally, when wolves kill new calves there is little left of the carcass; possibly a few small bones or a piece of the skull but usually just a bloody place on the ground is all that remains because the calf is totally consumed. Even if wolves are confirmed in the immediate area, the cause of death of the livestock is unknown unless the kill can be positively documented based on the strict depredation conformation standards.

What makes confirmation of a wolf kill very difficult includes:

1. Missing livestock with no remains resulting from wolves eating the whole carcass of calves including skull, hooves, bones, and hair;
2. Coyotes or other scavengers consuming remainder of calf carcasses;
3. Calves/yearlings/cows not being found in rough remote terrain;
4. Advanced decomposition, rapid and severe in summer weather;
5. Insect infestation;
6. Scavenging birds;
7. Other scavenging carnivores;
8. Weather condition;
9. Rocky, hard ground conditions limit impressions; and/or
10. Untimely carcass detection.

If a larger calf is killed, there are remains left much of the time, but there may not be capture bite marks. The reason is the calf is bedded and the wolf pins the calf down and the feeding begins. The wolf does not have to bite the calf to capture it.

Wolves kill by consumption; they eat their victims alive. Those who perish, die from stress, and tissue and blood loss. In 266 wolf-depredation investigations, I have never documented a lethal bite site on cattle carcasses. Some cattle are stressed down and the wolves eat twenty pounds from the victim and the injured cow, calf, or yearling, and then leave. Most cattle die at the feeding site.

Some survive after the wolves have eaten their fill. Such animals are not dead and walk around with its rear end eaten out. Still, the victim with massive tissue loss has to be put down by the resource owner.

Research on Five Ranches

As part of my job, I conducted a study of the impact of wolves on five ranches located within the Blue Range Wolf Recovery area in Catron County. The study included data gathered during a six-year period from 2005 to 2010—both onsite investigations and research on records. These ranches all were identified as having wolves denning in or near calf-yearling cattle core areas. When I came on board, the relationship between high calf loss and proximity to denning Mexican wolves was not well understood.

On one ranch wolves denned in 2005, 2006, and 2007. On the other ranches denning occurred in 2008 and 2009. The study compared confirmed wolf-livestock depredations to actual losses, and identified other monetary losses to the resource owner. The study was finalized on March 1, 2011.

This study compares the following factors on the five ranches:

1. Historic pre-wolf normal calf losses;
2. Confirmed and probable wolf-livestock depredations;
3. Actual livestock losses;
4. Compensation paid by Defenders of Wildlife.

One of the first discoveries of this study showed that coyotes swarm to areas where Mexican wolves are continually killing livestock, contributing to the removal and destruction of evidence of the predations by wolves, as well as increasing depredation. This may not be consistent with wolf-coyote inter-actions in other parts of North America, but it is true for Mexican wolves. Perhaps this relatively peaceful co-existence is associated with wolf habituation and/or hybridization.

The Magic of a Clean Slate

Prior to this study, the focus of predation studies was on irrefutably proving that wolves had solely initiated and killed livestock because of the funds for reimbursing ranchers whose cattle are killed by wolves; if it can be proven that wolves are killing livestock, then wolves can be removed. In the past the "Three Strike Rule" governed removing depredating wolves. But, when the wolves started to kill large numbers of livestock the rules were changed to protect the wolves, and there is no "Three Strike Rule" any longer.

One year after a committed confirmed depredation, that depredation is removed from the offending wolf. Each year the depredating wolves start with a clean slate. The Middle Fork Pack has a history of sixteen confirmed livestock depredation and is still on the landscape. We now know that such proof is often difficult or impossible to determine, especially if the kill site is not immediately investigated as many scavengers quickly convene on wolf kills, and some wounded animals wander away and are never found. This is consistent with a 2003 USFWS study by John Oakleaf, field coordinator of the Mexican Wolf Recovery Program, which concluded that only one in seven wolf kills of livestock can be conclusively proven to be due to wolves.[4]

Pathological Fatigue

We now know that wolves kill cattle by stress. Wolves will run, attacking and biting grown cattle for a long period until they are exhausted, resulting in "pathological fatigue." Pathological fatigue interferes with the activity of every gland in the cow's system; its principle effect is to destroy the capacity of muscles and nerves to perform the work natural to them. A chemical change takes place in the muscles, releasing toxic substances including 1. lactic acid, 2. creatine, and 3. carbon dioxide. These toxic substances are acids and cause a state of fatigue in the cow's muscles and system. During rest following fatigue, these acids are neutralized by alkaline of the blood and internal secretions, which restores freshness, strength, and tone of the muscle. I feel that once a cow's system has been saturated to a certain point, "beyond recovery" from these toxic substances, there is no ability for the cow's system of neutralization (alkalinity) and the cow shuts down and dies. I have seen healthy cows in prime condition just seem to fall over dead. Lying on their sides, there is no indication of leg movement, no sign that the hooves disturb the ground, or ground liter at all. Also noted were no noxious plants in the area.

An Example of Wolf-Caused Pathological Fatigue: Case #AP-030

This cow was black and heavy, approximately twelve hundred pounds. I documented the torn-up ground around the dead cow. The ground looked like a small circle racetrack, and the cow's hooves gouged the ground, uprooting the vegetation. The cow was bitten at the root of the tail and on down the tail about twelve inches. Upon necropsy there were bite sites with corresponding hemorrhage; the canine spread was consistent with a wolf. The bitten tail did not kill the cow. There were no lethal bite sites. This was a stress death caused by continued harassment chasing and was confirmed. Further, the wolf was located up on a nearby mountain and was documented by aerial telemetry. If there were no bite sites, hemorrhage, or a canine spread, this cow's death would have been "unknown." I feel there have been numerous depredation cases that were stress death, but they were classified "unknown" based on the best available evidence at the scene. In twelve confirmed wolf-killed yearling calves on one ranch, five did not die at the attack and feeding site. They traveled for some distance after being fed upon by wolves. Four yearlings were found alive and walking around with massive tissue loss. One yearling was found dead and the scene lacked evidence of an attack and feeding site. Dried blood found on the (hind) legs indicated the yearling was bleeding while standing upright and walking.

It's not just that wolves are killing livestock. Wolf-caused chronic stress in cattle produces non-compensated economic losses. Wolf-caused chronic stress disrupts the cows' breeding cycle, causing them to abort calves, birth weak calves, undergo weight loss, and become more susceptible to disease; disfigured livestock also brings less money at the sale and providing vet care for wolf injured animals is costly. All of this is incurred at the expense of the resource owner without compensation. Catron County has documented on one ranch that 36 percent of the depredated yearlings that were confirmed as having been attacked and fed upon by the Middle Fork pack were still alive after the initial attack/feeding. All these animals have had to be put down.

Procedure: Investigating a Livestock Carcass

Once notification is given by the resource owner or others that may have found a suspected victim predator depredation in Catron County, USDA Wildlife Services and I respond to the scene to perform an investigation to determine the cause of death of the animal.

Arriving at the scene is similar to being at a crime scene. Canine tracks can be destroyed by people walking on the scene. Other livestock and scavenging birds can also destroy tracks. Humans too can destroy tracks if necessary precautions aren't taken. The best procedure when entering the attack scene is to protect the evidence such as canine tracks as you find them; cover these tracks to prevent other livestock, animals, and people from trampling them. Cover the carcass with a tarp rocked around the edges to prevent scavenging canines and birds from feeding on it. Cover blood trails or droplets of blood leading to the carcass if rain is eminent.

Timely carcass detection and notification is key to determining cause of death. Lost or destroyed evidence can result in a non-confirmation. Calf carcasses left uncovered in the field will disappear during the night. If you do not have a tarp, hang the calf high up in tree; if there is no tree, mark the area, bring the calf in, and store it so dogs cannot get to it.

All scene evidence is photographed. The carcass is photographed at head, back, rear, and belly. Further, all injuries are catalogued, including attack sites on the carcass, bite sites, feeding sites, and impact injuries. Scavenging canines and birds are also noted. Measurements are taken to document predator tracks and scats. A diagram is drawn to reflect attack and feeding site, drag marks, carcass site, blood trails, predator/victim track location, and direction of travel.

Barbwire fences are also checked, especially bottom and second for hair caught in the barbs when predators pass under or through them. A predator's identification can be made with this transfer evidence (hair).

Dirt roads are checked for predator tracks, scats, and any sign of predators as you near the area of the carcass. If tracks are located on the roadway they are marked and protected so no one drives over them.

Other cattle nearby are observed for unusual behavior such as calling and alert, defensive, and frightened behavior, injury bite sites, and impact wounds like running into barriers or barbwire fences.

The area is also checked for a wolf collar signals using a ground telemetry receiver. If a signal or signals are picked up the corresponding wolf number is noted.

The scene around the carcass is searched to identify the attack site, feeding site, drag marks, tracks, scats, blood trails, trampled/uprooted vegetation, torn up ground, broken fences. The scene could be less than fifty yards to several hundred yards in size.

Once everything is documented, the investigation focuses on the carcass and a necropsy is performed. The percentage of carcass remains is noted, as well as disarticulation of limbs and bones. Some carcass remains are just dried skin and bones; these have to be soaked in water three to five days to soften the skin, yet compression bite sites on the skin still remain. A compression bite site can only be confirmed if the victim was bitten while alive.

First the hair is clipped from the skin of the carcass to detect bite sites and rake marks. Without clipping the hair you cannot see the bite and rake marks. Photographed measurements of all canine spreads are documented. The skin is removed to document a bite site's corresponding hemorrhage, and deep hemorrhage in the muscle tissue and injuries. Much of the time, there are no internal organs left inside the carcass for assessment. The skin is held up to the sun and photographed to document bites sites and rake marks with hemorrhage in the skin.

The Five Ranches

Ranch A
Ranch A is a cow/calf operation. Records of average annual pre-wolf introduction losses were 16 percent. The herd consisted of 300 head. Herd makeup: 20 bulls, 25 replacement heifers (not expected to calve), 0 steers, and 255 production cows.

255 production cow numbers X 16 percent average pre-wolf annual calf losses = a 41 head loss. 255–41 = 214 fall calf crop number, representing an 84 percent calf crop.

Losses pre-wolf were attributed to calving, open-ranging cows, coyote predation, and winter weather. In 2008, the San Mateo pack denned in calf core areas on Ranch A.

- Pre-wolf annual losses = $24,600
- 2008 losses = $71,400
- 2009 losses = $60,000
- Defenders of Wildlife compensation paid = $600
- Rancher sold off remaining herd and went out of business

Ranch B

Ranch B is a cow/calf operation that adjoins Ranch A. Records of average annual pre-wolf introduction calf losses were 2.5 percent for three years running with an average annual loss of 4 to 6 head of calves per annum. The herd consisted of 256 head. Herd makeup: 18 bulls, 30 replacement heifers (not expected to calve), 5 steers, and 203 production cows. Average calf crop = 97.5 percent. Losses pre-wolf were attributed to calving, open range cows, coyote predation, and winter weather. In 2008, the San Mateo Pack denned near calf core areas on Ranch B.

- Pre-wolf losses—5 head/year = $3,300
- 2008 losses—27 head = $16,200
- 2009 losses—58 head = $34,800
- Total losses 2008 and 2009 = $51,000
- Defenders of Wildlife compensation paid = $1500

Ranch C

Ranch C is located approximately 35 miles in a southerly direction from Ranches A and B. Records show that Ranch C had a 3 percent average annual pre-wolf introduction loss. Total herd is 330 head. Herd makeup: 18 bulls, 0 steers, 30 replacement heifers (not expected to calve), and 282 production cattle. Average annual pre-wolf losses of 9 head per annum were noted. Losses were attributed to birthing, coyote depredations, open range cows, and winter weather. In 2005, the Luna Pack denned in calf core areas on Ranch C.

- Pre-wolf losses—9 head = $5,076
- 2005 losses—42 head = $25,200
- 2006 losses—82 head = $49,200
- 2007 losses—69 head = $41,400
- Defenders of Wildlife compensation paid = $0
- Ranch went out of business in 2007

Ranch D

Ranch D is located to the west of Ranch C. When the livestock were removed from Ranch C the wolves immediately left the vicinity of Ranch C and dispersed to Ranch D where there were livestock. Records show Ranch D had an 11 percent annual pre-wolf introduction loss. Total herd is 205 head. Herd makeup: 15 bulls, 0 steers, 10 replacement heifers (not expected to calve), and 180 production cattle. Average annual pre-wolf losses of 20 head per annum were noted. Losses were attributed to birthing, coyote, bear depredations, open range cows, and winter weather. In 2008 the Luna Pack denned in calf core areas on Ranch D.

- Pre-wolf losses—20 = $8,100
- 2008 losses—35 = $21,000
- 2009 losses—23 = $13,800
- Defenders of Wildlife compensation paid = $0

Ranch E

Ranch E is located north east of Ranch C and runs yearlings. The Allotment consisted of three pastures. There were 300 yearlings in excellent condition in Pasture A and B, and 287 yearlings in pasture C. In 2009, the Middle Fork pack denned in yearling core areas on Ranch E.

- Pre-wolf annual losses—5 = $2,827.50
- 2009 confirmed kills—11 = $6,307
- 2009 carcasses—14 = $7,917.00
- 2009 missing animals—73 = $41,281.00
- Losses confirmed wolf kills carcasses missing
- Defenders of Wildlife compensation paid = $6,307

The final analyses indicate that annual post-wolf introduction livestock losses are higher than the average annual pre-wolf losses for the five study ranches:

- Total combined livestock losses = 651 head
- Total combined dollar value losses = $ 82,198.50

Two of the five ranches in this study went out of business, one selling the ranch and the second was on the market as of this writing. A third ranch sold off their livestock in the fall of 2009 and did not re-stock cattle in 2010.

In addition, confirmed and probable findings do not reflect the true number of livestock losses. Wolf-caused stress disrupts a cow's breeding cycle, the resulting calf loss must be measured in monetary value as if the wolf depredated a calf.

The findings for these five ranches of confirmed and actual losses and overall damages are consistent with other ranches across Catron County where wolves den in calf- and yearling-core areas.

Many ranchers have cooperated with wolf recovery agencies utilizing recommended non-lethal schemes to prevent wolf-livestock interactions that result in livestock depredation. These ranches have added additional range riders, moved livestock to other pastures, penned livestock, fed hay, and worked multiple additional hours to prevent wolves from killing their livestock. All these measures add costs to operation. Still the wolves depredate their livestock. The ongoing added effort, plus stress and expense, is a high loss cost factor far beyond pre-wolf introduction.

Wolf-Human Interactions

The USFWS has changed their terminology from "supplemental feeding" to "diversionary feeding" of wolves. This gives the public the false sense that the wolves are able to hunt and make it on their own. This diversionary feeding contributes to food conditioning, resulting in habituation. Habituated wolves are bold, fearless, and are a threat to our children. Wolves come to homes and yards where children play and I am often called in to investigate. Since the release of Mexican wolves in 1998, unusual wolf behavior has been documented by Catron residents numerous times including:

- Territorial scrapes at one residence where wolves were documented at the home twenty-three times.
- Mexican wolves urinating on vehicle tires and on an ice chest located outside an occupied camp trailer.
- A wolf defecated on the front of an ATV vehicle located in a front yard.
- Wolves defecated on porches and yards at door entrances of occupied homes.
- Territorial scrapes at occupied residences where the wolves were claiming the residence as part of their territory.
- Wolves playing and/or mating with domestic dogs.
- Wolves following young children as they walked to and from the school bus stop.

Just like cattle who are harassed by wolves, causing stress, we have documented cases of psychological trauma (post-traumatic stress disorder) in children and families of Catron who were subject to fearless wolves approaching them. One case really sticks out. Wolves were regularly coming to the home of a family with a fourteen-year-old daughter. Wolves were documented at this home a total of twenty-three times. The county was in the process of trapping the wolves. Sitting in the living room with the family, the father had a ground telemetry unit on. Little was said as the signal began to beep, first faintly, then a little louder and louder and louder as a wolf made it way to the residence from the nearby hills. No one said a word. Everyone's breathing was shallow for some time. I remember the look on their faces—it was one of sheer terror.

Conclusions

From April 2006 until February 2012, there were 420 reported wolf-animal, wolf-human interactions and forty-six information reports. Two hundred fourteen were on private property and 206 were on non-private property. The fact that approximately 50 percent of all wolf interactions occurred on private property clearly indicates the severe habituation of wolves towards humans and human-used areas. Mexican wolves seek out humans and human-used areas due to habituation, and their lack of an avoidance response towards humans. Therefore, designation of critical wolf habitat is useless when the wolf's definition of that habitat includes homes, communities, and people.

To alleviate the taking of private property (livestock) without compensation by the Federal Government, confirmation standards and the compensation scheme as a whole must be re-evaluated. In-depth studies must be conducted to evaluate the negative impacts of wolves' denning in calf/yearling core areas and the effects of wolf-related stress on livestock. Evaluation of data must include the wide spectrum of negative impacts to livestock and livestock producers, rather than the current focus solely on benefits to wolves. Recommended areas of study include:

1. Pre-wolf introduction historic annual losses;
2. Post-wolf introduction annual livestock losses;
3. Wolves denning in calf/yearling core areas;
4. Wolves denning near calf/yearling core areas;
5. Wolf rendezvous sites located in calf/yearling core areas;
6. Wolf-claimed territory overlapping livestock core areas; and
7. Wolf-caused chronic stress and effects on livestock.

Defenders of Wildlife, a pro-wolf organization, has had a Compensation Fund to reimburse ranchers for livestock lost to wolf confirmed kills. Several ranches received no compensation on livestock depredation investigations conducted by Wildlife Services for documented for confirmed or probable losses. Defenders of Wildlife no longer pays compensation.

Wolf recovery causes negative effects that can only be described as collateral damage. The mentality of the USFWS wolf recovery team to date is: so what if our wolves are habituated and come to homes, communities, confront children, and kill family pets, farm animals, and livestock? You need to accept the Mexican wolf for the way he is and coexist by changing your daily lives to accommodate the true "New Age Mexican Wolf"—lacking wild wolf characteristics and flawed to the extreme by habituation.

Endnotes:

1. http://www.fws.gov/southwest/es/mexicanwolf/cap_manage.shtml
2. http://www.krqe.com/dpp/news/politics/new-mexico-abandoning-wolf-program
3. http://www.azgfd.gov/w_c/es/mexican_wolf.shtml
4. http://www.fws.gov/news/NewsReleases/showNews.cfm?newsId=3878CF69 -1F55-4B23-AF326A3A9BA7F930

Photo credit: LittleMiss/Shutterstock.com

CHAPTER 12

• • •

The Wolf in the Great Lakes Region

By Ted B. Lyon

Protecting wolves in the Midwest will cease to be practical or even possible according to the Endangered Species Act, as species integrity is disappearing.

Before 1974

The nationwide campaign to eradicate wolves did just that by the 1960s, except in the upper Midwest where a small group of wolves (three hundred or less) retreated into wilderness. Wolves were declared close to extinct and given the equivalent of endangered species protection in Wisconsin in 1957 and Michigan in 1965. Until the mid-1970s, gray wolves in the Great Lakes region were almost entirely found in northern Minnesota in the Superior National Forest. This was the largest population in the United States, numbering about one thousand. It was kept in check by hunting and trapping.

In Minnesota, there was a bounty on all predators, including wolves, until 1965. Between 1965 and 1974, Minnesota had an open season on wolves and a Directed Predator Control Program that removed about 250 wolves per year. Anti-wolf sentiment in that region ran fairly high.[1]

After 1974

Following the passage of the Endangered Species Act in 1973, wolves all across the United States were listed as endangered in 1974. Almost immediately, the wolf population in the Midwest began to increase and expand, aided by a succession of mild winters and a large whitetail deer population. According to the Michigan Department of Natural Resources, wolves were breeding in Wisconsin by 1975 and in Michigan's Upper Peninsula by 1989.[2] There was also a small population of wolves on Isle Royale that got to the island by crossing the ice in winter; incidentally, these wolves, which fed on moose, were perhaps the most studied wolves in the world.

Due to their population increase, wolves in Minnesota were reclassified to "threatened" in 1978. As the numbers of wolves in the Great Lakes area began to rise, people became more aware of them, and pro-wolf campaigns were launched by environmental and animal rights groups. Early public opinion surveys, after listing wolves under the Endangered Species Act, were largely positive. Christening Minnesota's NBA team the "Timberwolves" no doubt aided early positive public opinion of wolves.

Rising Population

Protection aided wolf population growth. According to former Michigan Department of Natural Resources wildlife biologist James Hammill, "Michigan and Wisconsin have typically had a 15 percent annual rate of increase in the number of wolves since 1977. The wolves in Minnesota have also continued to increase but at a slower rate of roughly 4 percent annually."[3]

Just how many wolves there are in this region is controversial. The US Fish and Wildlife Service estimates that the current Midwest wolf population is about 4,400.[4] This is a conservative number. The Michigan Department of Natural Resources (DNR) says there are nearly seven hundred wolves currently in Michigan's Upper Peninsula.[5] Wisconsin DNR says it has at least as many.[6] And Minnesota DNR says that it has over three thousand wolves.[7]

Hammill, who lives in Michigan's Upper Peninsula and is active in studying wolf behavior, believes that the Midwest wolf population of 2012 will actually approach six thousand animals, about four thousand of which are in Minnesota.[8] Wolves also have been reported in North and South Dakota, Iowa, and northern Illinois. And, of course, wolves are found in neighboring Canadian provinces.

There have been multiple reports of wolves in Michigan's Lower Peninsula.[9] Presumably, these wolves reached Michigan's Lower Peninsula by crossing Lake Michigan or Lake Huron in the winter, but as in other states, some people raise wolves and wolf-dogs in captivity, and some escape or people purposefully release them into the wild. In 2000, Michigan took steps to reduce this problem by passing the Wolf-Dog Cross Act, which requires wolf-dog owners to prove the hybrid has been sterilized, to keep the animal in a fenced area, and to display a sign reading: "A potentially dangerous wolf-dog cross is kept on this property."[10]

As Dr. Matthew Cronin states in his chapter on wolf genetics, the identification of specific species and/or subspecies of wolf is not uniformly agreed upon among scientists. Basically, there are two wolf species: the gray wolf (*Canis lupus*) and the red wolf (*Canis rufus*). At one time, there were as many as twenty-four subspecies of wolf recognized in the United States. Today, most scientists agree that there are five subspecies of gray wolf in North America: the Arctic wolf *Canis lupus arctos*, the eastern timber wolf *Canis lupus lycaon*, the Great Plains wolf *Canis lupus nubilus*, the Rocky Mountain wolf *Canis lupus occidentalis*, and the Mexican wolf *Canis lupus baileyi*. Of these five, at least two subspecies—the Great Plains wolf and the eastern timber wolf—interbreed in the Great Lakes region, and Ontario also considers the Arctic wolf territory along Hudson Bay to overlap with eastern timber wolf territory.

Today there are only forty-five to sixty wild red wolves, a population which was bred in captivity and released into the wild. They all live in the mountains of North Carolina. As this chapter is being written, the US Fish and Wildlife Service is proposing that starting in late 2017 they plan to limit wild red wolves to a federal wildlife refuge and some adjacent land in eastern North Carolina's Dare County, rather than the five-county area where they currently roam. For this discussion of wolves in the Midwest, the red wolf needs to be noted as, at one time, it may have been found as far west as Texas and Missouri and north-ward to southern Ontario. Therefore, some Midwestern wolves may well also contain DNA of red wolves.[11]

The point simply is that, sooner or later, if it is not already the case, all wolves in the Midwest will be hybrids—either mixtures of wolf sub-species or wolf-coyote-dog hybrids.

"I can tell you this: The adage that a wolf only kills the old animal or the sick or wounded is total bull. I had them kill twelve sheep in one night."
—Tony Cornish, retired MN game warden[12]

Wolves and People in the Great Lakes States

Gray wolves in Minnesota have normal ranges of twenty-five to 150 square miles, which is much smaller than wolves in the Northern Rockies. As food availability influences the territorial behavior of wolves, presumably the smaller range area in Minnesota is due to the abundance of whitetail deer. These wolves are also smaller than the Canadian wolves that were transplanted into the Northern Rockies. Minnesota gray wolf females run fifty to eighty-five pounds, and males, seventy to 110 pounds.

Reviewing each state's website about wolf management, one finds fasci-nating differences of opinion about the number of deer per year that an adult wolf consumes. The Michigan Department of Natural Resources estimates that an adult wolf kills thirty to fifty deer per year.[13] The Minnesota Department of Natural Resources estimates only about twenty deer per year are taken by an adult wolf.[14] The two states share a border and are so similar in habitat that it's hard to imagine that the Minnesota estimation is lower because of some dramatic difference in habitat.

210 •• THE REAL WOLF

As the wolf population in Minnesota, Wisconsin, and Michigan has grown, wolf-hunting territory expanded, both to find new land not already occupied by a wolf pack and to find food. The Midwest has National Forests, but extensive wilderness habitat like in the Northern Rockies is not found in the Great Lakes region, except in the far north. So Midwest wolves increasingly approach farms and urban areas where hunting has not been permitted. As wolf populations increase, they move closer to population centers including Minneapolis-St. Paul and Duluth, Minnesota; Green Bay, Wisconsin; and even Chicago, Illinois—and wolf-human contact increases. According to James Hammill, wolves have attacked and killed pets in the immediate vicinity of homes and within city limits in all three states.[15]

As predation on pets and livestock has increased, according to the USDA Wildlife Services, over three thousand wolves in the Great Lakes area have been killed due to conflicts with human settlements. Wolf attacks on pets and livestock hit record levels in 2010. In Minnesota, fifteen dogs were killed by wolves, up from an average of just two dogs per year from 2006 to 2008, according to the federal agencies. Minnesota officials verified 130 of 272 complaints—both records—involving 139 livestock and poultry and twenty-three dogs. The verified complaints were 31 percent above the five-year average. One person's safety was threatened by a wolf.[16]

In Wisconsin in 2010, wolves attacked livestock on forty-seven farms, fifteen more than the previous high, killing sixty-nine animals.[17] Also, wolves killed twenty-four dogs and injured fourteen more, the most ever.[18]

In May 2012, the Michigan DNR killed eight wolves that had moved into the Upper Peninsula town of Ironwood, where they hunted pets and consumed garbage.[19]

Minnesota, Wisconsin, and Michigan have a livestock compensation program for animals killed or injured by wolf predation, and Wisconsin has a pet owners' compensation program. In 2011, Minnesota paid out $154,136 compensation to farmers for livestock—mostly for cattle but also for sheep, turkeys, pet dogs, a horse, and a llama.[20] The state paid 128 claims totaling $102,230 in fiscal 2011 and 104 claims for $106,615 in fiscal 2010. Contrast this with $72,895 paid for 71 claims in 2006.

In Minnesota, trappers with the USDA's Wildlife Services killed 192 problem wolves in 2011, down slightly from the 196 killed in 2010. The state Department of Agriculture paid about $96,000 in 2010 to people who lost livestock to wolves.[21]

Wolf Politics
Like the Northern Rockies and the Southwest wolf population centers, attempts to manage wolf numbers by hunting and trapping in the Midwest have been

met by legal action from pro-wolf organizations. In 2003, the US Fish and Wildlife Service (USFWS) reclassified Midwestern wolves from endangered to threatened, but this was judicially overturned. In April 2005, the USFWS issued sub-permits to allow the Michigan and Wisconsin Departments of Natural Resources (DNR) to kill depredating wolves. But in September of the same year, the sub-permits were stopped by Federal District Court in Washington, DC, due to inadequate public notice of the states' applications.

In May 2006, a permit to kill depredating wolves was issued to the Wisconsin DNR and the Michigan DNR by the USFWS. The Humane Society of the United States (HSUS) and others filed suit against the Department of the Interior (DOI) and USFWS for issuance of those permits. On August 9, 2006, a US District Court judge ruled against DOI, and the permits were no longer in effect.

In 2007, the USFWS delisted the wolf in the Midwest, but again HSUS sued to block it. Finally, Endangered Species Act protection for gray wolves in the Western Great Lakes Distinct Population Segment (DPS) was published in the *Federal Register* on December 28, 2011.[22] However, in 2012, suits were filed to prevent hunting wolves with dogs in Wisconsin and to block the Minnesota management wolf hunt. These lawsuits have cost the federal government considerable money to defend.

Ontario: A Model for Wolf Management?

On the eastern side of the Great Lakes, the US gray wolf population comes into contact with gray wolves from Ontario. Ontario presently has eight to ten thousand wolves, about the same as Alaska. Provincial parks, such as Algonquin Provincial Park, are known to be wolf population centers. As the landscape there is heavily forested and viewing wolves is less easy than in the more open western United States, "howl-ins" are a popular tourist activity in some Ontario parks.[23]

The Ontario Ministry of Natural Resources (MNR) considers there to be two species of wolf in Ontario: the gray wolf in northern boreal and tundra areas and the Eastern wolf in the central and northern hardwoods and conifers. MNR also states:

> Coyotes, a close cousin to the wolf, co-exist well with humans and are common in the developed and agricultural areas of southern and northern Ontario. Where the ranges of Eastern wolves and coyotes overlap, interbreeding makes it difficult to distinguish between Eastern wolves and coyotes, and they can easily be confused.

Other researchers say there are four subspecies of wolf in Ontario: 1. *Canis lupus hudsonicus* that inhabits the subarctic tundra; 2. a race (Ontario type) of the eastern timber wolf (*Canis lupus lycaon*) that inhabits the boreal forests; 3. a second race (Algonquin type) of *Canis lupus lycaon* that inhabits the deciduous forests of the upper Great Lakes; and 4. a small wolf (Tweed type) in central Ontario that may be a hybrid between the Algonquin-type wolf and expand-ing coyotes, *Canis latrans*.[24] The hybridization between wolves and coyotes is a major reason why some scientists are opposed to establishing a wolf population in the Adirondack Mountains, as it would not be able to be protected under the Endangered Species Act.[25]

Regardless of the species or subspecies of wolf, Ontario has some impor-tant lessons for US wolf management. For example, Ontario has long had wolf hunting and trapping, and their wolf population is robust—more wolves than in the entire lower forty-eight, and the population has been stable for about thirty years.[26] The size of the Ontario wolf population in itself should answer concerns about hunting and trapping alone driving wolves to extinction. Science-based management is the key.

The Ontario Ministry of Natural Resources has an Enhanced Wolf Management Plan (EWMP) that seeks to ensure that wolves occupy 85 per-cent of their historic range in the province. It states that hunting and trap-ping should play an integral role in controlling wolf populations within the province.[27]

What we can also learn from Ontario's experience is that, as wolf popula-tions grow and expand, more and more conflicts between wolves and people will occur. This has translated into several recent attacks on people:

- In Algonquin Provincial Park, there have been five attacks on humans, mostly on children, between 1987 and 1998 by four different wolves.[31]
- September 2006. A lone black wolf attacked and seriously injured six people, including three children, at the popular Katherine's Cove Beach in Lake Superior Provincial Park. The wolf was killed by park staff and tested negative for rabies.[28]
- December 31, 2011. A lone wolf attacked two ice fishermen and their Border Collie on a lake adjacent to Algonquin National Park, where hunting and trapping wolves had been curtailed.[29]
- Summer 2012. There were two wolf attacks on people with dogs in the Thunder Bay, Ontario area.[30]

Conclusion

Before wolves were placed on the Endangered Species Act protected list, public opinion was not too favorable towards wolves. As soon as they were listed, envi-ronmental groups sought to exploit wolves for fund-raising, and people who did not have daily contact with wolves became more positive about them. A 1990 study by Steven Kellert found that 80 percent of Michigan's deer hunters favored Michigan's wolf population being replenished.[32] At that time, there were a handful of wolves in the Upper Peninsula. Fifteen years later, when the wolf population had reached about four hundred, 812 people attended a series of meetings about wolves in the Upper Peninsula—22 percent said they wanted no wolves in the Upper Peninsula, and 36 percent said they wanted fewer than existed at that time.[33] Another study conducted in 2002 found that residents of the Lower Peninsula (where there were no wolves) were much more supportive of having wolves than were residents of the Upper Peninsula.[34] Elsewhere, as Midwestern wolf populations have increased, conflicts with farmers and dog owners have risen. As conflicts have increased, public acceptance of wolves has declined.[35]

Reviewing these studies and several others, wildlife biologist James Hammill concluded: "The current trajectory of public attitudes, especially in Michigan and Wisconsin, is not favorable to sustaining wolves in those states."[36] This conclusion leads Hammill to suggest that—in addition to the biological carrying capacity of an area determining the desired population for a species, especially species like wolves where contact with humans may lead to conflicts—a "social carrying capacity" should be developed for a realistic management plan.

As Midwestern states grapple with wolf management, much could be learned from Ontario about the practical sensibility of managing wolves in modern times.

For some dramatic perspective on what it's like to find yourself living with habituated wolves, Bruce Mahler, police chief of Marenisco Township in Michigan's Upper Peninsula, has provided Ted Lyon with details on the situation in that area for the last two years.

In Michigan's Upper Peninsula, Wolves Come Into Town

Marenisco Township is a scenic outdoor recreation area in Gogebic County in the copper country of the far western Upper Peninsula of Michigan. With approximately 330 square miles of numerous lakes, streams, and densely wooded areas, it's the second largest township in the state. The area is a mecca for fish and game, as much of the township is located in the Ottawa National Forest and Commercial Forest Reserve. Approximately 1,704 people live in the township, many residents residing on the shores of Lake Gogebic, the largest inland lake in the Upper Peninsula.

Wolves were originally native to Michigan, but vanished from the Upper Peninsula by the early 1960s due to hunter bounties and declining local white-tailed deer populations. Ironically, the Marenisco police department patch for their officers' uniforms features a wolf's head—the design coming the drawing of a high school student who won a contest for a new patch design in 2006. Bruce Mahler, chief of the township police, says that inspiration probably came from the appearance of wolves near the town of Marenisco, population three hundred, at that time.

Since the winter of 1993–94, combined wolf numbers in Michigan and Wisconsin have risen to and surpassed a hundred, meeting federally established goals for population recovery. The Michigan goal of a minimum sustainable population of two hundred wolves for five consecutive years also was achieved. From 1994 to 2003, the Upper Peninsula wolf population saw an average annual growth rate of 19 percent. Growth then shrank to 12 percent as the wolf population neared the maximum biological carrying capacity of the area.

In January 2012, citing wolf recovery in the region, the US Fish and Wildlife Service took gray wolves off the federal endangered species list in Michigan and Wisconsin and the threatened species list in Minnesota, as the USFWS said wolves were thriving in that area.

In 2013, a total quota of forty-three wolves was set for the Upper Peninsula, with all twelve hundred available licenses sold. Trapping was not allowed. Hunters killed a total of twenty-two wolves during the hunt, which lasted from November 15 to December 31, 2013.

Nonetheless, the Humane Society of the United States filed a formal complaint, and in December 2014, a federal judge ordered the wolves relisted in those three states under the Endangered Species Act. Michigan's laws on wolf depredation and the ability of wildlife managers to use lethal means, including hunting, to control wolves was suspended in December 2014, after a ruling from the US District Court in Washington, DC.

The Michigan DNR's most recent minimum estimate of the Upper Peninsula wolf population in the spring of 2014 was 636, and current population estimates are that there are at least that many wolves roaming the UP.

There were no wolves in Marenisco Township for several decades, but recently they've been returning. There are now three to four packs of wolves in the township, including several habituated wolves that have decided to make the town their home in winter months.

Because wolves have been relisted in Michigan, the wolves of Marenisco cannot be harmed, unless they are actually attacking a person. This is causing a serious problem for a Chief Mahler, who wears a wolf patch on his shirt every day when he goes to work.

What's been happening in Marenisco Township clearly follows Valerius Geist's habituation model. The Michigan Department of Natural Resources reports that at least twenty-six cattle and seventeen hunting dogs were killed by wolves in the Upper Peninsula in 2014. Only twenty animals were attacked in 2013.[37]

The attacks on dogs continued in 2016, including dogs that were not in the woods with hunters.[38]

According to Chief Mahler, since the fall of 2015, several wolves have moved into town, following a herd of fifty to sixty deer. The deer were originally encouraged to move into town because people were feeding them, but it soon became clear that the deer had also moved into town to avoid the wolves, which prey heavily on deer in winter; so the wolves fearlessly followed the deer.

On February 25, 2016, Chief Mahler sent a letter to the Michigan DNR documenting forty-one reports of wolf interactions with humans close to their homes in 2015 and 2016. The report suggested between two to four wolves have been involved. This is only a sample of what's been happening.

The sightings started on January 9, 2015, when two wolves, including a large black male, were seen in the early morning hours on Presque Isle Street between and within twenty feet of two homes that were built right up to the street.

The next day, two wolves, including a large black male, were seen in the early morning hours in the yard of a home on another street. Nearly a week later, two wolves—again, including a large black male—were sighted at the end of a Main Street driveway in the early morning hours by a woman letting her dog out. She immediately brought the dog inside.

Almost a month passed before, on February 14, two wolves, again including a large black male, were seen in the same Main Street driveway. That person saw the wolves in the early evening and called the homeowner to urge her not to let her children outside.

Six days later, things became a lot more tangible when a corrections officer on her way to work at the Ojibway Correctional Facility hit a wolf in town with her car. According to the officer, the wolf appeared to be chasing a deer.

"There was substantial damage to her car but the wolf ran off into the woods," Mahler said. "I was unable to locate that wolf. This incident occurred early in the morning."

As the weather warmed, the deer moved out of town, and so did the wolves. When winter came in 2015, the wolves and deer returned to Marenisco.

- March 4, 2015, a deer was killed by wolves in town in a yard between two houses.
- March 18, 2015, a resident reported a wolf walking down the main street toward Dutch's Old Bar. Shortly thereafter, a resident saw a wolf staking his dog who was inside a fence. The wolf retreated when the resident opened his door and yelled at the wolf.

- March 19, 2015, a resident living on the shore of Lake Gogebic reported that a wolf had killed a deer in her yard and was eating the deer.
- April 2, 2015, a resident reported a black wolf was in her yard, about twenty feet from the house. She was concerned about her five-year-old granddaughter who lives with her.
- The black wolf was reported in town again on April 2, 3, and 8, and on April 14 it killed a domestic rabbit in the yard of a woman who lived in town.
- April 15, 2015, a resident reported that a pack of four wolves had been seen in the state park on Lake Gogebic. Sightings continued in April and May, and then the wolves disappeared until October, when they returned to town. On November 25, one of the wolves appeared on the back porch of a resident.
- In February 2016, two ice fishermen on Lake Gogebic reported that wolf was circling their fishing shanty. A week later wolves were seen near the school in the afternoon as children were heading for the school buses. There were several more reports of wolves that month including wolves walking down streets, in people's yards, and appearing near children waiting for school buses or returning home from school.

Since Chief Mahler's report to the DNR in February of 2016, wolf encounters in town have continued. For example, on March 11, 2016, a homeowner reported a wolf killing a deer by near his home. Mahler said, "He stated he was awakened to the cries of the deer, which seemed to be right outside his window. In the morning, the homeowner found the kill and reported it."

Five days later, on March 16, a resident walking his dog during daylight hours encountered a wolf that came out of the wood-line, stopped about thirty feet away, looked at the man and his dog for several seconds, then turned and walked back into the woods.

Another sighting occurred on March 18. Mahler says, "At approximately 2000 hours I was called at home by a resident who saw a large black wolf walking down the middle of the road in front of her house . . . she said the wolf was walking toward Dutch's Bar."

Finally, on March 18, a resident reported that his dog, chained outside, was barking wildly. The man said he looked out—the yard has a backyard flood light on when the dog is outside—to see a wolf "stalking" the dog. He yelled, and brought his dog in the house.

"The above instances are all within the town limits of Marenisco," Mahler says. "In addition, I have reports of wolves near homes in the Lake Gogebic area and this doesn't include the usual sightings of wolves in the outlying areas of the township."

The incidents that Mahler reports, forty-two total and thirty nine in town, took place during winter months, when the deer herd had moved into town. "My concern is not environmental, it is simply public safety. People are afraid to walk their dogs, let their children and grandchildren outdoors to play, and some are even afraid to go outside and shovel snow."

Chief Mahler said that he has three major concerns about the wolves. The first is that they seem relatively fearless. Wolves kill deer in people's front yards, watch kids get off school buses, attack dogs, and people now are carrying firearms when they walk their dogs. Mahler says that since legally he has to wait until someone is physically attacked before shooting the wolf, he asks, "Just when and where does a case of 'imminent threat' exist?"

"It's not good enough anymore to say the wolves are not a threat or that there is nothing we can do because they're an endangered species," Mahler says. "I now have people arming themselves and I foresee it won't be long before they take matters into their own hands. If I get a call and there's a wolf and it's a threat to a child or someone in my community I will shoot that wolf. Plain and simple."

The second concern is that one of the major winter recreational activities in the area is snowmobiling. Abandoned railway lines now form a network of ORV and snowmobile trails, which provide many residents and visitors access to miles of scenic areas of the township. The hundreds of inches of snow that fall on the area every winter make the area a preferred destination for numerous snowmobilers. There are several instances of wolves attacking snowmobilers in Alaska. Mahler says that wolves quickly seem to learn that loud noises and other repellant methods mean no harm, and so they begin approaching people.

"I keep thinking about what will happen if someone is out there alone in the evening on their snowmobile and it breaks down or runs out of gas. We find carcasses of deer that are killed by wolves on the snowmobile trails. There is no way that we can be sure that a person who has to walk back into town from one of trails won't be a target for wolves."

A third concern is that the deer herd in that area is virtually gone—except for the deer that are trying to hide in town. If the wolves remain in that area, the town deer herd will decline and the wolves will inevitably be driven to prey on livestock and pets, and feed on garbage and roadkill.

Mahler says that even if the wolves haven't attacked a person, the presence of a pack of habituated wolves in the area has serious economic implications for tourism, livestock, and hunting in Marenisco Township.

There may not have been any attacks on humans in Marenisco Township yet, but Mahler called attention to an incident on January 2, 2015, in Presque Isle, Wisconsin, which is located about eighteen miles from Marenisco, where the body of sixty-one year-old Corrine Gerster's was found outside the Towne House store.

Gerster's neighbor has two retrievers. One of them tried to spend the night on her porch, and so she put it on a leash and walked the dog over to the neighbor's house. That was the last that anyone saw of her until the next morning when her body was found. The woman's clothing was torn from her body, her arm was partially amputated below the elbow, and there were signs of a real struggle.

An autopsy found that Corrine Gerster had died of a heart attack, but just what caused her to have a heart attack is the real question. Initially law enforcement in Presque Isle and the Wisconsin DNR said that animal tracks were discovered near the body, but couldn't say what animal it was. Wisconsin DNR then later said some of the tracks belonged to domestic dogs and led to two different homes.

Chief Mahler says the woman had deep puncture bite wounds on her head and in the heart-lung area of her body. It seems likely that the woman died because of the attack. The dog she was returning was not hurt.

Charged with protecting the people of Marenisco Township, in his letter to the DNR Chief Mahler says, "I will take whatever action 'lawfully' necessary to safeguard the people of my township—including lethal force if a wolf is a threat to either myself or another. I strongly urge you to do whatever you can to help overturn this judicial order so an adequate management plan can be adopted and put in place."

Marensico is less than ten miles from Wisconsin, which has a growing problem with wolf attacks on dogs. Between March 25, 2016, and September 18, 2016, the Wisconsin DNR reports thirty-six wolf attacks on dogs in the state. All but two of these attacks resulted in one or more dogs being killed.[40]

The significance of this has already been explained by Dr. Geist as a part of the sequential wolf habituation that brings them closer and closer to attacking people.

In September 2016 Paul and Judy Canik, who own a farm in Butternut, Wisconsin (about fifty miles from Marenisco), with 350 sheep and dogs to guard them, reported at a "wolf summit" held in that area in early September 2016 that in the last year and a half, wolves had killed two dogs and seventeen exotic sheep. The ewes they lost in late May were all pregnant with twins, tripling their losses. The wolves didn't eat any of the sheep.[41]

Wyoming, Colorado, Idaho, Kansas, Montana, Minnesota, New Hampshire, North Dakota, and Utah—all have filed briefs in support of the US Fish and Wildlife Service's delisting of the wolf in the Great Lakes region, calling for Congress to pass legislation overturning a 2014 federal judge's decision to place gray wolves in the Great Lakes states of Wisconsin, Michigan, and Minnesota back on the endangered species list after they had been delisted in 2012.

While the wolf population in Michigan's Upper Peninsula is stable, there have been reports of wolves moving into the Lower Peninsula, presumably the wolves walked across the ice on Lakes Michigan and Huron in winter. According to the Michigan DNR, in March 2014, biologists with the Little Traverse Band of Odawa Native Americans discovered wolf tracks and collected scat from what was presumed to be a wolf in Emmet County. In September 2015, confirmation was received from Trent University in Peterborough, Ontario, that the Emmet County scat submitted for DNR analysis was from a male gray wolf.

This was the second confirmation of wolf presence in the Lower Peninsula since 1910. The first occurred in 2004 when a wolf, collared by the DNR in Mackinac County, was accidentally killed by a coyote trapper.[42]

For more information about wolves in Michigan, including statistics of attacks on dogs, see: http://www.michigan.gov/dnr/0,1607,7-153-10370_12145_12205-32569—,00.html.

Ted Lyon: Update as of October 2016

In July of 2016 in the American Association for the Advancement of Science magazine, Science Advances, a group of scientists from Princeton, UCLA, UC Santa Cruz, UC San Francisco, SiChuan University from the Republic of China, and Efi Arazi School of Computer Science in Israel published an extensive study of the wolf genome sequence that revealed that the eastern wolves of the Great Lake states, and the red wolf of the southeastern United States are in fact wolf-coyote hybrids. The red wolf showed 75 percent coyote ancestry while the eastern wolf showed 25 percent coyote ancestry.[43]

This scientific study is the beginning of what I predict will eventually lead to the delisting of eastern wolves in the Great Lake states and the red wolf in North Carolina. The simple reason for this is that the ESA is designed to protect only species not hybrids. It's never been used to my knowledge to protect hybrids and probably never will be in the future.

It's my position that protecting hybrids will require Congress to rewrite the Endangered Species Act to include hybrids, and that's something that probably won't happen. Delisting lawsuits challenging the listing of the eastern wolf and the listing of the red wolf will probably be filed in the near future and both, if done properly with the proper evidence, will probably be successful. This study did not sequence the genetics of the Mexican wolf which has been highly controversial but if that testing is done in the near future, it will probably reveal the same thing. This is very controversial right now. There is strong evidence that the original Mexican wolves were wolf-dog hybrids.[44]

There is also research that indicates Mexican wolves in captivity may be pure wolves,[45] but there are other studies that conclude that Mexican wolves are hybrids.[46]

Regardless which study is correct, as soon as they are released into the wild, as chapters by Laura Schneberger and Jess Carey demonstrate, Mexican wolves show habituated behavior that includes breeding with domestic dogs, and wolves waiting along roadsides for USFWS trucks to come by and dump food for them, such as roadkill.[47] The 2016 study reported in *Science Advances* even revealed that the Yellowstone wolves show an 8.5 percent coyote ancestry. This study should have profound implications for coyote/wolf management in the future and is even more reason to turn over all management of wolves to the states.

Endnotes:

1. Weiss, et. al., "An Experimental Translocation of the Eastern Timber Wolf," *Audubon Conservation Report 5*, 1975, Twin Cities, MN: NAS.
2. Michigan Department of Natural Resources, *Michigan Gray Wolf Recovery and Management Plan*, Lansing, MI, 1997.
3. James Hammill, "Policy Issues Regarding Wolves in the Great Lakes Region," Transactions of the 72nd North American Wildlife and Natural Resources Conference, Wildlife Management Institute, 2007, pg. 378.
4. http://www.fws.gov/midwest/wolf/aboutwolves/WolfPopUS.htm
5. http://www.michigan.gov/dnr/0,1607,7-153-10370_12145_12205-32569-,00.html #recovery
6. http://www.jsonline.com/news/wisconsin/45452492.html
7. http://www.dnr.state.mn.us/rsg/profile.html?action=elementDetail&select edElement=AMAJA01030
8. Hammill, *Op. Cit.*, p. 387.
9. http://greatlakesecho.org/2010/02/20/wolf-count-raises-questions-about-mich-igan-range-threats/
10. http://www.animallaw.info/statutes/stusmi287_1001.htm
11. http://www.wolf.org/wolves/learn/wow/regions/United_States/North_Carolina .aspandhttp://www.valleymorningstar.com/news/us_news/article_90da8565-be47-5bd2-8f96-21e2d959c8b5.html
12. http://www.startribune.com/sports/outdoors/169262076.html
13. http://www.dnr.state.mn.us/mammals/wolves/mgmt.html
14. http://www.dnr.state.mn.us/mammals/wolves/mgmt.html
15. Hammill, *Op. Cit.*, p. 383.
16. http://www.dnr.state.mn.us/mammals/wolves/mgmt.html
17. http://www.startribune.com/sports/outdoors/116787898.html; and http://dnr.wi.gov/topic/WildlifeHabitat/wolf/
18. http://dnr.wi.gov/topic/wildlifehabitat/wolf/dogdeps.html
19. http://www.mlive.com/outdoors/index.ssf/2012/05/government_killing of_8_wolves .html
20. http://www.twincities.com/localnews/ci_21518716/minnesota-wolfs-recovery-seen-higher-livestock-loss-payouts
21. http://www.dnr.state.mn.us/mammals/wolves/mgmt.html

22. http://www.thewildlifenews.com/category/wolves/wisconsin-wolves/
23. http://www.mnr.gov.on.ca/en/Business/SORR/2ColumnSubPage/STEL02_163450 .html
24. http://jhered.oxfordjournals.org/content/100/suppl_1/S80.abstract
25. http://consbio.org/
26. http://www.ofah.org/news/Research-needed-on-Algonquin-wolves
27. http://www.ofah.org/news/Government-announces-Enhanced-Wolf- Management-Plan
28. http://www.propertyrightsresearch.org/2006/articles09/six_injured_in_rare_ wolf_attack.htm
29. http://www.ontariooutofdoors.com/News/?ID=203&a=read
30. http://www.ontariooutofdoors.com/news/?ID=153&a=read
31. http://www.fws.gov/mountain-prairie/species/mammals/wolf/ FinalWolfHabPlan foremail.pdf
32. Kellert, S.R., *Public Attitudes and Beliefs about the Wolf and its Restoration in Michigan,* Wisconsin: HBRS, Inc., 1990.
33. Hammill, *Op. Cit.,* p. 383.
34. Mertig, A.G., "Attitudes about Wolves in Michigan," *Report to the Michigan Department of Natural Resources-Wildlife Division,* 2004, East Lansing, MI.
35. Mech, L. D. 2001. *Managing Minnesota's Recovered Wolves,* Wildlife Society Bulletin 29:70–77.
36. Hammill, *Op. Cit.,* p. 387.
37. http://www.9and10news.com/story/27877103/number-of-wolf-attacks -double-in-michigan
38. http://abc10up.com/warning-graphic-wolves-attack-dog-dnr-confirms/
39. http://www.rivernewsonline.com/main.asp?SectionID=6&SubSectionID=59&A rticleID=68395
40. http://dnr.wi.gov/topic/wildlifehabitat/wolf/dogdeps.html
41. http://www.wausaudailyherald.com/story/news/local/2016/09/15/ great-lakes-wolves-spil-blood-stir-fear/90315186/
42. http://www.michigan.gov/dnr/0,4570,7-153-10366_46403_63473-378496—,00 .html
43. http://www.sciencemag.org/news/2016/07/how-do-you-save-wolf-s-not-really-wolf
44. https://arizonadailyindependent.com/2015/01/12/are-mexican-wolves-in-arizona-actually-wolf-dog-hybrids/
45. http://mexicanwolves.heritageparkzoo.org/Genetic%20Management%20of%20 the%20Mexican%20Wolf%20Captive%20Population.pdf
46. http://www.earthtouchnews.com/natural-world/animal-behaviour/is-this-coyote -wolf-hybrid-taking-over-north-america
47. http://www.azgfd.gov/w_c/wolf/faq.shtml#4

Photo credit: William Larrison/Shutterstock.com

CHAPTER 13

. . .

The Wolf as a Disease Carrier

By Will N. Graves

*"The only way the E. granulosus parasite will be completely eliminated is
elimination of wild canids, including the wolf."*

—Will N. Graves

Wolves in Finland

There's a long history of wolves attacking domestic livestock, semi-wild reindeer, and people in Finland that traces back to before Rome occupied the
Scandinavian northland. In response, the Finns would periodically launch campaigns to reduce wolf numbers. In the early 1700s, for example, while some
game species were only hunted by noblemen, anyone could kill a wolf on any
land, including the private property of another person. Additionally, at least
one male from every family had to participate in hunts to control predators.[1]

In the 1960s, Finland conducted a war on wolves in northern Lapland.
Planes were sent into the air with soldiers carrying sub-machine guns to kill
the enemy wolf. Wildlife biologist Dr. Karloo Nygren remembers the time
well. He says there was "a radio program where the local Game Chief (we have
fifteen such in Finland) gave instructions on how to shoot a wolf through the
window—what gun and cartridges one should use and how close to the glass
one should keep the barrel when the beast is watching through it."[2]

The True Concern

The war against wolves in the 1960s and 1970s in northern Finland was not
so much about predation but rather a tapeworm carried by wolves that was
infecting reindeer and the people who herd them (*Echinococcus granulosus*) and
the horrible illness (*Hydatidosis* or hydatid disease) created by the tapeworms.

Russian scientists have identified fifty different diseases that wolves may
carry—including rabies, hoof and mouth disease, anthrax, and brucellosis—that

can affect livestock and/or people. All canids, wild and domestic, can carry many diseases, including those caused by tapeworms. Every dog owner knows that his/her dogs should be periodically wormed, but relatively few realize that some tapeworms carried by canines can infect humans, as well as livestock and wild game, and that these diseases can be deadly.

After the wolf-culling period in the 1960s and 1970s in Lapland, when wolf numbers were severely reduced, shooting and trapping wolves was greatly reduced. Soon, their populations began to rebound. Dr. Nygren wrote in 2010 that, as wolf populations increase, hydatid disease is again on the rise:

> It appears to be spreading in my own home area, Karelia, on both sides of the Fenno-Russian border. I am afraid it will not only affect our staple food and essential part of our heritage, moose, but also us directly. Hunters, dog owners, forest workers, and berry and mushroom pickers will indeed be in danger . . . none of it is exaggeration.

The Wolf Disease Problem in the United States

The possibility that wolves transplanted into the United States from Canada by the 1995–96 relocations might carry *Echinococcus granulosus* was recognized by the US Fish and Wildlife Service. When the 1993 Draft Wolf Environmental Impact Statement to Congress on wolf relocation was made public, it stated that "wolf recovery is unlikely to have any measurable impact on disease or parasite transmission."[3]

Veterinarian Jacob Wustner, who developed and coordinated the veterinary part of the 1995–96 wolf relocation program, stated:

> We treated every wolf (at least twice) with three paraciticides, including Droncit, which is essentially 100 percent effective against *Echinococcus* in wolves with a single treatment. It is extremely unlikely that any reintroduced wolf from Canada could have carried *Echinococcus* tapeworms into the United States.

However, Wustner then conceded that *Echinococcus granulosus* was already present in the Northern Rockies, and so the wolves could have quickly picked it up as they became established.[4]

There is considerable disagreement with this statement from experts in North America and around the world. Parasitologist Dr. Delane C. Kritsky, professor emeritus, Idaho State University, writes:

> I conducted research for seven years on *E. multilocularis* in North Dakota during the 1970s, and it is a very dangerous parasite to human beings.

However, the species of *Echinococcus* occurring in wolves and ungulates in Idaho is *Echinococcus granulosus*, a close relative of *E. multilocularis*.

E. granulosus is, in my opinion, more dangerous than the strain of *E. multilocularis* that occurs in the upper Midwest (North Dakota, eastern Montana, South Dakota, and points southeast). The strain of *E. multilocularis* in the north-central states appears to be relatively non-infective to human beings. However, *E. granulosus* is more dangerous because it is highly infective to man and, as a parasite of sheep and domestic dogs, it is much more easily brought into homes in Idaho, Montana, and Wyoming where human beings can be exposed.

Utah had a focus of *E. granulosus* during the 1970s and 1980s, during which time people were dying or undergoing dangerous surgery for the parasite cyst. The Utah focus occurred primarily in rural areas where sheep were raised. My friend and colleague, Dr. Ferron Anderson at Brigham Young University, was conducting research on *E. granulosus* in Utah. He developed an educational program that primarily endorsed burying sheep carcasses and deworming dogs, which eventually eliminated the parasite in central Utah.

The parasite in Idaho will not be dealt with as easily—and I doubt that it can ever be eliminated as long as wolves are present—because wolves and ungulates (deer and elk) will maintain a sylvatic (wild) cycle, which did not occur in Utah during the 1970s and 1980s. Thus, elimination of the parasite from sheep and dogs, as occurred in Utah, will not be successful as it was in Utah because the wild cycle will continuously provide eggs of the parasite for infection of man and his domestic animals in the future. The only way that the parasite will be eliminated from our area is elimination of the wolf. I have examined coyotes, which can carry both species of *Echinococcus*, and foxes from southeastern Idaho since 1974 and never found either *Echinococcus multilocularis* or *E. granulosus*. Ferron Anderson never found the latter species in Idaho either when he examined canines in Idaho during the 1970s and 1980s; that is, the *E. granulosus* was never in Idaho until the introduction of the wolf.

Finally, I asked Fish and Wildlife, during one of their public meetings concerning introduction of the wolf (prior to wolf introduction), and was "brushed off" by their "promise" that the wolves introduced to Idaho would be "dewormed," which action everyone (and especially they) should have known is never 100 percent effective.[5]

Hydatid Disease

Hydatid disease or *hydatidosis* is caused by the *Echinococcus granulosus* tapeworm. According to the US Army, there are six species of *Echinococcus* that are currently recognized, and all are capable of infecting man. However, *E. granulosus* is the most likely of the six species to cause serious or lethal illness.[6] *Echinococcus granulosus* is a three-millimeter tapeworm that requires two hosts to complete its life cycle: the final host, which harbors the adult parasite, and the intermediate host, which harbors the larval stage. The adult parasite occurs in the small intestine of carnivores, canines (wolves, coyotes, dogs, and foxes), felines (lion, puma, and jaguarondi cat), raccoons, and their domestic counterparts (dogs and cats). Ungulates—deer, elk, moose, sheep, goats, caribou, reindeer, and antelope in the wild, and domestic sheep, cattle, buffalo, and goats—are the main intermediate host-group. Humans also can be an intermediate host.

According to the Center for Disease Control, the larval stage of *Echinococcus granulosus* is transmitted as dogs and other canines eat the organs of animals that contain hydatid cysts. Canines pass the parasite to sheep, cattle, goats, and pigs from eggs released in canine feces. *Echinococcus* eggs can stay viable for up to a year.[7]

Humans typically are exposed to eggs through handling carcasses and skins, touching infected animals, and contact with feces of infected animals. Those photos you may have seen of people hugging or kissing wolf-dogs or tame wolves potentially spell trouble, unless the animals have been wormed, as wolves are the most common wild carrier of *Echinococcus granulosus*. All around the world, the people most likely to contract hydatid disease are those who have close contact with dogs that have had contact with either livestock that have the intermediate larval stage or wild, infected canines. This is why the incidence of infection in many areas is highest among women who clean houses, and children who play with dogs. Anyone who tends to sheep and other livestock, and hunters or ranchers who butcher animals also are at risk from contact with the hair of the animals. Thoroughly sanitizing your hands after contact with dogs that may have rolled in wolf scat or are infected themselves is imperative if you live in areas with wolves. When eggs are ingested through contaminated food or water, they hatch in the small intestine, releasing a small larva (known as an oncosphere) that then penetrates the gut wall and enters the circulatory system. The oncosphere is then carried via the circulatory system first to the liver. Typically, 60 to 75 percent localize there as cysts. *Echinoccocus granulosus* is the most common cause of liver cysts in the world.[8]

If the larvae pass through the liver, they are carried into the right auricle of the heart and then to the lungs, where they develop into the larval cyst stage called a hydatid cyst. Cysts may become so large and sometimes numerous that organ failures occur. In wild elk, deer, sheep, goats, caribou, reindeer, and moose, nearly all larvae are found in the lungs. Predators that catch and eat infected animals contract the disease from eating the cysts.

While the lungs and liver are the most common places to find hydatid cysts in ungulates, in human beings, it is not uncommon to find them in other organs of the body, including the brain and reproductive organs. Human beings are not the natural host for *E. granulosus*; therefore, the oncosphere (the larva of the tapeworm), in a sense, becomes lost when entering the blood stream of a human being and can lodge in almost any organ in its body.

After the larvae are established in an organ, growth begins quickly, resulting in a solid mass in which a central cavity soon appears—the beginnings of the hydatid cyst. A membrane soon lines the entire cavity. This membrane is in direct contact with a thin connective tissue capsule separating it from normal host tissue. The hydatid cyst swells and fills with a milky fluid. The larvae grow steadily and are produced in great numbers. Secondary germinal vesicles may develop inside the first cyst, increasing the infected area and the overall size of the cyst. Cysts in the liver may not be detected until discovered accidentally, depending on their size. In the lungs, difficulty breathing and inflammation are common. If a cyst breaks open, then more and more cysts are formed, and they occupy more space and draw vital fluids from the body, resulting in a wide variety of symptoms as they spread through—out the organs and body cavity.

Hydatid disease in the human kidney, spleen, or brain is generally serious. Seizures and paralysis may occur. Osteoporosis of the skull bones is possible. If larvae reach bone marrow, the bones are weakened, resulting in fractures. Spontaneous or traumatic (including surgical) rupture of the cysts may cause severe anaphylactic shock, which can be fatal. Treatment options include medication and surgery. Surgical removal of cysts must be done carefully, without cutting the cyst open. If a cyst is cut, the liquids can spread the disease throughout the entire body.

Detection of *hydatidosis* is not easy. While the parasites quickly become established in the body, actual symptoms may not appear for some time, as long as ten to twenty years. Often the cysts are not recognized until other illness or body trauma occurs. New ultrasound and serology tests aid detection, but these are not always able to detect hydatid disease. Often infections are ultimately found in autopsies of people who died of other causes.

Prevalence

There is no question that hydatid disease is serious and can be deadly. It is also well established that, among wildlife, wolves are a major carrier. They may ingest larvae by feeding on infected wild or domesticated ungulates, rolling in feces, and licking each other or themselves. The real question is just how common is hydatid disease among people?

An international survey of hydatid disease published in 1977 in the Bulletin of the World Health Organization (WHO) entitled "Hydatidosis: A

Global Problem of Increasing Importance" finds that hydatid disease is global and increasing not only where it is endemic but also in countries where it was not previously found.[9] The authors—professors of bacteriology and parasitology from Lebanon, Australia, and England—report that, during the nineteenth century, Iceland had the highest prevalence of human *hydatidosis* in the world. In 1900, it was 25 percent. They attribute this to the prevalence of sheep and dogs, and they note that an aggressive public health effort to control the disease resulted in a significant decline in human cases during the twentieth century. Other areas with high levels of *hydatidosis* were northern Scandinavia; throughout the Middle East (Israel once reported an average of one hundred deaths a year from human *hydatidosis*); the Indian subcontinent, where hydatid disease was the most important helminth (parasitic) health problem in Afghanistan,[10] and where 22 percent of the adult humans in one village in India tested positive for hydatid disease; the northern part of Russia; North Africa; Alaska and Canada, where the disease is frequently found in Native Americans and Inuits; southern South America, especially Argentina, Chile, Peru, and Uruguay; and Australia, where the disease is most prevalent in the south and west.

The authors conclude: "Prevention is better than cure . . . Unless urgent action is taken to redress this situation, the natural pattern of the disease is likely to change for the worse in the next few years." Some may argue that, as the WHO article was published in 1977, it does not reflect contemporary times. In response, following are some brief quotes from other articles:

China

According to Weiping Wu of China's Institute of Parasitic Diseases, quoted in *Scientific American* in 2005, at least six hundred thousand Chinese are currently infected by the deadly disease, and an additional sixty million are at risk. "It is an epidemic," he says."[11]

East Africa

In 1988, 18,565 nomadic pastoralists, from twelve different groups—living in the vast, semi-desert regions of Kenya, Sudan, Ethiopia, and Tanzania—were screened for hydatid cysts using a portable ultrasound scanner. High prevalence of *hydatidosis* were recorded among the Die northwestern (5.6 percent) and northeastern (2.1 percent); the Turkana of northwest Kenya; the Toposa (3.2 percent) of southern Sudan; the Nyangatom (2.2 percent), Hamar (0.5 percent), and Boran (1.8 percent) of southwest Ethiopia and northern Kenya; and the Maasai (1.0 percent) of Tanzania.[12]

Uruguay
A study of 1,149 people in the village of LaPaloma in central Uruguay in 1998 found 5.6 percent had *hydatidosis*. Almost 20 percent of the dogs in the village were infected.[13]

Wales
Echinococcus granulosus in sheep and dogs has been known to be endemic for many decades. An analysis of national hospital records for 1974 to 1983 found that the incidence of human cystic *echinococcosis* was 0.2 cases per million in England, two cases per million in Wales, and the highest rates (5.6 cases per million) occurring in southern Powys County.[14]

Australia
Scientists believe that hydatid disease came to Australia with sheep and dogs, and today it is most prevalent in sheepherding areas. It was most common in the late 1800s, but between 1987 and 1992, 321 people were diagnosed with hydatid disease in New South Wales and the Australian Capital Territory.[15]

Nepal
Hydatid disease is of considerable economic and public health significance in Nepal. One study found that 25 percent of the residents of one region showed symptoms of *hydatidosis*. Testing of blood samples of patients admitted to different hospitals of Kathmandu Valley showed that the disease had slightly higher prevalence among the males (53 percent) and considerably higher prevalence among the thirty-five-or-older age group (76 percent).[16]

Finland
In northeastern Finland, 25 percent of the wolves today are carrying hydatid disease. The incidence of the disease is increasing, although it is not yet commonly found among dogs.[17]

Kazakhstan
Kazakhstan has the most wolves per capita of any country in the world. In 2002, researchers reported that human cystic *Echinococcus* rates increased fourfold within ten years after post-Soviet independence due to dismantling collectives and changes in organized livestock and farming practices.[18]

Russia
Over five hundred cases of hydatid disease are reported in Russia every year, and the number is increasing. Hydatid disease is on the rise in the Bashkiria

region of Russia where fifty-three cases were identified in 2008, which is 1.7 times the number of cases reported the year before.[19]

Hydatid Disease in Northern Rockies

The relocated wolves came from Canada, where hydatid disease is well known. For example, between 1991 and 2001 in a clinic in Edmonton, Alberta, Canada, forty-two cases of hydatid disease were identified in people ranging in age from five to eighty-seven years—77 percent were female, 41 percent were native (women and children tend to most often contract the disease), 40 percent of the patients had cysts in their lungs, and 55 percent had cysts in their liver. The researchers report that the most common intermediate hosts are barren ground caribou and moose. Studies have shown that 50 percent of the moose in Ontario and British Columbia have hydatid disease.[20] Between 2006 and 2008, a team of federal and state veterinarians and biologists—Foreyt, Atkinson, and McCauley—evaluated the small intestines of 123 gray wolves collected in Idaho and Montana, looking for *Echinococcus granulosus* tapeworms. Reporting their findings in the *Journal of Wildlife Diseases* in 2009, they stated, "The tapeworm was detected in 39 of 68 wolves (62 percent) in Idaho, and 38 of 60 wolves (63 percent) in Montana. The detection of thousands of tapeworms per wolf was a common finding."

Intermediate-form hydatid cysts were found in Idaho elk, mule deer, and mountain goats and in Montana elk. The researchers concluded, "To our knowledge, this is the first report of adult *E. granulosus* in Idaho or Montana . . . Based on our results, the parasite is now well established in wolves in these states and is documented in elk, mule deer, and a mountain goat as intermediate hosts."[21]

The extent of human infection in the United States may not be known for years; however, in 2011 a woman in Idaho, who lives in an area where wolves are common, was operated on to remove a large hydatid cyst on her liver.[22]

What Can Be Done?

Dr. Valerius Geist, who had a relative die of hydatid disease, advises hunters and those with domestic dogs where wolves are present to take the following steps to protect themselves and their pets:

1. To prevent infection of dogs, do not consume or allow your dog to consume uncooked meat or organs of wild or domestic ungulates. If your dog does have access to carcasses, talk to your veterinarian about an appropriate deworming treatment. Deworming your dog and preventing access to carcasses of dead sheep, which should be buried immediately, are important preventive steps.

2. Hunters should not touch or disturb wolf, coyote, or fox scat. Wear gloves when field dressing a canid carcass, and wash any body part that may have been exposed to feces or contaminated fur.

Neospora Caninum

Wolves also may carry a prozoan parasitic disease, *Neospora caninum*, which so far has not been shown to affect people, but it does cause cattle, sheep, and goats to abort, resulting in considerable economic loss to ranchers and farmers. *N. caninum* was not identified until 1988, but has since been found worldwide.[23] It is a major cause of abortions in ruminants. Adult infected cattle appear healthy, but they will abort at least 20 percent of the time during their lifetime,[24] causing millions of dollars in damages every year.[25]

Like *Echinococcus*, *N. caninum* has a life cycle with two hosts—canids as adult hosts and ruminants as the intermediate host. *Neospora caninum* may be transmitted by dogs, wolves, coyotes, and foxes. Wild, free-ranging wolves make it very difficult to control the disease in areas where there are cattle, sheep, and goats. Ruminants contract the infectious eggs when they feed on grass where there are wolf, fox, and coyote feces.[26]

When introduced into a herd, up to 90 percent of the unborn calves can be infected. When the brain and nervous system of cows, as well as sheep and goats, are infected, it results in abortions. When canines eat the aborted fetuses and placenta, the disease is spread, as their feces contain oocysts (hardy, thick-walled spores able to survive for lengthy periods outside a host). When ruminants feed where the canines have defecated, they ingest the oocysts, which then infect the dogs, wolves, foxes, and coyotes; the life cycle then begins again. *Neospora caninum* is now recognized as being worldwide and one of the most damaging of all parasites transmitted to cattle.[27] *Neospora caninum* does not appear to be infectious to people, but in dogs, it has been found that infection can lead to neurological damage.[28]

In 2011, a team of researchers from the USDA Animal Parasitic Research Laboratory found *Neospora* oocysts in three of seventy-three necropsied wolves in Minnesota.[29] One study in 2004 found 39 percent of 164 wolves from Minnesota were infected with *Neospora caninum*.[30] A study of wolves in Yellowstone National Park published in 2009 found that 50 percent of the wolves tested positive for *Neospora caninum*, with the oldest wolves most likely to be infected.[31]

Researchers in Wisconsin also report wolves testing positive for *Neospora caninum*. As both Minnesota and Wisconsin have a significant dairy industry and wolf populations are increasing in both states, one would expect that *Neospora caninum* would also be increasing.[32] In a 2011 study reported in the

Journal of Veterinary Parasitology, *Neospora*-like oocysts were found in the feces of three of seventy-three wolves in Minnesota examined by necropsy.[33]

N. caninum has been shown to be a large economic loss to the dairy and beef industry with infected animals being three to thirteen times more likely to abort than non-infected cattle. The researchers also note that the presence of wolves near cattle and other livestock causes the livestock to run about in fear, which adds further stress that also increases abortions.[34]

Considering there are wild and domestic intermediate hosts in areas where wolves are now found, eradication is very difficult. However, a vaccine has recently been developed that yields partial protection to ruminants.[35]

Rabies

Rabies is a viral disease normally spread by the saliva of infected mammals that bite others. After being bitten by a rabid animal, the disease attacks the central nervous system, resulting in swelling, which progresses to the brain. Symptoms may appear as early as seven days after a bite to as long as several years. The average incubation time is three to seven weeks. Early-stage symptoms of rabies include malaise and fever, progressing to acute pain, violent movements, uncontrolled excitement, depression, hydrophobia, hallucinations, and delirium. As rabies progresses, violent mood swings may occur before the victim lapses into a coma. If not treated with a vaccine before symptoms begin, rabies is usually fatal. Death typically occurs two to ten days after symptoms appear. Modern vaccines seldom have any side effects beyond those from common influenza shots.

Rabies vaccines for dogs have severely curtailed bites by dogs as a cause of rabies in the United States. In underdeveloped countries abroad, however, dogs are a major vector for rabies. In North America, rabies is most often spread by wild animals, especially raccoons, bats, skunks, foxes, coyotes, and, where they are found, wolves. When animals become rabid, they seem to lose all fear. Thus, a rabid animal becomes easy prey for a predator, and conversely, rabid animals are not afraid of anything. Rabies is present in almost all countries of the world, and especially in Asia, attacks on humans by rabid wolves are not uncommon. There are an estimated fifty-five thousand human deaths from rabies worldwide every year—about thirty-one thousand in Asia, and twenty-four thousand in Africa.[36] There were forty-five cases of rabies in the United States between 1995 and 2010; of these, nine are thought to have been acquired abroad. Bats are the most common carriers of rabies in North America.[37]

Wild wolves can catch rabies. In 2007 in the Alaskan Inuit village of Marshall in western Alaska, a pack of wolves attacked sled dogs, killing three adults and three puppies. A wolf killed by villagers as they drove off the wolves was found to be rabid by the Alaska Department of Fish and Game.[38]

Conclusion

Wolves are wild ancestors of dogs, but unlike pet dogs, wild wolves do not receive treatment from veterinarians. As a result, the fifty-some diseases they may carry double their negative impact on human settlements. This reality amplifies the need to keep a safe distance between man and *Canis lupus.*

Endnotes:

1. http://www.ekoi.lt/info/azl/2003/AZL percent2013_15-20.pdf
2. http://www.savewesternwildlife.org/killer-wolves.html
3. http://huntingnewsdaily.com/2011/07/08/idaho-fg-perpetuates-ignorance-with-misinformation/
4. http://missoulian.com/mobile/article_cfd5615e-77ce-11df-bcde-001cc4c03286.html
5. Personal correspondence with Dr. Delane C. Kritsky, Professor Emeritus, Idaho State University, 2011.
6. http://tmcr.usuhs.mil/tmcr/chapter3/intro.htm
7. http://www.cdc.gov/parasites/echinococcosis/epi.html; for detailed life- cycle information, see: www.dpd.cdc.gov/DPDx/html/Echinococcosis.htm
8. http://www.ispub.com/journal/the-internet-journal-of-pulmonary-medicine/volume-10-number-1/unusual-presentation-of-hydatid-disease.html
9. Matossian, R.M., M.D. Rickard, and J.D. Smythe, "Hydatidosis: A Global Problem of Increasing Importance," *Bulletin of the World Health Organization,* 1977, 55 (4. 499–507.
10. Pertaining to intestinal worms, such as the tapeworm and the roundworm.
11. *Scientific American*, July 2005 issue, page 22; http://www.khamaid.org/ about_kham/news/hydatidosis.htm
12. http://www.tropicalmedandhygienejrnl.net/ article/0035–9203%2889%2990664–0/abstract
13. http://www.ncbi.nlm.nih.gov/pubmed/9790441
14. http://findarticles.com/p/articles/mi_m0GVK/is_4_11/ai_n13609429/
15. http://www.science.org.au/nova/056/056print.htm
16. http://homepage.usask.ca/~shb292/hydatid.pdf
17. http://www.sciencedirect.com/science/article/pii/S0304401702003813
18. Togerson, P.R., B.S. Shaikenov, K.K. Baitursinov, and A.M. Abdybekova, "The Emergent Epidemic of Echinococcus in Kazakhastan," Translated, *Society of Tropical Medical Hygiene,* 2002; 96: 124–8.
19. http://www.wormsandgermsblog.com/2009/01/articles/animals/dogs/ echinococcus-on-the-rise-in-bashkiria-russia/
20. http://www.biomedcentral.com/1471–2334/5/34/prepub
21. Foreyt, William J., Mark L. Drew, Mark Atkinson, and Deborah McCauley, "*Echinococcus Granulosus* in Gray Wolves and Ungulates in Idaho and Montana, USA," *Journal of Wildlife Diseases,* 2009: 1208–1212.
22. http://www.sciencedirect.com/science/article/pii/S0304401711003566
23. Dubey, J.P., "Neosporosis—The First Decade of Research," *International Journal for Parasitology.* 29 (10), 1485–8.

24. http://www.wvdl.wisc.edu/PDFpercent5CWVDL.Info.Recognizing_and_Preventing _Neosporosis_Infections.pdf

25. http://cmr.asm.org/content/20/2/323.abstract

26. http://www.ars.usda.gov/Main/docs.htm?docid=11007

27. http://www.merck-animal-health-usa.com/binaries/NeoGuard_Fact_Sheet _tcm130- 126564.pdf

28. Barber, J.S., C.E. Payne-Johnson, and A.J. Trees, "Distribution of Neospora Caninum Within the Central Nervous System and Other Tissues of Six Dogs with Clinical Neosporosis," *Journal of Small Animal Practice* 37 (12): 568–74.

29. "I am a Woman with a Story to Tell," *The Outdoorsman*, March–May, 2011: 7–9.

30. http://dnr.wi.gov/org/land/er/publications/pdfs/wolf_impact.pdf

31. http://www.ncbi.nlm.nih.gov/pmc/articles/PMC2738425/

32. http://www.wvdl.wisc.edu/PDF%5CWVDL.Info.Recognizing_and_ Preventing _Neosporosis_Infections.pdf

33. J.P. Dubey, M.C. Jenkins, C. Rajendran, K. Miska, L.R. Ferreira, J. Martins, O.C.H. Kwok, and S. Choudhary, "Gray Wolf *(Canis lupus)* Is A Natural Definitive Host for *Neospora Caninum*," Veterinary Parasitology, 2011.

34. http://dnr.wi.gov/org/land/er/publications/pdfs/wolf_impact.pdf

35. http://www.merck-animal-health-usa.com/binaries/NeoGuard_Fact_Sheet_ tcm130-126564.pdf

36. "Rabies," World Health Organization, September 2011. Retrieved 31 December 2011.

37. "Rabies Surveillance Data in the United States," Centers for Disease Control and Prevention; http://www.cdc.gov/rabies/

38. http://usatoday30.usatoday.com/news/nation/2007-11-03-798767073_x.htm

Photo credit: Dennis Donahue/Shutterstock.com

CHAPTER 14

• • •

The Wolf as a Cash Cow

By Karen Budd-Falen and Joshua Tolin

The Economics of Litigation

Environmental groups have profited immensely by "defending wolves" in federal litigation. Interestingly, in their fundraising appeals, environmental groups explain their need for funds as essential to supporting education, administration, lobbying, and litigation. What they fail to explain is that the litigation itself is also a significant source of income and the wolf is but one of many species that these groups use to raise money so they can keep on litigating, which hampers federal and state agencies in doing their conservation work, and that in turn makes them easier targets for more litigation.

Even more troubling is that the attorney's fees paid to these groups by the federal government are not properly tracked, so the public cannot know for sure exactly how much money the environmental groups are taking in by their constant filing of litigation.

Additionally, the funding source of the attorney's fees payments made to the environmental groups is not always clear. Generally, payments are made either from the Department of Treasury's Judgment Fund (an open-ended budget item) by way of the Endangered Species Act of 1973's citizen-suit provision or from the responsible agency's budget pursuant to the Equal Access to Justice Act by way of the Administrative Procedure Act (APA). Oftentimes, environmental groups file their complaints mixing both ESA citizen-suit claims and APA claims; therefore, even a review of the claims cannot provide a clear answer.

Environmental Litigation Gravy Train

There are two major sources of attorney fees that can be paid to plaintiffs that "prevail" in litigation either by winning a case on the merits or by the Justice Department agreeing that the group "prevailed" in a settlement by achieving the purpose of the litigation.

One source is called the "Judgment Fund." The Judgment Fund is a Congressional line-item appropriation and is used for Endangered Species Act cases, Clean Water Act cases, and other statutes that directly allow a plaintiff to recover attorney fees. There is no central database for tracking the payment of these fees; thus neither the taxpayers nor members of Congress nor the federal government knows the total amount of taxpayer dollars spent from the Judgment Fund on individual cases.

The second major source of payments to "winning" litigants against the federal government is the Equal Access to Justice Act (EAJA). EAJA funds are taken from the "losing" federal agencies' budget. Within the federal government, there is no central data system or tracking of these payments from the agencies' budgets.

The Judgment Fund

Prior to 1956, most judgments against the United States could not be paid from existing appropriations, but required specific congressional appropriations. In 1956, Congress enacted a permanent, indefinite appropriation called the Judgment Fund for the payment of final judgments which were "not otherwise provided for" by another source of funds.

EAJA

In 1980, Congress embarked upon an experiment in "curbing excessive regulation and the unreasonable exercise of government authority" by directing that attorney fees be awarded in favor of private parties who resist unjustifiable government conduct in litigation. Signed into law by President Carter on October 21, 1980, the EAJA contained a sunset provision and expired by its own terms on October 1, 1984. In 1985, Congress passed new legislation that reinstated EAJA retroactively to October 1, 1984. On August 5, 1985, President Reagan signed the bill, making the EAJA a permanent statute.

The EAJA awards costs of litigation and attorney fees (up to $125/hour) when a citizen, non-profit organization, or small business wins a case involving the federal government and can show that the federal government's position was not "substantially justified." The EAJA's purpose is to help those who had been locked out of the decision-making process by virtue of their income, race, economic scale, or educational limitations.

Thirty Years of Profits

While the time span of wolf litigation actually covers almost thirty years of profits, the ridiculousness of the attorney's fees awards under EAJA has ballooned in the last ten years, where the average case garnered over a quarter million dollars in attorney's fees. The following is a summary of major cases where environmentalists have won awards of attorney's fees by representing the interests of wolves over the interests of those people living nearby.

Sierra Club v. Clark

In 1985, the Sierra Club secured an appeals court win in the Eighth Circuit Court of Appeals in *Sierra Club v. Clark*, 755 F.2d 608 (8th Cir. 1985). In *Sierra Club*, the environmentalists challenged rules concerning the eastern timber wolves (gray wolves living in Minnesota). At the time, the wolves were listed as threatened species under the Endangered Species Act of 1973 ("ESA"), 16 USC. §§ 1531–1543. As the population of wolves in Minnesota spiraled out of control, the state requested transfer of control of the wolves to it so it could create a public sport season to control wolf numbers.

While the Fish and Wildlife Service ("FWS") did not transfer control of the wolves to the state, in 1983 the FWS promulgated regulations permitting trapping of the wolves in certain areas where repeat livestock predation had occurred. Shortly after publication of the new regulations, environmentalists filed suit. The Eighth Circuit held that the regulations violated the ESA because it found the ESA did not provide discretion to the Secretary of the Interior to allow for "taking" threatened species. The environmentalists won, and were awarded $55,369.45 in attorney's fees. That number included a 30 percent upward adjustment of the fees of the two attorneys for the "extraordinary circumstances and the importance of the case to the management of all threatened species." Little did anyone know that we would look back on awards of $55,000 of tax-payer funds as "the good ol' days."

Biodiversity Legal Foundation v. Babbitt

Ten years passed before the environmental groups again sued over wolves in *Biodiversity Legal Found. v. Babbitt*, 943 F. Supp. 23 (D.D.C. 1996). In *Biodiversity Legal*, the environmentalists petitioned FWS to list the Alexander Archipelago wolf, which is found in the Tongass National Forest in Alaska. For several years, the US Forest Service had been considering major revisions to the Tongass Land Management Plan ("TLMP"). Because the Forest Service had major changes planned for the TLMP, the Regional Director of FWS found that listing the Alexander Archipelago wolf was not warranted, as the Forest Service's revisions would reverse any declining population of the wolf. The District Court in Washington, DC, overruled the agency's action, holding that

238 •• THE REAL WOLF

the ESA only allowed consideration of current regulations and their effects on species, not upcoming changes. The environmentalists won and were awarded attorney's fees.[1] The exact amount is not available online, as the DC District Court does not keep digital records from twenty years ago. The award may be available through federal archives.

Defenders of Wildlife v. Department of the Interior

Another ten years passed before the environmentalists recognized their cash cow: ESA litigation. In *Defenders of Wildlife v. Department of Interior*, 354 F. Supp. 2d 1156 (D. Ore. 2005), the environmentalists cashed in on their first quarter-million-dollar wolf case. After numerous years of much higher than expected "recovery" of wolves throughout the country, the US Fish and Wildlife Service issued a rule creating three distinct population segments ("DPSs") for the gray wolf and down-listing the wolf from endangered to threatened in the Eastern and Western DPSs ("2003 Rule"). Wolves in the Southwestern DPS remained listed as endangered, and the experimental populations in the Northern Rocky Mountains remained unchanged. The environmentalists succeeded in convincing the court that the Fish and Wildlife Service had violated the ESA by creating the DPSs and down-listing wolves in major parts of the country. In doing so, the environmentalists found what they were looking for: attorney's fees paid by the federal government to the tune of $272,710.54.

National Wildlife Federation v. Norton

In the same year as *Defenders of Wildlife*, environmentalists double dipped on the same issue. Nine months after the United States District Court for the District of Oregon held the 2003 Final Rule promulgated by the US Fish and Wildlife Service violated the ESA, another environmentalist group scored another quarter of a million dollars from challenging the same rule. In *National Wildlife Federation v. Norton*, 386 F. Supp. 2d 553 (D. Ver. 2005), environmentalists challenged the FWS's final rule because it had combined two DPSs listed in its proposed rule into one DPS in its final rule. The environmentalists won their argument, and even though the final rule had already been invalidated more than six months prior, secured more funds for their coffers: $255,500.

Humane Society of U.S. v. Kempthorne

After paying the environmentalists half a million dollars to litigate against its 2003 final rule, the US Fish and Wildlife Service issued a new rule in 2007 creating a Western Great Lakes DPS and simultaneously delisting that DPS, as the wolf numbers had over the years exploded in Minnesota. In *Humane Society of U.S. v. Kempthorne* ("*HSUS*"), 579 F.Supp.2d 7 (D.D.C. 2008), the animal rights group challenged FWS for creating a DPS based on geographical boundaries and at the same time removing that DPS from ESA protection by delisting the DPS. HSUS was able to successfully argue that FWS was incorrect in considering the ESA clear on its face as for providing authority to create a DPS and delist it at the same time. The court did not hold that FWS's interpretation of the statute was unreasonable—only that the statute was not clear on its face. The court vacated the 2007 rule and remanded it back to the FWS to explain its interpretation. For that, the HSUS cashed a check for $280,000.

Defenders of Wildlife v. Hall

The final case environmentalists used to make wolf litigation a million-dollar industry was *Defenders of Wildlife v. Hall*, 565 F.Supp.2d 1160 (D. Mont. 2008). In *Hall*, the environmentalists challenged a rule similar to that in *HSUS*, only the rule in *Hall* concerned wolves in the Northern Rocky Mountain DPS instead of in Minnesota. In 2008, FWS created this DPS and delisted all the wolves in the DPS. Judge Don Molloy held that FWS acted arbitrarily by delisting these wolves without proof of genetic exchange between the sub-populations and without explaining its reversal concerning Wyoming's wolf management plan.[2] Judge Molloy granted a preliminary injunction, finding that the environmentalists would likely succeed on the merits of their case. As a result, the wolves in Idaho, Montana, and Wyoming were relisted, and the environmentalists took in an additional $263,099.66. (The environmental groups had actually requested $388,370 in attorney fees.)

USFWS attempted to comply with Judge Molloy's ruling in *Defenders I*. In 2009, FWS issued another final rule, this time delisting the wolves in Montana and Idaho. With respect to the wolves in Wyoming, FWS stated that the proposed management plan for Wyoming was deficient, and therefore the wolves in Wyoming would remain listed until the issues were sufficiently addressed.

Again, the environmentalists challenged this rule in *Defenders of Wildlife v. Salazar* ("*Defenders II*"), 729 F. Supp. 2d 1207 (D. Mont. 2010). And again, Judge Molloy ruled in favor of the environmentalists, this time on the basis that ESA does not allow for designations smaller than DPSs, which the court held FWS had done by delisting the wolves from the same DPS in Idaho and Montana, but not delisting them in Wyoming.

No Matter What

The United States Congress eventually took notice that no matter what USFWS would do with wolves, the environmentalists would challenge those decisions in court. In response to Judge Molloy's decision in *Defenders II* and while that case was on appeal to the Ninth Circuit, Congress passed H.R. 1473, the Department of Defense and Full Year Continuing Appropriations Act of 2011. Congress, in § 1713 of that Act, directed FWS to reissue the 2009 rule. Moreover, Congress mandated that the 2009 rule could not be subject to judicial review, thus shielding it from the environmentalists' never-ending challenges.

Yet again, the environmentalists filed another challenge in court, this time to Congress's passage of § 1713 in *Alliance for the Wild Rockies v. Salazar*, 800 F. Supp. 2d 1123 (D. Mont. 2011). The environmentalists claimed that Congress overstepped its authority and unconstitutionally violated the separation of powers doctrine. The court sided with the government, and the environmentalists then appealed that decision, but the Ninth Circuit affirmed the constitutionality of § 1713 in *Alliance for the Wild Rockies v. Salazar*, 672 F.3d 1170 (9th Cir. 2012).

Finally, more than fifteen years after FWS introduced Canadian wolves to the Rockies and after paying more than one million dollars to environmentalists' attorneys, wolves in Idaho and Montana have been removed from ESA protection.

Conclusion

Environmentalists have taken one animal, the wolf, and used it and the ESA to develop a profit-generating business opportunity for its lawyers. In a span of only a few years, environmentalists have been able to promote the interests of wolves over the interests of those people living near the animals, and other species of wildlife, and in doing so, cashed in on over one million dollars of taxpayer funds. These funds, taken from taxpayer dollars and sportsmen contributions, could have been used to conduct many other wildlife management programs, and the man-hours spent on the case by the federal government—both attorneys and staff—could also have been directed to other purposes.

As if this was not bad enough, the wolf is just the tip of the iceberg in environmentalist and animal rights litigation against the federal government.

According to Pulitzer Prize–winning *Sacramento Bee* journalist Tom Knudsen, during the 1990s the US government paid out $31.6 million in attorney fees for 434 cases brought against federal agencies by environmental organization attorneys.[1]

Studies released independently in 2012 by the Notre Dame Law School, the Federal Government Accountability Office, and the Boone and Crockett Club find ever-increasing sums of money being paid by the government to environmental and animal rights groups, and although the original EAJA law calls for the government to keep track of money paid out in EAJA cases, the GAO report finds that accounting stopped in 1995, so the government does not know how much money is actually being paid to environmental litigants. The GAO report, dated April 12, 2012, finds: "Most USDA and Interior agencies did not have readily available information on attorney fee claims and payments made under EAJA and other fee-shifting statutes for fiscal years 2000 through 2010. As a result, there was no way to readily determine who made claims, the total amount each department paid or awarded in attorney fees, who received the payments, or the statutes under which the cases were brought for the claims over the 11-year period."[2]

Some of the highlights of that report include:

- When the GAO asked seventy-five bureaus and agencies at USDA and the Department of the Interior (DOI) for records on payments, only ten could provide data on cases and attorney fee reimbursements; and the records provided were incomplete and unreliable, based on manual calculations from older files, and the memory of career employees. Also, some records may overlap, so the GAO is not even certain of their totals.
- The GAO identified $4.4 million per year of EAJA payments to environmental groups during the period of 2000–10 from suits against the ten units of USDA and DOI that had any records at all.
- The GAO's minimum numbers do not add up to totals available from public court records and tax returns over the same period. Public federal court records from just thirteen federal courts revealed $5.2 million in legal fees per year, compared to GAO's estimate of $4.4 million, as tabulated by legal staff for the Western Legacy Alliance. A broader analysis including additional federal court records and public tax returns from just twenty environmental organizations showed $9.1 million reimbursed during FY2010 alone, as demonstrated by attorneys for the Boone and Crockett Club.[3]

Boone and Crockett Club research shows EAJA actual costs are exceeding fifty million dollars per year from litigation by the top twenty environmental litigants.[4] Additionally, a careful review of the cases finds that many are about challenging "minor procedural decisions"—red tape—rather than substantive issues. Thus, what emerges is that not only does the government not know just how much money is being spent on EAJA cases (their accounting does not include fees for government attorneys), but the adversarial actions of the environmental and animal rights groups are causing more stress on the federal agencies, which is burying them in paperwork and making them more prone to overlook bureaucratic details, which makes them an easier target for more legal actions.

In short, the wolf is a cash cow for environmental litigants, and the successes of wolf litigation cases has helped establish a sizeable source of revenue for environmental and animal rights groups, all at the expense of the taxpayer, and wildlife in general. And unfortunately, the wolf is only one example of how environmental and animal rights groups are using the Endangered Species Act and the EAJA to drain economic resources from taxpayers and reduce the effectiveness of federal agencies in practicing effective wildlife management.

> "Here is your country. Cherish these natural wonders, cherish the natural resources, cherish the history and romance as a sacred heritage, for your children and your children's children. Do not let selfish men or greedy interests skin your country of its beauty, its riches or its romance." —Theodore Roosevelt

Endnotes:

1. icecap.us/images/uploads/ENVIRONMENTINC.doc
2. http://www.gao.gov/products/GAO-12-417R
3. http://www.boone-crockett.org/news/featured_story.asp?area=news&ID=137
4. Lowell E. Baier, Reforming the Equal Access to Justice Act, *Journal of Legislation*, Notre Dame Law School. Vol. 38, No. 1, 2012, pgs. 1–70; or http://www.boone-crockett.org/news/featured_story.asp?area=news&ID=114

Photo credit: Holly Kuchera/Shutterstock.com

CHAPTER 15

. . .

How the First Political Wolf War Was Won

By Ted B. Lyon

"I knew that, if we could get the sportsmen groups together and tell the truth about what was happening, the law would eventually be changed."

—Ted B. Lyon

Democracy

As Americans, we are used to the fairness of democracy and seeing the will of the people acknowledged and their choices made clear through legislation—slowly perhaps but inexorably. It became painfully clear that this matter of wolf management had to come to Washington, DC, and sooner, not later. It was also clear that an extraordinary effort was going to be needed to undo the damage done by the introduction of Canadian wolves, and I wanted to be part of that effort.

The real beauty of democracy is the potential for individual concerned citizens to reach out to their legislators to enact change, and I felt strongly that I should add my voice to the clamor of the stakeholders affected by the wolves. Because of my background in politics and my relationships with so many sportsmen and conservationists, I already knew many who might be willing to take some time to see what plans could be laid to either change the Endangered Species Act (ESA) or, failing that, simply find a way to get the wolves delisted and under individual state's management.

As a country, we are the victims, as well as its beneficiaries, of free speech. Just about every point of view is available to be heard, and it takes an educated and determined listener to winnow through the chaff to find the wheat. What myself and others who have contributed to this book were dedicated to do was to provide very busy men and women the kernels of truth not currently

243

available to the general public in such a way that they would be able to reassess their perceptions of the wolf management issue and then work together to make changes.

Once a group of us started the dialogue with legislators about getting things changed, it created an energy of its own, and we found stakeholders all across the country, angered by the continued listing of the wolves, who were contacting their congressmen for action. We started with small meetings—here and there across the country with concerned organizations, governors, senators, and representatives—to establish the basic groundwork of presenting the hard cold facts of the situation at hand. And those facts spoke for themselves. They spoke so well, in fact, that some of the legislators were stunned by what they learned.

Allies

The first step in any political battle is to find allies. In early 2010, I called Ray Anderson and asked him, as a member of the Rocky Mountain Elk Foundation (RMEF), to reach out to their headquarters in Missoula, Montana, and ask them to call me. RMEF is one of the premier wildlife conservation groups in North America. They have almost two hundred thousand members and have conserved over six million acres of land in conservation easements for wildlife. RMEF had never really taken a stand on wolf issues because some of their former executives thought it might be too controversial. Ray must have been effective because I was asked to call Chief Operating Officer Rod Triepke the next day.

Triepke suggested that I talk to the RMEF President, David Allen. Allen is active on a national basis and knows many of the big-name players in the conservation movement. Additionally, RMEF has tremendous credibility throughout the West and across the conservation community.

Networking

That was in March of 2010. No one involved in the wildlife conservation community really knew who I was. I was simply some guy from Texas, who had a home in Montana and who was complaining about wolves. I was certain that, if we could get the sportsmen groups together and tell the truth about what was happening, the law would eventually be changed.

Bombshells and Shock Waves

On March 30, 2010, Mike Leahy, Director of the Rocky Mountain Region of the Defenders of Wildlife, and Kirk Robinson, Executive Director of the Western Wildlife Conservancy in Salt Lake City, sent Allen an adversarial letter in which they accused the Rocky Mountain Elk Foundation of polarizing the important conservation issue of wolves and elk. They claimed that the wolves introduced into Yellowstone were native (not true), and they then called for "a scientific review of wolf recovery criteria to incorporate the best available science, followed by a regional stakeholder process to guide development of state plans that meet wolves' biological needs while addressing the legitimate concerns of affected people and communities." In an effort to intimidate Allen, they sent a copy to every Rocky Mountain Elk Foundation board member and published it.

Allen sent me a copy of the letter. He responded to it immediately, and he released his letter to the press. His reply letter—covered throughout the West and dated April 8, 2010—was a bombshell. He pointed out that Defenders of Wildlife and the Western Wildlife Conservancy were "contributing to perhaps one of the worst wildlife management disasters since the destruction of bison herds in the 19th century."

David went on to tell them that—until the federal lawsuit they had brought, which stopped the states of Montana and Idaho from managing the wolves, was dropped—there really was no need to meet. David went on to point out that groups such as Defenders had continued to move the "goal line" of wolf introduction, which had gone from three hundred total wolves in Montana, Idaho, and Wyoming under the original introduction to a current estimate of a minimum of seventeen hundred wolves with many respected biologists considering there to be many more. David pointed out that, therefore, the wolves were not endangered at all. He further pointed out that the states and the federal government had, through their wildlife agencies, already called for delisting of wolves but that the wolf support groups opposed those scientists. At the end, he offered to meet with groups to try to work the issues out. They eventually met, but Leahy could provide no new science to support the Defenders' position.

David's letter, which was widely published, sent shock waves throughout the political world. If an organization that was as well respected in wildlife conservation circles as RMEF felt so strongly about the Endangered Species Act and wolves, then surely something must be wrong with current policy.

Help from Congressman Edwards

In May 2010, I called then-Congressman Chet Edwards. Chet and I served together in the Texas Senate for ten years, and he and I are extremely good friends. While serving in the Senate, Chet always looked to me for advice on hunting and fishing issues since I was such an avid outdoorsman and served for a number of years as the vice chairman of the Senate Natural Resources Committee. I asked Chet if he would introduce a bill that would change the Endangered Species Act and take wolves out of the act. Chet knew that I had a home in Montana, and after I explained to him what was happening, he agreed that something needed to be done.

Chet directed his staff to go to legislative counsel and draft a bill. The first bill that was sent to me was so complex that I could see it being litigated for years—just like the Endangered Species Act. I called Chet's aide up and dictated to him over the phone how I thought the bill should be drafted. The next draft was simple. Chet filed the bill, and from there, things started to happen very fast. This gained me the advantage of instant credibility with many of the sportsmen groups.

I met with David Allen, RMEF President, in Missoula, Montana, and outlined my plan to get all of the other sportsmen groups to support Chet's bill to take the wolves out of the Endangered Species Act and let the states manage them.

Politics

Also in May of 2010, I reached out to Don Peay, Founder of Sportsmen for Fish and Wildlife (see chapter 6). Don is from Utah and is truly a human dynamo. He has done more for wildlife conservation in Utah and the West than most other people. I asked him to fly to Dallas and meet Congressman Edwards, a Democrat. Don is a moderate, who supports both Democrats and Republicans that support wildlife. He is very close to a number of Republican senators and understands politics better than anyone I know in the conservation movement.

Don, Chet, and I met that evening and formulated a plan. Don would work with the Republicans, and I would work with the Democrats. We knew going in that the end result would not be the bill that Chet had introduced, but a bill that would probably have to be phased in. Chet was able, within a few weeks, to get former Congressman Denny Rehberg, a Republican from Montana, to sign on as a co-sponsor of the bill, and by that fall we ended up with close to sixty co-sponsors in the House, both Democrats and Republicans.

In June of 2010, I drove to Missoula, Montana, and met with David Allen and former Montana Governor Brian Schweitzer. A rancher himself, Governor Schweitzer was fed up with wolves and what they were doing to the wildlife and people of Montana, Idaho, and Wyoming. He agreed to support the concept of the bill and, over the next few months, was quite helpful in lobbying former Secretary of the Interior Ken Salazar, who comes from a ranching family in Colorado, to help get the wolf delisted.

The summer and fall of 2010, I attended a small event with then-President Obama and other events in Dallas with Senators Claire McCaskill, Sherrod Brown, and Charles Schumer. At every one of those events, I talked about wolves to the Senators and their staff and to close advisors to the President. While the initial reaction was skepticism when I talked about the loss of elk in Yellowstone, the other elk herds that had been lost, the livestock killed by wolves, and the way the federal government treated the people that suffered livestock losses, these politicians came to realize the situation was critical. The clincher for most of the Democrats was when I told them of polling that showed that over 80 percent of the people in the United States wanted wolves managed by the states and not the federal government. In the meantime, I also had lunch in Dallas with my longtime friend, US Trade Representative Ron Kirk, and relayed to him the problem of wolves. He was well aware of the problem.

Politics and Intrigue

As time went on that summer, more and more people became aware of Chet's bill, but the pro-wolf groups did not believe it had a prayer of passing.

Don Peay, who has worked for years with Senator Orrin Hatch, was told very clearly by Senator Hatch that, in order to pass the bill, it would be necessary to get the support of the then two Democratic senators from Montana—Max Baucus and Jon Tester. The offices of Senators Tester and Baucus were flooded with emails, phone calls, and personal visits as dozens of private citizens across the West began to pressure their senators and congressmen to get something done.

I personally talked on the phone with Idaho Governor Butch Otter after my old friend, Brent Hill, called Governor Otter and asked him to support Chet Edwards' legislation.

One Republican senator called me personally and told me that, while I may have passed hundreds of bills in the Texas Legislature, passage of legislation of this magnitude, changing the Endangered Species Act, in these partisan times was simply not going to happen. While thinking to myself, not so, I was extremely polite and told him that if he could get the Republicans to cooperate, I would somehow get the Democrats.

Don Peay told me that, with my Democratic connections, he felt strongly that we could get wolves delisted and asked me to fly to Washington, DC, to meet with industry leaders about Chet Edwards's wolf delisting bill.

I got on a plane and flew to DC the next day and attended a luncheon meeting with a number of professional folks that worked for Safari Club International, the National Rifle Association, the Congressional Sportsmen's Foundation, and several other organizations. The consensus was that Congress would not tackle anything that controversial in an election year. This meeting occurred on September 15, 2010. One participant was particularly adamant that nothing could get done. I told him, quite frankly, that we didn't have a choice, that elk were being slaughtered by the thousands as the records showed, that this was a wedge issue, and that, if it wasn't taken care of, congressmen and senators from both parties would be defeated in the next election.

L to R: Ted Lyon, former Senator Harry Reid, Don Peay, Clint
Bentley, and Miles Moretti.

Breakfast Meeting

In September of 2010, I was able to set up a breakfast meeting with Senator Reid
in Washington. In attendance at the meeting were Don Peay with Sportsmen
for Fish and Wildlife; Miles Moretti, president of the Mule Deer Foundation;
and Clint Bentley, former president of Nevada Bighorns Unlimited and former member of the Nevada Wildlife Commission. Clint, who had just had
open heart surgery two weeks before, was there on behalf of the Wild Sheep
Foundation at the request of Don Peay.

As we sat down to breakfast, we had a plan. We had met the night before,
and it was decided that no one else would talk until I made my presentation to
Senator Reid. This fifteen-minute presentation was one of the most important
presentations of my life. I knew that if we could persuade Senator Reid, a powerful member of the US Senate, then other Democratic senators would follow.
The whole ballgame to me was the US Senate. I had stayed up half the night
writing and rehearsing what I was about to say. I knew that if I could convince
the senator, then we would have a good chance of passing the legislation in the
Senate. The House, I always felt, with sixty co-sponsors, was doable.

At the beginning of the meeting, I asked Senator Reid if he wanted the
science or the politics first. Without hesitation, he replied science first so I gave
him the facts. Almost all of the information had come from the 2009 Montana
Wolf-Ungulate Report and from RMEF biologists.

Miles Moretti and Don Peay both made very brief but heartfelt pitches
on behalf of their groups. The fifteen-minute meeting stretched to forty-five
minutes. I finally wrapped it up by telling the senator that—under the law as it
existed in Montana, Idaho, and Wyoming at that time—he could be walking
his dog down a country road and, if a wolf attacked it, there was nothing he
could legally do to stop the attack or to defend his dog. The senator could not
believe it. He was visibly upset and, without consulting his staff, said that he
would help us. He also said to me, "That was a hell of a presentation, lawyer."

Senator Reid had known in a minute what the right thing to do was, and he had committed to doing it: the states needed the right to be able to manage the wolves, and the endless stream of lawsuits by groups simply using the wolf as a fundraising tool had to stop.

Montana's Senator Tester

We still did not have the support for our bill from the Montana senators whose state was arguably hit the hardest, but at least we were making progress. I had reached out to Senator Tester's office through my good friend, former Lieutenant Governor Ben Barnes of Texas, and he gained me entry to Tester's Chief of Staff. Senator Tester's staff was well aware that former Representative Denny Rehberg, Tester's Republican opponent in 2012 for the Senate, was holding town hall meetings all across Montana about delisting the wolves. This was a smart move on Rehberg's part and enabled him to get one hundred or more people at several meetings across the state. Representative Rehberg's move to talk about the wolves put a lot of pressure on Montana Senators Tester and Baucus.

On October 13, 2010, Senator Tester and I had a meeting in Missoula, Montana, with his staff. Tester made it clear to me that he had no love for wolves and would do everything he could to pass legislation to delist them. My goal was to get total delisting of the wolves across the United States. He explained to me that a compromise was the only way we could get anything through the Senate; he was right. In 2013, the USFWS proposed delisting gray wolves in all states with the exception of the Mexican gray wolf in Arizona and New Mexico.

Senator Tester then filed Senate Bill 3864, which was drafted by the Senate's legislative counsel and Senator Tester and Senator Baucus's staffs. It was an extremely complicated bill that would have resulted, in my opinion, in an endless series of lawsuits. Senator Tester asked for my opinion, and I gave it to him on November 30, 2010, along with a legal memorandum addressing the consequences of S. 3864 as written.

Wins and Losses

At the same time this was going on, Senator Hatch was working on a rider to the Omnibus Spending bill that Congress had to pass by the end of 2010. Senator Hatch's staff was on the phone with me on a regular basis to talk about the rider, and Don Peay and I talked every day. I was also on the phone on a regular basis with Senator Tester's staff.

As the year 2010 came to a close, the Republican and Democratic parties were fighting, and our rider never got put on the bill. Many of our supporters were truly demoralized, but I pointed out to them that we had come a long way in a little less than a year. We had on our side all the US senators from Montana, Idaho, and Wyoming; the governors of Wyoming, Montana, and Idaho; the Majority Leader in the Senate; the Republican Vice Chairman of the Senate Finance Committee; and the Democratic Chairman of the Senate Finance Committee. Additionally, we had over sixty co-sponsors in the House of Representatives, and numerous other people in the wildlife conservation community supporting our cause.

Online Support
The wolf supporters, who still did not understand the forces that they were up against, sat idly by, not knowing that they were faced with folks who, armed with the truth, were making tremendous headway in Washington. The issue, with the help of a group of bloggers, caught the imagination of thousands of sportsmen across America. George Dovel of Idaho, who publishes the *Outdoorsman* magazine; Toby Bridges of Missoula, Montana, publisher of the *LOBO Watch* blog; and Tom Remington of the *Black Bear Blog*, on an almost daily basis, were on the internet alerting the US Fish and Wildlife Service, the Idaho Fish and Game, and the Montana Fish, Wildlife and Parks to what was happening to the elk, mule deer, and moose across the West. Steve Alder and Scott Rockholm from Idaho, along with Robert Fanning with the Friends of the Northern Yellowstone Elk Herd from Montana, were also active in criticizing the actions of the US Fish and Wildlife Service. They pointed out—with facts and statistics—what was really going on. Their combined efforts had a tremendous effect upon publicizing the issue.

Non-Partisan Support
The effort was non-partisan, and we needed every sportsmen group united. An NRA lobbyist had committed at our earlier meeting in DC, and having the NRA on our side was huge. With them would come most Republicans, and a few Democrats. Don and I went to the Congressional Sportsmen's Foundation meeting the night that Senator Jon Tester was sworn in as its new president.

By mid-November, Senators Tester and Baucus had filed a bill that took care of the delisting of the wolf in Montana and Idaho. Idaho Senators Risch and Crapo, Wyoming Senators Enzi and Barrasso, and former Wyoming Representative Lummis were also very involved.

Win Some, Lose Some
When the 2010 elections came, the Democrats lost many seats, and the rider to be attached to the spending bill failed because it was going to require unanimous consent, and Maryland Senator Ben Cardin objected. The message I gave to every Democratic senator that I talked to was that, in the West,

senators were in deep trouble in 2012 unless the wolf problem was solved. Again, a budget resolution was needed to keep the government running, and again our plan was to put a rider on the budget resolution that would delist the wolves.

Looking forward to the 2012 election, we shared an interesting data point with senators and congressmen: although Utah is a very Republican state, by a slim margin, traditional Republican voters in Utah had re-elected Democratic Congressman Jim Matheson, who had been a vocal and early supporter of returning wolf management to the states. Conversely in the Democratic state of New Mexico, Democratic Congressman Harry Teague was a big supporter of wolves, and the pro-wolf groups had poured hundreds of thousands of dollars against Teague's challenger, Steve Pierce. However, Pierce, a pro-wolf management Republican was elected.

We also pointed out that we have a political data point in the Western states where hunters and ranchers will not vote party lines; instead, they will vote for those who support the "right" wildlife management policies, and this group, when motivated, can be a difference maker in congressional elections.

Language, Lawsuits, and Law

The final language that came out from the group that included Senators Reid, Hatch, Tester, Crapo, and Risch delisted wolves in Montana, Idaho, and Wyoming while still preserving at least one hundred-fifty wolves and ten breeding pairs in Montana and Idaho.

In March of 2011, the US Department of Interior dropped its appeal in Wyoming Wolf Coalition v. US Department of Interior, Case No. 09-CV-118J, and Park County v. US Department of Interior, Case No. 09-CV-138J, which meant that Judge Alan Johnson's decision that Wyoming's proposal to designate wolves as trophy game animals in certain areas around the national parks and as predators in the rest of Wyoming was now law.

The absence of wolf breeding pairs in the 88 percent of Wyoming where the wolf is designated as a predatory animal has no impact on Wyoming's ability to maintain its share of the greater Yellowstone area meta-populations segment because that area is largely unsuitable for wolf habitat and it is not located between the core recovery areas so it cannot affect the rate of natural dispersal between the three core recovery areas.

Testimony by Ed Bangs, the top wolf biologist for the US Fish and Wildlife Service, stated for the record: "The 2007 Wyoming wolf plan is a solid, science-based conservation plan that will adequately conserve Wyoming's share of the greater Yellowstone area population so that the Northern Rocky Mountain wolf population will never be threatened again."

Actions and Reactions

Harriet Hageman—the Cheyenne, Wyoming, lawyer who won the lawsuit for Wyoming, which is the first lawsuit anyone ever won for a group not supporting the proliferation of wolves—was upset in March of 2011 when she found out about an amendment sponsored by US Representative Mike Simpson, a Republican from Idaho, and Senators Tester and Baucus. The amendment, as originally proposed, would have delisted wolves in Idaho and Montana and in parts of Oregon, Utah, and Washington, while Wyoming and the rest of the states would be left to fend for themselves. In a letter to her clients called "A Call To Action!" Hageman blasted the original amendment by Simpson, Tester, and Baucus that was also supported, as her letter notes, by the NRA, Safari Club International (SCI), Congressional Sportsmen's Foundation (CSF), and Boone and Crockett.

I am absolutely certain that neither Senators Tester or Baucus nor Representative Simpson knew that the amendment was flawed, nor do I believe that the NRA, SCI, CSF, or Boone and Crockett would have supported that kind of bill. But once Hageman sent her letter to her clients, all hell broke loose, and there were harsh words exchanged between the NRA, Big Game Forever, and Sportsmen for Fish and Wildlife, creating bad feelings.

Success at Last

Needless to say, after a flurry of action, Representative Simpson, to his full credit, amended the language in the budget rider to fix the problem. The key language was that the 2009 wolf delisting rules adopted by the US Fish and Wildlife Service "shall not be subject to judicial review." Ryan Benson of Big Game Forever, who is a Harvard-educated lawyer, helped draft the final language that concerned Wyoming. Congress also stated in that rider that the bill "shall not abrogate or otherwise have any effect on the order and judgment issued" in the Wyoming delisting lawsuit, which Wyoming had won in Judge Johnson's federal court earlier in 2011.

Don Peay called me to see if we should accept the language that was eventually agreed upon by all of the Western Senators, and I told him hell yes!

The appropriations bill with the rider passed the Senate by a vote of 81 to 19 on April 14, 2011. "This wolf fix isn't about one party's agenda. It's about what's right for Montana and the West," Senator Tester said in a statement. President Obama signed the bill into law the next day, and the Endangered Species Act, for only the second time in its history, was amended. Sportsmen groups had all worked together in a nonpartisan way, and as a result, wolves will now be managed by the states up to certain levels. Wyoming will have a predator zone outside of the national parks where wolves are treated the same as coyotes, and Oregon, Washington, and a small part of Utah will be managed by the individual states.

Was this legislation everything we wanted? Absolutely not, but it was a start. As Confucius once said, "The journey of a thousand miles begins with a single step."

Conclusion

Although much more needs to be done, this first and critical victory in wolf management proves the effectiveness of sportsmen groups working together. When stakeholders make their voices heard to their legislators, with hard evidence supporting their claims, laws can be changed.

Across the Western states of New Mexico, Arizona, Washington, and Oregon, as well as states where wolves are just beginning to show up—California, Utah, Colorado, Texas, and North and South Dakota, wolf management issues remain to be resolved, and stakeholders need to vigilantly continue their efforts. Additionally, there are growing wolf populations in Michigan and Wisconsin, dispersals into neighboring states, and proposed plans for wolves in New York, and all these populations will need proper management—and soon.

In order to ensure that wolves are managed responsibly, people need to know the truth, and citizens need to be willing to do what it takes to communicate with their legislators. This first victory in wolf management sets the stage for the future.

Afterword:

Ted Lyon's Note: From my viewpoint, the legislative heroes who stood up for their states include Senators Orrin Hatch, Jon Tester, Harry Reid, Mike Crapo, James E. Reich, Mark Pryor, Mike Enzi, and John Barrasso along with Representatives Chet Edwards, Bob Simpson, Denny Rehberg, Cynthia Lummis, and Jim Matheson. There are many people and groups that deserve credit for their involvement in the Montana, Wyoming, and Idaho delisting of the wolves under the Endangered Species Act. It took all of our efforts—as individuals and as groups—to achieve this first step in wolf management, and it will take the same nationwide involvement to achieve further delisting. This is one person's perspective of how it happened.

Photo credit: Holly Kuchera/Shutterstock.com

CHAPTER 16

. . .

How the Second Political Wolf War Can Be Won

By Ted B. Lyon

Photo credit: stayer/Shutterstock.com

*"The biggest culprit that I found in my research was not the human beings
trying to implement government policy because it was their job,
but the law they were charged to implement."*

—Ted B. Lyon

Just Doing Their Jobs

At this point I want to say that in the 1990s, the governmental employees for the US Fish and Wildlife Service (USFWS) and all of the state wildlife agencies were doing their jobs, introduction of the Canadian gray wolf throughout the West. There are many people whose lives have been impacted by this introduction, and therefore who dislike or even despise those government employees, but I'm not one of them. All the people in federal and state agencies that I've interviewed for this book who worked for wolf introduction were extremely nice people who love wildlife and who thought then, and in many cases still believe, that wolves are a positive force in nature. I think they are wrong in their interpretation of this idea, and I believe that history will show them to be in the wrong, but I still respect their opinion. I can't believe that people who work for fish and wildlife agencies want to destroy the wildlife that they have spent their lives protecting, but in some cases this is exactly what's happening.

Photo credit: Nolodymyr Burdiak/Shutterstock.com

Declining Moose Populations

Dramatic declines in elk were the first significant impact of the wolf re-introduction program. A second large ungulate native to areas where wolves are found is the moose, *Alces alces*.

All across northern North America, and the world for that matter, one may find moose (they are called "elk" in areas of Europe), plentiful in some areas, and rare in others. Moose are native in nearly all Northern forest zones: Northern Europe, North America, the Baltic region, and Siberia, as well as in the colder regions of Asia. They are the largest of ungulates, the largest subspecies standing six feet tall and weighing as much as 1,600 pounds.

Ideal moose habitat is dense forests for food and cover and bodies of water that provide an abundance of aquatic vegetation.

In Europe, moose are found in large numbers throughout Norway, Sweden, Finland, and the Baltic States. They are also widespread through Russia. Small populations remain in Poland, specifically Biebrza National Park, Belarus, and the Czech Republic. In 2007 there were approximately 120,000 moose in Norway, 730,000 in the Russian Federation, 264,000 in Finland and 400,000 in Sweden.

In Canada there are approximately one million moose in nine provinces and three territories. The Provinces where moose are found are: Alberta, British Columbia, Manitoba, Newfoundland and Labrador, New Brunswick, Nova Scotia, Ontario, Quebec and Saskatchewan. The Territories are Nunavut, Yukon, and Northwest Territories.

In the United States moose are found in fifteen states: Alaska, Colorado, Idaho, Maine, Minnesota, Montana, New Hampshire, New York, North Dakota, Utah, Vermont, Washington, Wisconsin, Michigan, and Wyoming, with the largest population and the largest subspecies found in Alaska which has 170,000 to 220,000 moose. South of the Canada-US border, Maine has most of the population with a 2012 headcount of about seventy-six thousand moose. Other New England and Midwest states have another twenty-five thousand. There are at least twenty-five thousand more moose in the western United States.

Since the 1990s, moose populations have declined dramatically in much of temperate North America, although they remain stable in arctic and subarctic regions. The exact cause of the die-off is not determined, but appears to be a combination of factors, including a change in habitat and heat stress caused by global warming, poaching, the reintroduction of wolves, and the northward migration of warmer-weather parasites to which moose have not developed a natural defense, such as liver flukes, brain worms, and winter tick infestations.

What role do wolves play in the declining moose population? According to The US National Park Service: "Elk are the primary prey for wolves . . . Wolves have not reduced mule deer or bison populations . . . Moose represent less than 4 percent of wolf diets in winter and only twenty-six instances of wolf predation on moose were recorded in Yellowstone during 1995–2003."[1]

Often biologists seem like they cannot see the forest for the trees. Wherever wolves exist, we are seeing alarming rates of declines for moose. From Yellowstone National Park to Minnesota, Montana, Idaho, and Wyoming, the moose are disappearing, and the one common ingredient that all four states and Yellowstone National Park have is the presence of wolves. At this point I would refer the skeptical reader back to the chapter in this book by Dr. Arthur Bergerud, where he says that in the areas where they reduced wolves, recruitment was greatly increased: for moose, recruitment averaged forty calves per one hundred females; for elk, forty-nine per one hundred females; for Stone's sheep, forty-one per one hundred females; and for caribou, thirty-nine per one hundred females. For those herds with no wolves removed, the recruitment for moose was eleven calves per one hundred females; sheep, twenty per one hundred; elk, twenty-six per one hundred; and caribou, seven per one hundred. The elk and moose populations in this study that had wolf management increased from eighteen thousand to thirty-three thousand. In all removals, wolves were harvested in the spring before wolf denning, and in all years, young wolves were already dispersing into the removal area but insufficiently organized for denning so that the four prey species secured positive recruitment. At the end of the ten years, the wolf population was a healthy twenty-plus wolves per one thousand kilometers.

The results of Bergerud's research are confirmed by an increasing number of others. In 2013, Wyoming Game and Fish biologist Doug Brimeyer counted 239 moose, the second lowest count in twenty-eight years. Wildlife Conservation Society biologist Joel Berger was quoted in the *Jackson Hole Wyoming News & Guide* on March 2, 2013, as stating that wolf predation had played a role in the moose's decline.

Minnesota, which has approximately four thousand wolves, announced on February 5, 2013, that it had canceled the state's 2013 and foreseeable future moose hunting seasons, citing a "precipitous decline in moose population" as reported by the Associated Press. "The population has dropped 35 percent over the past year and 52 percent from 2010 to an estimated 2,760 moose left in northeastern Minnesota, according to the Department of Natural Resources in January 2013." The Associated Press story noted that some scientists suspected global warming, parasites, disease, and changes in vegetation as the cause. The story did not, however, mention that four thousand wolves eat a lot of meat and that the likely culprits are the wolves.

No one claims that wolves do not kill and eat moose, especially calves, but there is a considerable amount of disagreement when it comes to what factors are responsible for declines in moose populations—climate change, diseases, ticks, bears, and possibly predation by wolves. The authors do not deny that environmental and disease factors may affect moose populations, but wolves definitely do prey on moose. In Alaska, research by Alaska Fish and Game finds bears and wolves attacking calves is key to otherwise healthy moose populations declining.[2]

Studies in Germany by Reinhold Eben-Ebenau also find that wolf predation can be a significant factor in moose populations, especially if there are heavy tick infestations, as the ticks weaken the moose.[3]

As of 2012, Colorado's moose population tops 2,300, up 35 percent during the previous two years, beyond the state's latest target maximum number.[4] Moose numbers have also increased in Utah and Washington, where wolves are very scarce.[5]

If moose populations are growing in Utah, Washington, and Colorado, then these are presumably states where wolves and grizzlies are not common. The consensus from academic articles and conversation with researchers is that there are many factors that lead to population decline, and while wolves may be one factor, they are not the major one. According to Dr. Michelle Carstensen, wolf packs are the second highest cause of Minnesota moose deaths, but many moose "killed by wolves had predisposing conditions that made them more susceptible," including parasites, disease, and injuries.[6]

One of the foremost experts on wolves in the world, Dr. David Mech, a senior research scientist with the US Geological Survey and director of the International Wolf Educational Center, has recently changed his position on the role of wolves in moose populations.

In 2014, Mech and John Fieberg of the University of Minnesota published in the *Journal of Wildlife Management* the results of a re-evaluation of the findings from Lenarz et al. that adult moose (*Alces alces*) survival in northeastern Minnesota was related to high January temperatures, while the predation by wolves (*Canis lupus*) played a minor role. Mech and Fieberg now report finding significant inverse relationships between annual wolf numbers in part of the moose range and various moose demographics from 2003 to 2013 that suggested a stronger role of wolves than heretofore believed. Based on this research and Mech's over forty years of research on wolves, they concluded that there's simply no evidence that climate change has contributed to the moose decline in the northeast.[7]

Dr. Geist agrees that wolves can have a significant effect on moose populations. He writes:

In 2006 I rode for a week, dusk to dawn, through Yellowstone National Park as well as south of the park, in some of the finest moose country on Earth, at least judging from my moose experiences from the Alberta to the Alaska boundary, but we saw no moose, no moose track, no feeding sign of moose. Before the wolves—and the big wild fires—this had been home to a dense moose population. And the old feeding signs were still there. The park blamed the decline of moose on the loss of sub-alpine wintering habitat to fire, but failed to notice that lower elevations in moist areas these fires created—massively—moose habitat! So much for 'Yellowstone ecology'.

I have no doubt that wolves and bears closely crop or eliminate moose populations, but that climatic and edaphic factors determine how easy or how tough—moose calves especially—are to catch, or how many eggs get fertilized in a moose cow. Weak calves through parasitic or climatic factors mean greater success by predators catching calves. And that allows one to blame global warming etc. for the decline in moose. Predation is the ultimate factor. That it was a huge factor, is shown by the anti-wolf tactics of wintering moose. Why anti-wolf? Because bears are hibernating in winter. For instance, when moose are surprised—if not, they flee rapidly long distance at once!—they back their rear end into a dense conifer for protection leaving them free to concentrate striking wolves with their front hooves. As an alternative, the moose moves—at once—to an area of low snow but with good footing, especially small lakes with a snow/ice crusted surface, or wind-blown hillsides with bare patches of hard footing. to reach such a spot, the moose will even run towards the danger (me!) and then demonstrate his readiness for action. The hair goes up on neck withers and croup as the moose faces the danger. On hard ground (or ice) the moose is free to exercise with great virtuosity his front and his hind legs as striking instruments.

And moose can spin around, and strike and kick with lightning speed and precise targeting. The broken bones in Alaska wolves speak of that.

Also, wolves avoid moose (and horses) if there is anything easier to catch. But note: to succeed, the moose must understand how snow and hard ground influence their behavior and safety. And the way these moose acted when confronting me in winter made it clear they knew what they were doing, although defense was clearly second choice to flight. They understood hard lake surfaces as well as wind-blown ridge-tops, and they refused to run—and flounder—in the surrounding deep snow. All this is wolf specific. Also moose can utter under these conditions a bone chilling roar.

All this tells is that wolves and moose have confronted each other in winter for a very long time in order that such anti-predator strategies and tactics could evolve. These are old adversaries. The extinction of moose in Yellowstone with a burgeoning wolf population was most likely inevitable. Ditto for mule deer.

I recall an elegant study in which grizzlies were removed and relocated from a moose calving ground. Calf survival shot up. Next year the bears were all back and calf survival was very low. I have also no doubt that with heavy tick infections, cow moose produce weak calves that fall prey easily to any predator. I am well aware that wolves shun moose if any easier prey is available, such as abundant rabbits during a rabbit high. In such years wolves were observed hunting rabbit among moose and the moose ignored the wolves. In my field observations of moose I found that moose panicked at the howling of wolves. Secondly, moose have wolf specific anti-predator winter adaptations which could only have evolved under wolf predation as bears hibernate in winter. And nothing evolves unless predation pressure is severe.

The bottom line here is that one must always remember that while the scientific method seeks to find the truth in an unbiased way, scientists themselves are not unbiased. It takes courage for a scientist to not consider if the results of their research may not be politically correct, or likely to make it easier to get funding for their next project.

The Role of the ESA
The biggest culprit of moose declining that I found in my research was not the human beings who were trying to implement government policy because it was their job, but the flawed laws that they were charged to implement. The Endangered Species Act (ESA) itself is a large part of the reason wolves were introduced into Yellowstone National Park and Idaho and why states were unable to control wolves until 2011.

To that end, I asked a number of experts who deal with the ESA to convene in Bozeman, Montana, on January 12, 2013, to brainstorm ideas on what could be done to improve the ESA and legislative solutions that would make the ESA more effective in achieving its mission of recovering species threatened with extinction. The attendees were as follows: Arnie Dood, endangered species biologist for Montana Fish, Wildlife, and Parks; Miles Moretti, president and CEO of the Mule Deer Foundation, who has formerly served as acting director of the Utah Division of Wildlife Resources; Karen Budd-Falen, an attorney recognized as an expert on environmental policy issues; Dr. Matthew Cronin, associate professor of Animal Genetics at the University of Alaska; Ryan Benson, attorney, graduate of Harvard Law School, and co-founder and national director of Big Game Forever; and Ben Barmore, an associate at my law firm who graduated cum laude in 2011 from SMU Dedman School of Law and who worked closely with me in defending the amendment to the Endangered Species Act in the US Court of Appeals for the Ninth Circuit.

After a full day of debate, this high-powered group came up with the following problems and proposed solutions:

Problems Identified

- The failure to include state wildlife agencies and local authorities in listing and wildlife management decisions undermines support for recovery efforts under the Endangered Species Act (ESA).
- The ESA lacks a clear methodology and practice for the delisting of recovered species.
- The governmental funds spent on wildlife recovery efforts under the ESA are disproportionately spent on "charismatic" species that are not threatened at a global level.
- "Animals of greatest concern" are determined based on litigation from non-governmental organizations instead of on objective, science-based analyses.

Proposed Resolutions

After vigorous discussion and debate, the meeting's participants agreed that legislative changes are needed to enable the ESA to achieve its original purpose of efficiently and effectively recovering species threatened with extinction. Consistent with this belief, the participants agree that the following legislative changes are necessary:

- The states affected by a listing determination should be statutorily authorized as equal partners in management decisions affecting wildlife populations within their states. States possess broad police powers and primacy over fish and wildlife within their borders and will ultimately be responsible for management of the recovered species populations upon a successful delisting of the species. As such, they should be partners in wildlife management to ensure support for recovery at a local level and a smooth transition to state-only management upon delisting of the species. The cooperation between federal, state, and local governmental agencies in the Memorandum of Understanding on the management of the Mexican gray wolf experimental population that was recently dissolved in response to attacks from the Defenders of Wildlife is a good example of how an equal partnership between federal and state agencies can be achieved.

- State wildlife agencies must have an opportunity to participate in listing decisions on species with populations that are located in the state or that, with reasonable probability, will be located in the state after recovery. At a minimum, this opportunity should include a formal "vote" from the affected state on whether listing is necessary to preserve the species and consideration of any scientific data provided by the state's wildlife agencies without any special deference to the scientific evidence produced by the US Fish and Wildlife Service (USFWS). If USFWS determines that listing is necessary against the state's recommendations, a reasoned opinion supporting the decision should be required to be published by the USFWS.

- The Recovery Plan for a listed species should be required to include quantifiable recovery goals that are developed in coordination with the wildlife agencies of the affected states. Additionally, there should be statutorily mandated deadlines for preparing recovery plans, and states should be partners in development of those plans. The statute should also provide the necessary funding for state participation. The ESA should then provide a regulatory mechanism for states and non-governmental organizations to initiate delisting once the published recovery goals have been met.

- Listing determinations should be made based on threat to the species on a global level. The current distinction between "endangered" and "threatened" is meaningless and allows for the listing of species that are not in danger of extinction. The ESA authorizes the federal government to usurp the states' inherent primacy in the management of wildlife resources within their borders and so should be used sparingly in a manner consistent with our nation's founding principles. To more effectively distinguish the level of threat to a species, so as to justify listing under the ESA, there should be additional categories of "threatened" to more accurately reflect the species' danger of extinction.

- Note: Our discussion centered upon the IUCN (International Union for Conservation of Nature) categories, which include:
  - Extinct: No free-living, natural population.
  - Extinct in the wild: Captive individuals survive, but there is no free-living, natural population.
  - Critically endangered: Faces an extremely high risk of extinction in the immediate future.
  - Endangered: Faces a very high risk of extinction in the near future.
  - Vulnerable: Faces a high risk of extinction in the medium-term.
  - Near threatened: May be considered threatened in the near future.
  - Least concern: No immediate threat to the survival of the species.
- When developing these categories of "threatened," the vitality of the species should be assessed on a global level without consideration of international borders unless there is an affirmative finding that the foreign nation's policy actually threatens the species with the danger of extinction.
- A species that is not currently threatened with extinction should not be listed based solely on predictive models indicating that the species may become threatened or endangered in the future. Listing decisions should be based solely on the current status of a species.
- Listings under the ESA should be made exclusively at a species level. The determination of "subspecies" and "distinct population segments" is too subjective and controversial to constitute a meaningful, science-based determination.
- The US Fish and Wildlife Service should be required to provide an annual accounting to Congress detailing the amount of spending on recovery efforts for each species listed under the ESA.
- There should be an annual accounting of attorneys' fees awarded for actions brought under alleged violations of the ESA or NEPA (National Environmental Policy Act). This accounting should be available, and it should list each entity that received an award and the total amount of fees awarded in the calendar year. This should be reported to the GAO (Government Accounting Office).
- A litigant who wins a legal proceeding based solely on "process" under the ESA or NEPA should not be entitled to attorney's fees.
- It should be specifically provided that private property may not be listed as "critical habitat" under the ESA; instead, the protections for private landowners that enter into Conservation Reserve Program (CRP) contracts for the purpose of wildlife conservation should

be strengthened. This is necessary to provide incentives for private landowners to voluntarily cooperate in conservation efforts.

Conclusion

There are times where critical habitat designation may appropriately include private lands. There are significant numbers of endangered species that are only found on private lands. There certainly needs to be additional discussion about critical habitat and how it is applied. It has always been my belief that the statement "habitat is critical" is true and accurate; however, "critical habitat," as currently designated often rewards those who get out of the various conservation programs early and penalizes those who have stayed in the programs, done a good job, and maintained species on their lands. As far as protections for private landowners, there should be additional discussion within government agencies about validating programs such as the Safe Harbor Program, the Candidate Conservation Agreements with Assurances, and so on by securing needed flexibility within the statute.

Endnotes:

1. http://www.yellowstonepark.com/gray-wolves-impact-elk/
2. http://www.adfg.alaska.gov/index.cfm?adfg=wildlifenews.view_article&articles_id=266
3. Rheinhold Eben-Ebenau "Goldgelbes Herbstlaub" https://www.amazon.de/Goldgelbes-Herbstlaub-Reinhold-Eben-Ebenau/dp/B0000BHLLC
4. http://www.durangoherald.com/article/20130703/NEWS06/130709840&template=mobileart
5. http://wdfw.wa.gov/publications/01699/wdfw01699.pdf
6. https://www.researchgate.net/profile/Michelle_Carstensenhttp://news.nationalgeographic.com/2016/09/animal-biology-science-moose-population-disease-canada-united-states/
7. http://wolf.org/wp-content/uploads/2013/08/Re-evaluating-the-northeastern-MN-moose.pdf

Photo credit: Critterbiz/Shutterstock.com

CHAPTER 17

. . .

Canis Stew: The Inevitable Disappearance of the Pure Wolf

By Ted B. Lyon

Photo Credit: Holly Kuchera/Shutterstock.com

*"I was much struck how entirely vague and arbitrary is the
distinction between species and varieties."*

—Charles Darwin, *On the Origin of Species*

*"The domestic dog is an extremely close relative of the gray wolf, differing from it by at
most 0.2 percent of mtDNA sequence . . . In comparison, the gray wolf differs from its
closest wild relative, the coyote, by about 4 percent of mitochondrial DNA sequence.*

—Robert Wayne, PhD "Molecular evolution of the dog family"

As this book has the form of my presenting a case for managing wolves in
North America that will conserve purebred wolves and minimize conflicts with
people, I'd like to summarize some of what we've covered so far by beginning
with a story.

California once had a wolf population, but wolves were eradicated there at least a hundred years ago. Then, in 2011, a lone gray wolf, "OR7," left his pack in northeastern Oregon and journeyed 1,062 miles south, crossing into northern California. His path was monitored via radio collar. Becoming an overnight media celebrity, OR7 roamed about nine hundred miles in northern California, and then in early March of 2012, he turned north and headed back into Oregon, seemingly because he was unable to find any wolves to unite with in California. If only he would have gone a little farther south, he might not have returned to Oregon.

In August of 2015, a pack of seven wolves, two adults, and five offspring were filmed in northern California. All had coats that are very dark or black. OR7 was not one of the pack.

In December of 2015, another wolf wearing a radio collar entered California from Oregon. This canine known as OR25 was a male with a very dark coat and, was nearly three years old, according to the California Department of Fish and Wildlife.

These were reported in the news as the first appearances of wild wolves in California since the 1920s. Not quite.

In 2009, a sheep rancher in northern California contacted a USDA Wildlife Control Specialist for APHIS (Animal and Plant Health Inspection Service) as his livestock were being attacked and killed by an animal that was presumed to be a large coyote.

A large western coyote can weigh thirty pounds. The APHIS trapper put out snares and caught "two animals that weighed over a hundred pounds each that looked like wolves." A subsequent examination of the animals showed that they were wolf-dog hybrids. The trapper said that during the last ten years he had trapped several dozen similar animals. He said that people who grow marijuana frequently use wolf-dogs as guard dogs for legal and illegal crops. He added that there is also strong evidence that some people in northern California are raising and releasing wolves and wolf-dogs to create a California "wolf" population.

A US Fish and Wildlife Service Special Agent commented, "If they were 100 percent wolves, catching and killing them without a proper permit would have been breaking the law. Since they were wolf-dog hybrids, the trapper was following the local law that says that any dog or coyote attacking livestock can be shot on sight without needing a hunting license."

This incident foreshadows a problem that in the long run seems inevitable. Wild gray wolves will become ineligible for protection under the Endangered Species Act, regardless of their numbers, because they will breed with other canids, both wild and tame, resulting in the disappearance of purebred wolves.

CANIS STEW ·· 267

Tracing the Origins of *Canis lupus*

According to fossil record, wolf-like canids—wild dogs, or what we now call wolves—began to appear at least three million years ago. Originally, there were three species of canids in North America—the gray wolf, *Canis lupus*, the dire wolf, *Canis dirus* (a stocky carnivore with large teeth and jaws for crunching bones of megafauna and nearly twice the body size of the gray wolf) and the coyote, *Canis latrans*. For reasons we do not understand, the gray wolf, *Canis lupus*, migrated to Asia over the Bering Straits land bridge and spread throughout Eurasia. Ultimately, they became the most widely distributed predatory mammal on Earth. Later in the Paleolithic, gray wolves migrated back over the land bridge into North America.

The three wild dogs, *Canis lupus*, *Canis dirus*, and *Canis latrans*, lived together for about ten thousand years. The dire wolf ultimately went extinct about ten thousand years ago, seemingly along with the extinction of the megafauna.

After the recession of the Wisconsin Glacier some ten thousand years ago, smaller wild game became plentiful and gray wolves proliferated. There may have been as many as two million gray wolves in North America when the Europeans arrived in the 1500s. The gray wolf numbers were so large because of the abundance of big-game animals, especially the sixty million buffalo.

Lewis and Clark reported large packs of wolves following the buffalo herds of the Great Plains. As Meriwether Lewis wrote in his 1805 journal, "we scarcely see a gang of buffalo without observing a pack of these faithful shepherds on their skirts . . ." They also reported that wolves were relatively unafraid of people, and on occasion wolves attacked members of their expedition.[1]

As a wave of settlers pressed westward, the buffalo was nearly driven to extinction by market hunters and as a planned mass killing by the military to force Indians onto reservations. The European settlers replaced the dwindling herds of native buffalo, elk, and deer with sheep and cattle. Wolves switched their predation to domestic livestock, which resulted in a campaign to eradicate wolves, ultimately leading to the virtual extinction of wolves in the lower forty-eight states by the 1940s. While the wolves were nearly removed completely, their brother, the coyote, prospered.

Canis latrans, the coyote, which is normally one-third or less the size of an adult gray wolf, was originally found only west of the Rocky Mountains. Despite the eradication campaign that killed off wolves, and equal efforts to control coyotes, the coyotes proliferated, spreading eastward. They are now found in all states but Hawaii, as well as all Canadian provinces. Their population is estimated to be somewhere between ten million and one hundred million, even though as many as four hundred thousand coyotes are shot, trapped, and poisoned every year.

As a number of chapters have described, when wolves meet coyotes, or dogs, one of two things normally happens. The wolves attack the coyotes and wolves and drive them away, or they kill them. However, some wolves, especially beta wolves without mates, will breed with coyotes creating "coy-wolves," or dogs, resulting in "wolf-dogs" and quite possibly "wolf-coyote-dogs."

The word *Canis* means "dog" in Latin. As gray wolf populations are being restored in the Northern Rockies, Pacific Northwest, Great Lakes Area, New England, Southeast, and Southwest, this generation of wolves is encountering unprecedented numbers of coyotes as well as tens of millions of domestic dogs, tens of thousands that are feral. Interbreeding and hybridization will become the norm for wild canids.[2]

There are somewhere between seventy and eighty million domestic dogs in the United States; 39 percent of the households own at least one. The *Canis* genus contains seven to ten living species, including dogs, wolves, coyotes, and jackals, and many debatable subspecies. They are all capable of interbreeding and producing fertile offspring.

Never before in the United States have wolves had access to so many canids other than their own species to mate with, and that spells trouble for preserving wolves as a distinct species as well as problems for people trying to live with wolves.

The Evolution of Dogs from Wolves

The wolf has seventy-eight pairs of chromosomes, but there can be millions of variations. There is not a standard allele. Normally, through successive generations of breeding, mutations occur, which combined with adaptations for changing environmental conditions, result in new self-replicating species or sub-species forming. In contrast to earlier times, within the genus *Canis*, hybridization today is occurring very rapidly, especially in the Lower 48.

Domestic dogs are descended from wild wolves, *Canis lupus*. The Latin name for the dog is *Canis lupus familiaris*. The domestication process began somewhere between 60,000 and 135,000 years ago. Humans first used wolves for food and fur. Domestication most likely began when humans raised very young wolf puppies. Some animals were friendlier than others were, and could be tamed. This resulted in them performing tasks such as hauling loads, serving as watchdogs, and becoming hunting allies. Animals with desirable behavioral and morphological traits were bred with others with similar traits to create new unique breeds. This process of selective breeding has been demonstrated by Russian scientist Dmitry Belyaev in experiments to breed foxes in Siberia to form a new subspecies.[3]

Just what constitutes a "species" is hotly debated in scientific circles. John Wilkins has listed twenty-six different "species concepts" that scientists may use to determine what is a species of animal or plant.[4] Originally morphology, or the physical structure, was the key, but now with DNA analysis one must also look at genetics, and geneticists cannot agree on species identification either. The problem with canid species ID is that all canines can interbreed with other species of canines. Thus, the concept of preserving a "species" becomes difficult when one cannot agree on what it is, and its genetics are constantly changing from hybridization (this will be described in more detail in the following chapter by Dr. Matt Cronin).

The US Fish and Wildlife Service is currently monitoring several sub-species of gray wolves, *Canis lupus*, in a number of locations under the Endangered Species Act: *Canis lupus lycaon* (eastern timber wolf), *Canis lupus irremotus* (northern Rockies subspecies of gray wolf), *Canis lupus baileyi* (Mexican wolf), *Canis lupus monstrabilis* (Texas gray wolf), and the red wolf (*Canis rufus* or *Canis lupus rufus*), also known as the Florida wolf or Mississippi Valley wolf, which is found in the southeastern United States.

Some of these populations are breeding more successfully than others, especially the eastern timber wolf and the Northern Rockies wolf. There are also reports of Mexican wolves and Northern Rockies wolves connecting, as well as Mexican wolves breeding with dogs and coyotes. In the not-too-distant future, it seems very likely that there will be one genetic pool for all wild *Canis* species and subspecies, which will include genes from dogs, feral dogs, wolf-dogs, and coyotes as all members of *Canis* species can interbreed. There is also some evidence that wolves and foxes may interbreed in the United States as has been shown in Russia.

In 1978, USFWS classified the gray wolf as an endangered population at the species level (*Canis lupus*) throughout the lower forty-eight states and Mexico, except for the Minnesota gray wolf population, which was classified as threatened.[5] This classification based on location, not populations in the wild, has some real problems.

In northern Canada and Alaska, wolves are common. If the definition of what is "endangered" varies by region, and not the total actual size of the population of an animal, then could one claim that wolves should be "endangered species" in regions where they are not found, such as New York City, Los Angeles, or Washington, DC, except in zoos? Coyotes have already settled into parks in Chicago and Los Angeles, and have been sighted in Central Park in New York in 2015. Should coyotes be considered "endangered" in Central Park?

Species Conservation and the Endangered Species Act

The Endangered Species Act of 1973 (ESA) is the third in a series of acts designed to preserve wildlife species in danger of extinction.[6] The predecessors were the 1966 Endangered Species Preservation Act and the 1969 Endangered Species Conservation Act.

Richard Nixon signed the ESA into law on December 28, 1973. Its stated purpose is to protect species and "the ecosystems upon which they depend." ESA's primary goal is to prevent the extinction of imperiled plant and animal life, and to recover and maintain those populations by removing or lessening threats to their survival.[7]

According to ESA, "The term 'species' includes any subspecies of fish or wildlife or plants, and any distinct population segment of any species or vertebrate fish or wildlife which interbreeds when mature." The term "species" as used in the ESA has "vernacular, legal, and biological meanings."[8] The law defines species vaguely: "Species: includes any subspecies of fish or wildlife or plants, and any distinct population segment of any species of vertebrate fish or wildlife which interbreeds when mature . . . based solely on the best scientific and commercial data available."[9]

As per 16 USC. § 1532, an "endangered" species is one that is "in danger of extinction" throughout all or a significant portion of its range. A "threatened" species is one that is "likely to become endangered" within the foreseeable future.

To be considered for listing, the species must meet one of five criteria (section 4(a),

1. There is the present or threatened destruction, modification, or curtailment of its habitat or range.
2. An over-utilization for commercial, recreational, scientific, or educational purposes.
3. The species is declining due to disease or predation.
4. There is an inadequacy of existing regulatory mechanisms.
5. There are other natural or man-made factors affecting its continued existence.

The law now reads: "The Secretary may treat an unlisted species as listed if: it so closely resembles a listed species that enforcement personnel would have substantial difficulty in attempting to differentiate between the species; the effect of this difficulty is an additional threat to the listed species; this treatment will substantially facilitate enforcement and further the Act's policy. Regulations must be issued to provide for the conservation of threatened species."

One of the chief concerns of species preservation is whether the population number is too small, which will result in in-breeding, and lead to loss of health and ultimately extinction of the species. The gray wolf was placed on the Endangered Species List in 1974, and aside from a brief delisting in

2009 in Montana and Idaho, it remained there until delisting was done by amending the Endangered Species Act in 2011, despite numerous efforts to oppose delisting.

This concern is currently very much related to the red wolf, found in the wild only in North Carolina, where there are forty-five to sixty red wolves, and to the Mexican wolf, found only in the wild in Arizona and New Mexico, where there are about one hundred in the wild. (There are several hundred red wolves and Mexico wolves in captivity.)

Some wolf advocates worry about species extinction due to wolves being killed or dying from disease. The current North Rockies gray wolf population is at least five and a half times higher than the minimum population recovery goal and about three and a half times higher than the breeding pair recovery goal; and the eastern timber wolf population is no longer listed, according to the USFWS.[10] In reality, the greatest long-term threat to the preservation of the gray wolf may be hybridization.

Canid Hybridization—Wolf-Dogs

A Google search for "Wolves for Sale" in September of 2016 list, around 3,550,000 results. There are an estimated three hundred to five hundred thousand wolf-dogs in the United States in captivity, either as pets or in educational centers and sanctuaries. Adding wolf-coyote-dogs with wolves and wolf-dogs not in the wild, results in at least half a million canines in North America that have at least some wolf blood.

Anything less than a 100 percent wolf is considered a wolf-dog.[11] Currently there are only two states that have no laws regarding the ownership of a wolf or wolf hybrid, twenty-one states do not allow any ownership of a wolf or hybrid, and the remaining twenty-seven states require owners to obtain varying degrees of permits for the ownership of a wolf or hybrid.[12]

These restrictions are due to the hybrids being considered dangerous. They have the size and strength of wolves, seem to have less wild instinct, but are unpredictable. A number have been released into the wild by people who find that after two or three years the animal cannot be controlled.

According to Richard Polsky, a dog bite and wolf hybrid expert: "Wolf hybrids retain many wolf like characteristics which make their behavior unpredictable in the human setting. As a result, attacks upon humans likely happen at disproportionately high rates, and there are documented accounts of fatal attacks by wolf hybrids on people in the United States."

Polsky goes on to say that between 1979 and 1998 the top five dog breeds that were involved in dog bite-related fatalities were pit bulls, Rottweilers, German shepherds, Siberian huskies, and wolf-dogs. By compiling US and Canadian press accounts between 1982 and 2011, Animal People listed the following breeds and associated outcomes for dog bites. The combination of pit bulls, Rottweilers, their close mixes, and wolf hybrids yielded the following results: 77 percent of attacks that induce bodily harm; 73 percent of attacks to children; 81 percent of attacks to adults; 68 percent of attacks that result in fatalities; and 76 percent that result in maiming. Wolf-dogs topped the dog bite–related fatality list for the known crossbred dogs.[13]

An example of the dangers of wolf-dogs, the Wolf Hybrid Attack and Bite Statistics from 1982 through 2013 are as follows: bodily harm, eighty-five attacks; child victims, seventy attacks; adult victims, five attacks; deaths, nineteen deaths caused by a wolf hybrid; and maimings, forty-nine.[14]

Examples of some of these attacks include:

- June 29, 2015—Fredericksburg, Virginia—a four-month-old boy was bitten by wolf-dog hybrid. The child suffered "17 punctures to his rear abdomen and three to the front abdomen" plus "an abdominal evisceration," a medical term for disemboweling, according to the Inova Fairfax Hospital.[15]
- July 2010—Lincoln Park, Illinois—Five-year-old Kyle Holland was attacked while he slept by a German shepherd-wolf–hybrid. The boy died after he was "partially eaten" by the hybrid. The Wayne county prosecutor's office charged the dog's owner with manslaughter under "gross negligence theory," because he knew the dog was dangerous and could harm Kyle.[16]
- July 23, 2009—Lexington, Kentucky—A wolf-dog hybrid, Dakota, badly injured three-day-old Alexander James Smith. Michael Smith, Alexander's father, said Dakota was "a Native American Indian" breed and said the breeder told him the dog was "90 percent wolf." Ray Coppinger, a professor at Hampshire College who studies canine behavior, said, "Dogs like Dakota don't recognize infants as people. It's no more of an act of violence on the dog's part, than you eating a steak."[17]
- September 1, 2006—Pittsburgh, Pennsylvania—Wolf-dogs killed owner Sandra L. Piovesan who bled to death after being mauled by a pack of nine wolf-dogs that she had raised as pets. She did not have the required license from the Pennsylvania Game Commission required to own, breed, and sell wolves and wolf-dog hybrids.[18]
- April 2003—Boise Idaho—Child killed by wolf-dog hybrid. A thirteen-month-old boy died after a wolf-dog hybrid bit him and tore the baby's jugular vein at a Boise home. Andre Angel Thomas was bitten by the animal "hundreds" of times at the Frederic Street home, said Boise Police Detective Dale Rogers.[19]

- October 21, 2000—Cincinnati, Ohio—A five-year-old Cincinnati boy was killed Saturday morning when his grandmother's pet wolf hybrid attacked him at her home here. The animal was shot and killed after the attack.[20]
- August 21, 1999—Muskegon, Michigan—Four-year-old Cody Tyler Fairfield was playing in the backyard of his home in. Also occupying the yard was the Fairfield family's German shepherd–wolf hybrid, tethered to a chain. Left alone for a moment, Cody approached the family pet. Within minutes, the child was dead, his throat crushed, his trachea punctured.[21]
- September 18, 1991—Spokane, Washington—For the second time in a year, a five-year-old girl is attacked by a wolf–dog hybrid. The first attack required forty stitches to close the wounds in the upper chest and back. The second attack required eighty stitches to close all wounds to head, face, and back.[22]

Despite the warnings, some wolf advocates who own pet wolves and wolf-dogs will extoll about how wonderful their pets are. Recognizing the inherent dangers, some of our experts were contacted and they suggested the following books about how to live with socialized wolves based on solid scientific work including:

- *Wolf Park Guidelines for Keeping Wolves and Wolf-Hybrids* Hary Frank (ed.) 1987.
- *Man and Wolf.* Kluwer Academic Publishers, Dordrecht, The Netherlands.
- Raymond Coppinger and Lorna Coppinger 2001, *Dogs*, Scribner, New York. See Part I dealing with a contrast of wolf and dog characteristics.
- Jerome H. Woolpy and Benson E. Ginsburg. 1967. "Wolf Socialization. A study of Temperament in a Wild Species." *American Zoologist* 7:357–363.

When Wolf-Dogs Become Feral

Feral dogs of all types are a growing problem in many areas of the United States. When domestic dogs become wild, they lose their domestic behavior, form packs, and become secretive, surviving as opportunistic predators as well as scavengers, much like habituated wolves and coyotes.[23] They can be a danger to livestock and people.

In February of 2012, a rancher in Valencia County, New Mexico, reported that he and his employees have killed more than three hundred dogs over the past eighteen months. The feral dogs were attacking young calves and weak cattle for sport.[24]

The New Mexico rancher is not alone. In 1999, the National Agricultural Statistics Service reported that feral dogs were partly responsible for killing cows, sheep, and goats worth about thirty-seven million dollars a year.[25]

In 1997, the State of Colorado was faced with the task of how to handle a growing number of wolf-dogs in the state. The state veterinarian convened a special committee to study the problem and come up with suggestions. Based on advice from Dr. Ray Pierotti, a geneticist at University of Kansas, Dr. Nick Federoff, wildlife biologist, and Dr. Erick Klinghammer, ethologist from Wolf Park in Indiana, the report concluded: "All forms of wolf-dog identification are problematic. There is no genotype (the genetic constitution of an animal) or phenotype (the observable appearance of an animal) to distinguish between a dog, a wolf-dog cross, or a wolf. All DNA tests to differentiate wolf hybrids from domestic dogs are subject to challenge. There are no known DNA markers uniquely distinguishable in wolves that are not present in dogs. Blood tests, skull measurements, and skeletal measurements all have some merit but have not withstood legal challenge."[26]

Many wildlife forensic scientists agree that morphology alone cannot be used to determine whether an animal is a purebred or a hybrid. The only way that one can tell the species of an animal is with DNA analysis, which today has become sophisticated, but as the Colorado report states, with that complexity comes disagreement among scientists about what is a species.[27]

It has been known for well over two decades that Rocky Mountain gray wolves have interbred with domestic and wild dogs.[28] The DNA of Rocky Mountain gray wolves is currently being tested, and the scientific community is very aware that hybridization has biological, forensic, legal, and social implications.[29]

Genetic research from Stanford University and UCLA has already revealed that wolves with black pelts owe their distinctive coloration to a mutation that occurred through wolf-dog hybridization.[30]

Wolves normally will kill coyotes, but they also can breed with them, creating a coy-wolf. Geneticists say that eastern timber wolves, found in Ontario, Quebec, and Nova Scotia, are primarily coy-wolves, so just what you call these animals is a matter of choice.[31]

The reality, of course, is that the "eastern coyote" of New England and south-eastern Canada contains the DNA of western coyotes and eastern wolves, so you call it either an "eastern wolf" or an "eastern coyote." According to researchers Way, Rutledge, Wheeldon, and White, the "eastern coyote" should be called a "coy-wolf" as eastern wolves have at least 25 percent coyote genes.[32]

Several facts that further bolster this chapter's thesis:

- Some "wolves" in Oregon also have been found to be hybrids with coyotes and dogs.[33]
- The "red wolf" of the Southeast, *Canis rufus*, is already thought to be a coy-wolf by many scientists.[34]
- Similar research also finds hybridization taking place in the Northern Rockies.[35]
- Hybridization of wolves and dogs is also found in Europe.[36]
- It seems inevitable that in time the wolf species we know today will disappear into a hybridized gene pool that might best be simply called "wild canid hybrid."

"Wolves" and the Endangered Species Act

The Endangered Species Act is meant to protect species that can self-replicate. When one species breeds with another species, the resulting hybrid becomes problematic for Endangered Species Act protections. This is called the "hybrid policy," originally a rigid standard that states, "Protection of hybrids would not serve to recover listed species and would likely jeopardize that species' continued existence."[37] As DNA analysis has become more sophisticated, and species identification has become more complex, the US Fish and Wildlife Service have rescinded this position.

There is currently no official US policy to provide guidance for dealing with hybrids.[38] The questions of what a wolf really is can only be determined by a DNA analysis, and DNA analysis finds that 100 percent purebred wolves are growing increasingly rare due to hybridization.

A harbinger of the future, hybrid wild canids in general seem to be growing in size and number. In 2004, a wolf-dog was killed in Pennsylvania that weighed 105 pounds.[39] In 1996, an eighty-six-pound male wolf-like canid was killed in Maine that was later confirmed to be a wolf-coyote hybrid. In 1997, a seventy-two-pound coy-wolf was killed in Vermont.[40] And in 2010, a 104-pound "coyote" was killed in Missouri.[41]

As wolves co-exist with millions of dogs and coyotes, as they do today throughout North America, inter-species breeding will inevitably occur, and so hybridization of wolves will most certainly increase. It seems practical to not list wolves as an endangered species, for the only way to tell what is a wolf requires an expensive and time-consuming DNA analysis that can only be done when the animal is dead or captured. The number of "questionable wolves" being sent to the US Fish and Wildlife Service Forensic Laboratory in Ashland, Oregon, is growing. And as the Colorado report concluded, even with the best methods and equipment, the test may not always be completely conclusive.

A recent breakthrough in genetics, however, sheds light on the DNA makeup of wolves. An extensive investigation of the genetics of wolves was reported in the journal *Science*, in July 2016, that involved nine different scientists at nine different universities and research centers in the United States, China, and Israel. The researchers conclude that two endemic species of canids, the coyote and the American wolf, are pure species, both abundant in the United States and Canada, but the eastern wolf (found originally in the eastern United States and Canada) and the red wolf (found in the southeastern United States) are hybrids of coyote and wolf. Both species are smaller than true wolves and larger than coyotes. This genetic evidence has been confirmed by interspecies breeding in research facilities.

Additionally, it is the contention of the researchers, supported by research on the genetics of Great Lakes and western gray wolves, as well as Mexican wolves and coyotes, using five distinctly different methods, that there are few truly pure wolves in the wild today in the lower forty-eight states, and as existing wild wolf populations spread out throughout the United States, these animals will intermingle with feral dogs, domesticated dogs, and coyotes, resulting in what could be called "canis stew;" this is the genetic makeup of at least two, three, or four canid species. Such hybridization was also found to occur in Eurasia between the Eurasian gray wolf and dogs, and in Africa between wolves, dogs, and jackals. They also report finding that coyotes in several regions in the United States have gray wolf DNA. They also found that all North American wolves and coyotes have significant amounts of coyote ancestry.

Since the Endangered Species Act does not cover hybrids, these two hybrid subspecies of canids cannot be protected as Endangered Species, and there is doubt as to how many wild canids are truly purebred.

It seems inevitable that the "pure wolf" will become increasingly rare, especially as it spreads throughout the Lower 48, and management ultimately will shift from protecting "endangered species wolves" to controlling wolf-like apex predator hybrid wild dogs. Currently such hybrids can be shot on sight if they are chasing wildlife, threatening human safety, and preying on domestic livestock.

If purebred wolves do exist, and some scientists question their existence, it seems that the only way to preserve the genome of *Canis lupus* is to impede the contact of purebred wolves with dogs and coyotes, which is probably impossible. In his previous chapter, Dr. Valerius Geist suggests that perhaps the best way to preserve purebred wolves is to keep them on special enclosed areas, like fenced military reserves, or at special sanctuaries with strong fences and enough room to allow for natural predatory behavior on wildlife to take place.

As stated in previous chapters, if zoos and other small enclosures are used to preserve purebred wolf breeding stock, it is not likely that these wolves could be released into the wild, as they would be habituated. Their capacity to survive without preying on livestock and interacting with ranches and farms and their dogs, as well as with feral dogs, is questionable. In the spring of 2016, the USFWS took two newborn Mexican wolf puppies born in captivity and released in western New Mexico into a den with wolf pups—an effort that was opposed by New Mexico Game and Fish.[42] New Mexico Game and Fish has sued USFWS in response.[43]

The only way to know for certain if an animal is a pure wolf is through DNA analysis, and it seems only reasonable to expect that purebred wolves will soon, if not already, only exist in very remote regions of the far north or in enclosures that prevent interaction with all other canids.

It seems very clear that wolf politics and wolf biology do not hybridize very well. The next chapter describes the genetics of wild canines in more detail, again raising the question of what is a real wolf and how this in turn influences management of all canines in the wild, including which deserve protection under the Endangered Species Act.

"The wolf is neither a saint nor a sinner except to those who want to make it so."[44]

—David Mech

Endnotes:

1. http://fwp.mt.gov/mtoutdoors/HTML/articles/2006/LCmisadventures.htm
2. http://news.nationalgeographic.com/news/2003/08/0821_030821_straydogs.html
3. http://www.abc.net.au/animals/program1/factsheet5.htm
4. http://scienceblogs.com/evolvingthoughts/2006/10/a_list_of_26_species_concepts.php
5. http://ecos.fws.gov/speciesProfile/profile/speciesProfile.action?spcode=A00D
6. http://www.fws.gov/endangered/laws-policies/index.html
7. http://www.nmfs.noaa.gov/pr/pdfs/laws/esa_section3.pdf
8. Natural Research Council. "Science and the Endangered Species Act." *National Academy Press*, 1995:5.
9. http://wildlifelaw.unm.edu/fedbook/esa.html%205-YR%t20review%20PDF.pdf
10. http://ecos.fws.gov/docs/five_year_review/doc3978.%20lupus%205-YR%20review%20PDF.pdf
11. http://www.wolf-dogalliance.org/legislation/statelaws.html
12. http://www.bornfreeusa.org/b4a2_exotic_animals_summary.php
13. http://www.bradfordlicensing.com/documents/pets-fact-sheet.pdf; and https://www.dogexpert.com/wolf-hybrid-aggression-behavior/ and Sacks, J. *Breeds of dog involved in fatal human attacks in the United States between 1979–1998.* J. American Veterinary Medical Association, 2000, Vol. 217, 836–840.
14. https://pethelpful.com/dogs/Most-Dangerous-Dog-Breeds
15. http://www.fredericksburg.com/news/local/culpeper-baby-bitten-by-wolf-dog-hybrid-is-out-of/article_830b258a-1ea8-11e5-ac3e-6733f7144833.html

16. http://www.dailymail.co.uk/news/article-2009147/Kyle-Holland-5-died-partially-eaten-wolf-dog-Lincoln-Park.html
17. http://www.seattletimes.com/life/lifestyle/wolf-hybrids-scrutinized-after-pet-takes-baby-from-crib/
18. http://www.post-gazette.com/local/westmoreland/2006/07/19/Wolf-dogs-killed-owner-autopsy-determines/stories/200607190197
19. http://www.wolf-to-wolf-dog.org/boise.htm
20. http://www.enquirer.com/editions/2000/10/22/loc_wolf_hybrid_kills.html
21. https://www.petfinder.com/pet-adoption/exotic-pets/case-against-hybrids/
22. https://news.google.com/newspapers?nid=1314&dat=19910918&id=I1tWAAA AIBAJ&sjid=5-8DAAAAIBAJ&pg=2839,5322564
23. http://icwdm.org/handbook/carnivor/FeralDog.asp
24. http://www.cattlenetwork.com/cattle-news/New-Mexico-ranchers-protect-cattle-from-wild- dogs-139392968.html
25. http://icwdm.org/handbook/carnivor/FeralDog.asp
26. http://www.naiaonline.org/articles/archives/coloradotask.htm
27. http://onlinelibrary.wiley.com/doi/10.1046/j.1523–1739.1992.06040559.x/ abstract http://www.springerlink.com/content/kw1132542l0p2mgn/
28. http://www.rmrs.nau.edu/publications/pilgrim_et_al_1998/pilgrim_et_al_1998.pdf
29. http://www.jstor.org/pss/3783983
30. http://www.nytimes.com/2009/02/06/science/06wolves.html?_r=1&ref =todayspaper
31. Kyle, CJ. AR Johnson. BR Patterson. PJ Wilson. K Shami. SK Grewal. And BN White "Genetic Nature of Eastern Wolves: Past Present and Future." *Conservation Genetics*, 7: 273–287.
32. http://easterncoyoteresearch.com/downloads/GeneticsOfEasternCoywolf FinalInPrint.pdf and http://www.livescience.com/55586-wolves-only-one-species.html
33. http://www.defenders.org/publications/northern_california southwestern_oregon_gray _wolf_dps.pdf
34. http://www.scientificamerican.com/article.cfm?id=what-is-a-species; http://www.jstor.org /pss/2386371and http://www.nal.usda.gov/awic/newsletters/v5n4/5n4wille.htm
35. http://www.jstor.org/pss/3802344http://www.jstor.org/pss/3783983
36. http://www.nature.com/hdy/journal/v90/n1/full/6800175a.html
37. O'Brien, S. J. and E. Mayer. "Bureaucratic Mischief: Recognizing endangered species and subspecies". *Science*, March 8, 1991.
38. Ellstrand, Norman C., David Biggs, Andrea Kaus, Pesach Lubinsky, Lucinda A. McDade Kristine Preston, Linda M. Prince, Helen M. Regan, Veronique Roriver, Oliver A. Ryde, Kristina A. Schierenbeck. "Got Hybridization? A Multidisciplinary Approach for Informing Science Policy." *Bioscience.* May, 60:5.
39. http://www.snopes.com/photos/animals/coyote.asp
40. Adkins, Collette L. "The Big Bad Wolf Hybrid: How Molecular Genetics Research May Undermine Protection for Gray Wolves Under the Endangered Species Act." *Minnesota Journal of Science and Technology,* 6:2.
41. http://mdc.mo.gov/newsroom/hunter-shoots-unusually-large-coyote-northwest-missouri

42. http://www.abqjournal.com/765868/fish-and-wildlife-releases-2-wolf-pups-in-nm
    .html

43. http://www.abqjournal.com/759957/feds-to-release-wolves-in-nm-despite-state-
    opposition.html

44. http://www.sciencedirect.com/science/article/pii/S0006320712001462

Photo credit: critterbiz/Shutterstock.com

CHAPTER 18

· · ·

What Is a Wolf?: Classification of North American Wolf Species, Subspecies, and Populations

By Matthew A. Cronin, PhD

Photo credit: Holly Kuchera/Shutterstock.com

Introduction

The previous chapter describes the problems with hybridization of wolves and how this influences their management. In this chapter I describe how and why the scientific community is so divided on wolf taxonomy.

In order to know how to best manage wildlife, it's essential to know what you are managing. For most people that means taking out a field guide and comparing what you are seeing with what the book says.

The forerunners of today's field guides date back at least to Herodatus, who compiled a book of beasts in the fifth century BC. His work was expanded on about two hundred years later by Aristotle in *Historia Animalium*, which stood as the definitive work in the field the publication of a second-century text by Physiologus, a pen name. In the first century AD, Pliny the Elder wrote *Historia Naturalis*, which marked the founding of the science of natural history.

In the twelfth and thirteenth centuries, science and religion markedly separated and heatedly sparred for control of the central values of culture. Catholic priests working with the blessings of Albert the Great, Saint Thomas of Aquinas, and others sought to bring together the collected wisdom about the animal kingdom into voluminous, handwritten books called bestiaries. Drawing upon earlier works, folk tales, hearsay, personal accounts, and scholarly interpretations, bestiaries become monumental tomes, handwritten in Latin and often richly illustrated. Works of art as much as science, bestiaries were largely penned from secondhand sources rather than personal field studies of the authors.

For the Christian monks of the Middle Ages, recording the kingdom of nature for posterity was also way to guide new generations away from pagan religious beliefs about animals, for wild creatures were often the gods of the old religion. Thus the monks who penned bestiaries often advanced interpretations of animal behavior that supported Christian teachings as much or more than fact. Animals that had especially offensive qualities were made agents of the devil. The crocodile stood for the sin of pride, the viper for lust, and the partridge for covetousness, along with other familiars of Lucifer including the monkey, the wolf, the fox, the ass, the goat, the dragon, and the whale. (The whale was included as toothed whales could attack boats and tear them apart.)

A breakthrough for science in interpreting the meanings of animals came in 1732 through the fertile mind of twenty-five year-old Carolus Linnaeus. In 1735 Linnaeus published his thesis about categorizing animals according to their physical structure in *Systema Naturae*, forever influencing the nature of scientific thinking by a system of classification into kingdom, phylum, class, order, family, genus, species, and variety. In modern times "varieties" have been renamed "subspecies" for animals.

The wolf was first classified as a species (Latin name *Canis lupus*) by the Swedish naturalist Linnaeus in 1758. This included all of the wolves worldwide with a range that includes most of the Northern Hemisphere north of 30 degrees latitude, and south into India, Pakistan, the Middle East, and North Africa.

Since Linnaeus, classifications of wolves have included numerous sub-species, and recently new species and populations in North America. These classifications are important because species, subspecies, and populations (i.e., distinct population segments, or DPS) are all considered in the definition of "species" by the US Endangered Species Act (ESA).

The US Fish and Wildlife Service (FWS) and National Marine Fisheries Service (NMFS) implement the ESA and determine if a group qualifies as a species, subspecies, or population, and if it is threatened or endangered with extinction. The National Research Council has addressed taxonomic (i.e., classification) issues related to the ESA and notes that there is considerable debate among scientists as to what qualifies as a species, subspecies, or DPS.[1]

In this chapter, I review the current taxonomic status of wolves in North America beginning with a description of how scientists classify species, subspecies, and populations, and describe what has been happening on this topic, especially with regard to wolves. A key point I will emphasize is that designation of species, populations, and subspecies in particular, is not scientifically definitive.

Taxonomy

To understand wolf classification we must understand the science of biological classification, known as taxonomy. The basis of taxonomy is that organisms (plants and animals) are grouped according to phylogeny, which is simply the genealogy or ancestry of a group. In simple terms, we classify animals according to common ancestry, so two groups with recent common ancestry are classified together, and groups with more distant common ancestry are classified as different groups. This has been recognized since before Darwin.[2] Groups in taxonomy are called taxa (singular taxon). The basic groups of taxonomy are kingdom, phylum, class, order, family, genus, and species. For example, wolves are in the animal kingdom, chordate phylum, mammal class, carnivore order, *Canidae* family, *Canis* genus, and *lupus* species. There are sub-levels of each of these levels as well, including subspecies, so a wolf subspecies, such as Mexican wolves, is further classified with a third Latin name (*Canis lupus baileyi*). For many plants, a category "variety" has been substituted for subspecies.

There are other methods of classifying animals besides using ancestry. For example, one might classify animals as aquatic (living in water) or terrestrial (living on land). In this case, aquatic fish and whales would be classified together, but separate from land mammals. However, we know that all mammals, including whales, share a more recent common ancestry than any mammal does with fish, so in biology, aquatic whales are classified as mammals along with terrestrial mammals. Another example is aquatic sea otters and river otters. They are classified as mustelids along with the terrestrial weasels, mink, fisher, marten, skunks, badger, and wolverine, and not with the aquatic, but distantly related, seals, sea lions, and walruses.

Another non-taxonomic classification is the designation of "ecotypes," which are groups of populations of a species with similar characteristics due to local adaptations or environmental influence, not shared ancestry. Hence, ecotypes are not a part of taxonomic classification. An example is caribou (*Rangifer tarandus*), that include high arctic, mountain, forest, and barren-ground tundra ecotypes. These ecotypes have different characteristics in different ecological conditions but members of an ecotype do not necessarily share common ancestry and should not be classified as taxonomic groups.[3]

Wolves are now frequently classified as ecotypes, which sometimes confuses taxonomy.[4]

Species

A very important level of taxonomy is the species. Species are classified in two related, but subtly different, ways: by phylogeny (i.e., ancestry) and by reproductive isolation. There is a huge literature on what is known as the "species problem" in biology. The problem is that because closely related groups may interbreed, identification of species as distinct groups is not always possible. To identify species, it must be determined if two populations in different geographic areas share common ancestry and can interbreed. This is complicated because populations may interbreed, but the offspring may be inviable, infertile, or have reduced fitness. For example, horses (*Equus caballus*) and donkeys (*Equus asinus*) can interbreed but produce sterile mules as offspring. Horses and donkeys are considered different species because of this incompatibility. In some cases, wild populations have non-overlapping ranges and interbreeding is not testable. This leads to uncertain species designations. For example, if there are two wolf populations in different geographic areas that differ in size or coat color, should they be one species or two? The problem of uncertain species designations is succinctly described by natural historian J. C. Avise as: "In intermediate situations (and also in hybrid settings), educated nomenclatural judgments will remain necessary at species and subspecies levels."[5]

Species is an important taxonomic level because species are groups that are separate and will remain so. If two populations are isolated and then reconnect and interbreed freely, then they are not species but simply populations of the same species. Different species concepts are used by scientists, and include the "biological species concept" (BSC) in which reproductive isolation defines species. BSC defines species as groups of interbreeding natural populations that are reproductively isolated from other such groups.[6]

The BSC is the most commonly used species concept. Another species concept is known as the "phylogenetic species concept" (PSC) and considers a species to be a group with an independent ancestry, without consideration of reproductive isolation. Under the PSC, groups that share a most recent common ancestor are considered members of the same species. The PSC is most useful for very distinct groups (e.g., wolves and foxes), and not for closely related groups (e.g., wolves and coyotes). Regardless of how they are designated, species are considered groups that are now, and will remain, separate. Clear identification of species definition is important with regard to wolves because of the designations of the eastern wolf and red wolf as species in a recent review by FWS biologists[7] and subsequent reassessments.[8]

Subspecies

As noted above, subspecies are groups classified below the species level, and two definitions are commonly used. According to evolutionary biologist Ernst Mayr: "A subspecies is an aggregate of phenotypically similar populations of a species inhabiting a geographic subdivision of the range of the species and differing taxonomically from other populations of the species."[9]

Another interpretation by Avise and Ball: "Subspecies are groups of actually or potentially interbreeding populations, phylogenetically distinguishable from, but reproductively compatible with other such groups. Importantly, the evidence for phylogenetic distinction must normally come from the concordant distributions of multiple, independent, genetically based traits."[10]

These technical definitions simply mean that there must be genetically based traits that differ between populations for them to qualify as subspecies. However, the fact that subspecies are members of the same species and can interbreed results in subspecies that share genes and hence are not definitive groups. The scientific community recognizes the subjectivity of subspecies, as exemplified in Avise's quote above and the following, first by Mayr, who stated that "This concept of the subspecies is fallacious. . . . The better the geographic variation of a species is known, the more difficult it becomes to delimit subspecies and the more obvious it becomes that many such delimitations are quite arbitrary."[11]

In another view, scientist Paulo Vanzolini noted that " . . . present applications of the subspecies concept are uneven, frequently undocumented, and lead to no improvement of either evolutionary theory or practical taxonomy."[12]

Biologist Douglas Futuyma noted that there is so much variation among populations of most species that some combination of characters will distinguish each population from others and, consequently, there is no clear limit to the number of subspecies that can be recognized.[13]

Paul Ehrlich stated that "Widespread species thus can be divided into any number of different sets of 'subspecies' simply by selecting different characteristics on which to base them . . . as is the case with other species, geographic variation in human beings does not allow *Homo sapiens* to be divided into natural evolutionary units. That basic point . . . has subsequently been demonstrated in a variety of organisms . . . and use of the subspecies (or race) concept has essentially disappeared from the mainstream evolutionary literature."[14]

Bart O'Gara observed: "Classification below the species level often has been subjective because there are no standard criteria for naming subspecies or populations . . . subspecies designations should be based on phylogenetic relationships, the same for species. In practice, they seldom are."[15]

Robert Zink indicated that "Mitochondrial DNA sequence data reveal that 97 percent of . . . avian subspecies lack the population genetic structure indicative of a distinct evolutionary unit. . . . A massive reorganization of classifications is required so that the lowest ranks, be they species or subspecies, reflect evolutionary diversity. Until such reorganization is accomplished, the subspecies rank will continue to hinder progress in taxonomy, evolutionary studies and . . . conservation."[16]

And biologists Haig et al. stated: "Among taxonomists, definitions of subspecies are a source of considerable disagreement . . . In an extensive literature review, we found no universally accepted subspecies definition within or across taxa . . . the scientific community has some level of comfort with the subjective nature of subspecies classification."[17]

I provide these quotes to emphasize that the scientific community acknowledges that subspecies is not a rigorous scientific category and there are many different interpretations. However the subspecies concept can be useful to describe populations in geographic areas that differ from others, as long as the inherent subjectivity of the category is acknowledged. In addition, subspecies can be listed as "endangered species" under the ESA, so there is a practical need to deal with the category. Because subspecies are subjective, it is apparent that neither side in debates over subspecies designations is right or wrong in an absolute sense.[18] However, one of the primary problems with the ESA is the designation of subspecies by the FWS, without adequate acknowledgement of the inherent subjectivity of the category. This has resulted in many management and legal controversies.

Populations

The problems with taxonomic uncertainty and the ESA become even more apparent at the population level, which is also included in the Endangered Species Act definition of species. Scientifically, a population is considered a group of interbreeding animals in a specific geographic area. It's a general term and can be applied to a local area (e.g., the Yellowstone National Park population of wolves) or an entire continent (e.g., the North American population of wolves). However, the ESA includes in its species definition "distinct population segments" (DPS) which are defined as "populations that are discrete from other populations and significant to the species to which it belongs."

There are no truly empirical criteria on what constitutes a discrete or significant population, so like subspecies, DPS designations are quite subjective.

Hybrids

As the previous chapter has described, another category that is frequently encountered in wolf taxonomy is hybrids. Hybrids are the offspring of parents of different breeds, varieties, subspecies, or species. It's a general term, but the criterion for designation of a hybrid is that the parents are of different named groups. For example, a mule is an interspecies hybrid of horse and donkey parents. Wolf researchers have begun to consider both interspecies and intra-species hybrids. For example, interspecies wolf-dog and wolf-coyote hybrids and intra-species (inter-subspecies) Mexican wolf-northern wolf hybrids are current topics of research.[20] Because subspecies are indefinite as discussed above, hybrids of two subspecies will likewise be indefinite. A supporter of a subspecies designation will consider mixed offspring to be hybrids, while someone who doesn't support a subspecies designation will not consider mixed offspring to be hybrids, and a third group will simply consider them the products of dispersal and interbreeding among populations. Such differing perspectives can lead to political controversies regarding a litany of entities for the ESA.

Genetics, DNA, and Taxonomic Inflation

Traditionally, species and subspecies have been identified with morphology (i.e., anatomy), including body size, skull and bone proportions, coat color, or other characteristics. During the last half of the twentieth century, genetic technology (i.e., biotechnology) advanced to where DNA could be compared between animals. This is highly technical work, but the basic idea is that similarity of DNA reflects recent common ancestry, and hence can shed light on taxonomic relationships.

There are many technical approaches for using DNA to compare species, subspecies, and populations. One approach is to compare the proportions of gene variants (called "alleles") in populations. For example, if there are alleles for a gene controlling black and white coat color in wolf populations, the proportion of black and white alleles in each population can be compared and the level of interbreeding between the populations can be estimated. Another approach involves comparing actual DNA molecules between groups. In this approach, the number of differences (i.e., mutations) in the DNA sequence (which is composed of a series of four molecules called nucleotides or base pairs, designated A, G, C, and T) can be counted, and the relationship of the DNA sequences determined. This can then be used to infer the relationship, and even the time of divergence, of the groups. This is straightforward for very divergent groups (for example, dogs and cats) but for related groups it is complicated because they often share the same DNA sequences. Both of these approaches have been used in wolf taxonomy.

The use of genetics and DNA over the last fifty years has allowed better resolution of variation in wildlife populations. However, it has also been accompanied by a decided tendency to split groups into more species and subspecies. This has been termed "taxonomic inflation" and its impetus is often to elevate the status of groups for conservation.[22] For example, there has been an increase in elevating subspecies of wildlife to species status to enhance their chances of legal protection with laws such as the ESA.[23] The designation of the eastern gray wolf as a distinct species, *Canis lycaon* described below, is an example of taxonomic inflation that is relevant to the ESA status of wolves in North America. Taxonomic inflation reflects the current scientific tendency of extensive splitting of animals into species, subspecies, and populations. I believe this tendency springs from three primary factors: the recent availability of DNA technology that allows detailed comparison of populations, peoples' inherent tendency to classify animals into discrete groups, and the current focus on conservation and the use of the ESA to protect wildlife. This has led to acceptance of many new species, subspecies, and DPS without full acknowledgement of the scientific uncertainty of such designations, and has allowed science to be influenced by political and legal maneuvering.[24]

Wolf Species and Subspecies

Several scientific papers on wolf species and subspecies taxonomy have been published since the first edition of *The Real Wolf*. These studies have used new DNA technologies developed in dogs that allow assessment of large amounts of data covering the entire genome. The dog genome contains approximately 2.5 billion DNA base pairs (also known as nucleotide pairs) which is comparable in size to the genomes of other mammals, including humans. The dog is used as a model for genetic diseases, the genetics of behavior, and in pharmaceutical research.[25]

Analysis of the entire genome allows identification of differences in DNA sequences called single nucleotide polymorphisms (SNP). This technology is now used extensively in human medical genetics and dog and livestock breeding programs. Because of the close relationship of dogs and wolves this technology can be applied to study of wolf genetics. Here I will briefly review the status of wolf species and subspecies in light of these new data.

Wolf Species

In addition to wolves, the canid family (called the *Canidae*) also includes foxes, coyotes, jackals, dingoes, African wild dogs, and domestic dogs (*Canis familiaris*). There are wild canids on every continent except Antarctica. Both wolves and jackals were thought to be the progenitors of domestic dogs, but genetics has now shown that dogs were domesticated exclusively from wolves in Europe and Asia within the last fifteen thousand years or so. Recently, Fan and colleagues analyzed relationships of wolves and dogs with a large data set of thirty-four canine genome sequences that indicate that the dog was domesticated in Eurasia from ancestral wolves.[26] The same research team previously also have found extensive admixture between dogs and wolves, with up to 25 percent of Eurasian wolf genomes showing signs of dog ancestry.[27]

Fossils suggest that the genus *Canis* arose in North America more than three million years ago during the Pliocene Epoch. They spread to Eurasia and Africa and differentiated into jackals: side-striped jackal (*Canis adustus*), golden jackal (*Canis aureus*), black-backed jackal (*Canis mesomelas*), gray wolves (*Canis lupus*), and the Ethiopian wolf (*Canis simensis*). Another idea is that gray wolves originated in Africa and then spread to the northern hemisphere. Coyotes (*Canis latrans*) developed from the original *Canis* and have only occurred in North America.

Gray wolves are thought to have originated about five hundred thousand years ago and migrated from Eurasia to North America during the Pleistocene epoch. This was possible because the sea level dropped during glacial periods exposing the Bering land bridge between Asia and Alaska. It has been suggested that gray wolves entered North America at least three different times, with the new migrants mixing with or displacing the existing populations.

There are different views of wolf species in North America. In one view, there is one species of wolf in North America, and some geographic areas contain subspecies. In another, wolves have been classified as two species: the gray wolf that ranges across most of the continent, and the red wolf (*Canis rufus*) that inhabits the southeastern United States.[28]

Some taxonomists consider the red wolf a subspecies of gray wolf (*Canis lupus rufus*) and not a full species. Red wolves were extirpated in the wild, and now exist only in captive populations and an introduced population on an island in North Carolina. A third view developed amongst taxonomists recently identifies a third species that has been designated as the eastern wolf (*Canis lycaon*) in the Canadian provinces of southern Quebec and Ontario and in the northeastern United States as far west as Minnesota.[29]

The eastern wolf is also considered a subspecies of gray wolf (*Canis lupus lycaon*) (i.e., not a full species) by some taxonomists.[30]

The species designations of the eastern wolf and red wolf are based on their apparent common ancestry with coyotes in North America, separate from gray wolves in Eurasia.[31]

The assessment is uncertain because of the extirpation of wolves from much of their North American range and recent interbreeding between gray wolves, the proposed eastern wolf, and coyotes, as well as interbreeding between red wolves and coyotes. This greatly complicates the genetic patterns of living populations. Also, the species designations were based on a limited number of genes and an uncertain number of samples from pure red wolves and eastern wolves. In my opinion, this makes the species designation of an eastern wolf premature and unwarranted. A recent study of a large genetic data set described below supports this view and that there is only one species of wolf in North America, the gray wolf *Canis lupus*.[32]

The eastern wolf and red wolf species designations are based on the theory that these two groups and coyotes all evolved in North America from a common ancestor. It is proposed that gray wolves evolved independently in Eurasia and then entered North America where they came into contact with the coyote, eastern wolf, and red wolf. The fact that these four groups (gray wolf, eastern wolf, red wolf, and coyote) interbreed makes the historical reconstruction with DNA data very complex and speculative. Gray wolves and coyotes don't typically interbreed where they co-occur, but they may in areas where coyotes expanded their range, and where wolves are rare (like the northeast United States and southeast Canada). In these areas it is believed that coyotes have hybridized with wolves (red, gray, and eastern) within the last hundred years, making the genetic composition of each group mixed to varying degrees.

It's important to note that until recently the red wolf and eastern wolf species designations were based on limited data. The primary supporting data is from skull measurements and short sequences of mitochondrial DNA (mtDNA) and Y-chromosome DNA. These data are not in agreement with an extensive DNA assessment (forty-eight thousand single nucleotide polymorphisms, called SNP)[33] that show that the eastern wolf is genetically more similar to gray wolves than coyotes, and does not support the eastern wolf as a distinct species. However, a reanalysis of the data suggested that the eastern wolf may still be a legitimate species.[34]

The analyses of two studies are consistent in showing that red wolves are more similar to coyotes than to gray wolves. Because the genetic patterns of both red wolves and eastern wolves reflect recent interbreeding between various groups of wolves and coyotes, the taxonomy of the groups was still unclear following these studies.[35, 36]

More recent studies provide additional insights. A study by Rutledge and colleagues[37] using genomic data for 127,235 SNP found the eastern wolf is a distinct genetic group, and supported the recognition of three species of *Canis* in North America (gray wolf, eastern wolf, and coyote). This view considers the red wolf to be the same species as the eastern wolf. Rutledge et al. support recognition of the eastern wolf as a separate species, but note that they have hybridized with gray wolves and coyotes in some areas.

In contrast, vonHoldt and colleagues used twenty-eight genome sequences to derive 5,424,934 SNP with which to compare wolves and coyotes.[38] These extensive data showed that the red wolf and eastern wolf are hybrid populations of the gray wolf and coyote. This study showed what many have thought all along, that there is one species of wolf in North America (the gray wolf) and they have interbred with coyotes in some areas. My research involving 123,801 SNP also showed that coyotes in the northeast United States (Maine and Connecticut) have a degree of wolf ancestry, probably from hybridization.[39]

Further, vonHoldt and colleagues claim their data make removing the wolf from the ESA list invalid, and that the hybrid populations have unique genetics of wolves worthy of conservation and ecological processes (i.e., selection) that can be allowed to restore historical patterns of genetic variation.

The lack of consensus of the species status of wolves in North America indicates there is a need for a taxonomic assessment with full consideration of species concepts and definitions, additional genetic data, and the study of fossils. In regards to the ESA, there is a need to either determine consensus about taxonomy or clarify the law regarding the definition of species. In addition, comparisons should be made with other species groups for which DNA data are not concordant with species designations.[40]

This includes white-tailed deer (*Odocoileus virginianus*), mule deer (*Odocoileus hemionus hemionus*), black-tailed deer (*Odocoileus hemionus sitkensis*), brown bears (*Ursus arctos*), and polar bears (*Ursus maritimus*). In the case of the deer, mtDNA is more similar between white-tailed deer and mule deer (different species) than between mule deer and black-tailed deer (the same species). In the case of the bears, mtDNA is more similar between polar bears and one population of brown bears in southeast Alaska (different species) than between the southeast Alaska brown bears and other brown bears (the same species). It is well known among scientists that mtDNA relationships may not reflect species relationships.[41]

In both of these cases, other DNA data shows concordance with the species designations (i.e., white-tailed deer are different from mule deer and black-tailed deer, and all brown bears are different from polar bears). Comparisons of wolf genetics with others such as these will provide insights for wolf taxonomy.

Wolf Subspecies

Many subspecies of wolves were recognized during the 1900s based on morphology, particularly skull measurements. The primary classification recognized twenty-three subspecies of gray wolves in North America,[42] although up to twenty-seven subspecies were recognized.[43] This was later reduced to five subspecies including the Mexican wolf (*Canis lupus baileyi*), the northern timber wolf (*Canis lupus occidentalis*), the plains wolf (*Canis lupus nubilus*), the eastern wolf (*Canis lupus lycaon*), and the Arctic wolf (*Canis lupus arctos*) whose ranges are shown in Figure 1. The red wolf was designated a different species in these analyses.

Chambers and colleagues reviewed these five gray wolf subspecies with morphological and genetic data.[45] They found the existing data support designation of the Mexican wolf, northern timber wolf, and plains wolf subspecies. They also found that the data for the Arctic wolf is not definitive, and that the eastern wolf is a full species, not a subspecies. This work was prior to the recent recognition of the eastern wolf as a gray wolf-coyote hybrid by vonHoldt and colleagues, so their recognition of the eastern wolf as a species may change.[46]

The primary support of the three subspecies of gray wolf recognized by Chambers and colleagues. The Mexican wolf, northern timber wolf, and plains wolf include the skull measurements of Goldman[47] and Nowak[48] and genetic data showing limited interbreeding among them. The data indicate a degree of differentiation of the proposed subspecies, but the assessment is complex because data come from different studies with different sets of samples. Also, the subspecies range boundaries are uncertain so there was likely interbreeding across the original ranges of the different subspecies. The near extinction of the plains wolf and Mexican wolf subspecies make attempts to characterize their genetic makeup particularly problematic because the living populations from which samples can be obtained may not represent the original genetic patterns. North American wolf subspecies in general are questionable and have been described as arbitrary, typological, and an intergrading series of populations.[49]

Mexican Wolf

The case of the Mexican wolf (*C. l. baileyi*) warrants special discussion because it is listed as an endangered subspecies under the ESA.

My data[50] and that of vonHoldt and associates[51] show SNP differentiation of Mexican wolves from other North American wolves. However, living and historic samples show that Mexican wolves share mtDNA haplotypes with wolves in other areas and with coyotes and the living Mexican wolves came from only seven founders of a captive breeding population.[52] In addition, the living Mexican wolves may have included dog ancestry.[53] This finding is supported by a large genome study that showed 1 percent to 3 percent of the Mexican wolf genome has dog ancestry.[54]

I've pointed out that these factors indicate that designation of a Mexican wolf subspecies is questionable in my wolf genetic study published in 2015.[50] However, a study conducted by Fredrickson et al.[56] disagreed and claimed the Mexican wolf is a legitimate subspecies, despite additional arguments to the contrary.[57] Fan and colleagues also showed Mexican wolves as "a divergent form of gray wolf" compared with gray wolves from Yellowstone National Park (which originated with transplants from Alberta and British Columbia, Canada) and claimed their data contradicted my suggestion[58] that the Mexican wolf was not a distinct subspecies.[59]

In claiming subspecies status of the Mexican wolf, Fan and Fredrickson[60] missed the primary point of my colleagues and I made as indicated in the quotations about subspecies above[61] that the subspecies category is subjective. They also do not consider the importance of the point we made that living Mexican wolves are derived from only seven founding animals that are not representative of the native population of wolves in Mexico and the southwest United States.[62]

Other recent studies are relevant to Mexican wolf classification. Hendricks analyzed mtDNA and four autosome SNP loci in wolves in the southwest United States and Mexico including an historical sample (collected in 1922) in San Bernardino, California, and mtDNA in the Pacific Northwest United States.[63] The San Bernardino sample was deemed to have Mexican wolf ancestry based on the genetic data. They also modeled Mexican wolf habitat and potential distribution. The findings state that these data support expanding the historical range of the Mexican wolf subspecies and Pacific Northwest coast ecotype for recolonizing wolf populations.

However, this assessment has the potential problem of promoting typological thinking,[64] which is inherent in subspecies designations.[65] Typological thinking in biology assumes a morphological "type" for a species or subspecies, which is in conflict with "population thinking," and in which variation among individuals, not a fixed type, is the actual state of natural populations. In this case, an mtDNA haplotype and SNP alleles found in Mexican wolves were found in the San Bernardino wolf so it was considered of Mexican wolf ancestry.[66]

I think it is more appropriate to apply population thinking to these cases and consider the dynamic nature of populations' ranges expanding and contracting with dispersal and interbreeding among areas rather than associate a typological subspecies name with genetic markers.[67]

It should be obvious that there are many varying opinions on classification. Ecologists Wayne and Schaffer, and their associates, for example, discuss the issue of hybrids and ESA protection and provide a decision-tree framework for evaluating hybrid protection, including the processes that produced hybrids and the ecological impacts of hybrids.[68] They suggest that hybrid populations may have value and warrant protection. As examples, they discuss the Mexican wolf, red wolf, and eastern wolf, noting that Mexican wolves had historic admixture (i.e., hybridized) with wolves to the north and that such hybrids should be protected.

At this point, it's necessary to raise a basic question regarding subspecies in this case: If the Mexican wolf is a genetically distinct subspecies as Fan and Fredrickson maintain,[69] but historically mixed with wolves to the north, how can it be distinct?[70]

The Mexican wolf has been variously termed a subspecies, a population, an ecotype, an evolutionarily significant unit (ESU), a genetically distinct subspecies, and "the most endangered and . . . genetically distinct wolf subspecies in the New World," in the wolf genetics papers cited in this chapter, such as Hendricks and colleagues,[71] the Mexican wolf also had a "large zone of intergradation" with the subspecies to the north (*C. l. nubilus*),[72] and historic admixture with other subspecies, according to several researchers such as Vila, Frederickson, Hendricks, and Wayne and Schaffer, and their associates maintain.[73] How can the Mexican wolf be genetically distinct and also mix and intergrade with other wolves? Wolves in areas of admixture will not be as genetically distinct from the neighboring subspecies as are wolves in the center of the range. This apparent contradiction shows how typological subspecies confuse understanding of the reality of populations that are dynamic in space and time.

This issue doesn't need to be so complicated. A simple solution for conservation and wildlife management is to not use subspecies, but to refer to wolves and other wildlife as populations with geographic information as descriptors. This approach has been advocated by a number of researchers.[74] For example, Mexican wolves can be referred to as the population of wolves (*Canis lupus*) in Mexico and the southwest US management objectives can then be applied to the population without the uncertainties of subspecies.

Southeast Alaska Wolf

The subspecies classification proposed by Goldman and Hall include another subspecies of gray wolf in southeast Alaska and coastal British Columbia, the Alexander Archipelago wolf (*Canis lupus ligoni*).[75] However, both Nowak and Chambers and their respective colleagues[76] consider these wolves to be part of the plains wolf subspecies (*Canis lupus nubilus*). Figure 1 shows the range of the plains wolf subspecies including the coastal region of British Columbia and southeast Alaska. Indeed, wolves from this area share DNA genotypes with wolves from the late 1800s from Kansas and Nebraska (sampled from museum specimens). This group of wolves is thought to have colonized from the western United States northward to coastal British Columbia and southeast Alaska following the melting of the continental glaciers within the last fifteen thousand years.

My recent study assessed genome variation of dogs, wolves, and coyotes in several parts of North America, including southeast Alaska.[77] We determined genotypes of 305 wolves for 123,801 SNP loci, and found varying levels of differentiation, including low differentiation of wolves in interior

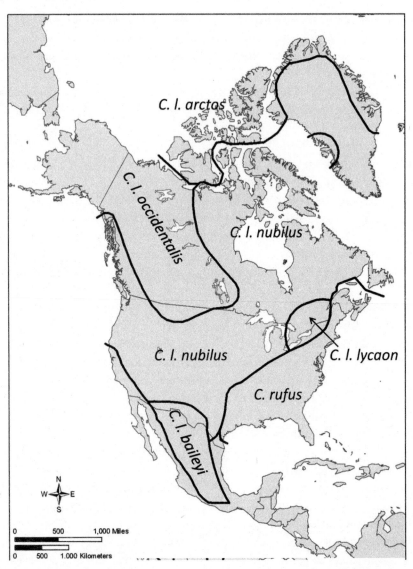

Figure 1. Ranges of the North American gray wolf (*Canis lupus*) and red wolf (*Canis rufus*) recognized by Nowak (1995, 2002). The subspecies are Mexican wolf (*Canis lupus baileyi*), northern timber wolf (*Canis lupus occidentalis*), plains wolf (*Canis lupus nubilus*), eastern wolf (*Canis lupus lycaon*), and Arctic wolf (*Canis lupus arctos*). The eastern wolf is recognized as a species (*Canis lycaon*) by some authors. This is Figure 3 in Chambers's paper, which is a publication of the US Fish and Wildlife Service and includes the following statement: All material appearing in North American Fauna is in the public domain and may be reproduced or copied without permission unless specifically noted with the copyright symbol ©.

Alaska, British Columbia, and the northern US Rocky Mountains. There is considerable differentiation of wolves in Southeast Alaska from wolves in other areas, but wolves in Southeast Alaska are not a genetically homogeneous group and there are comparable levels of genetic differentiation among areas within Southeast Alaska and between Southeast Alaska and other geographic areas. Considering these data, and the analyses of Chambers and associates, we suggested the wolves in southeast Alaska do not warrant subspecies designation.[78]

Weckworth and associates, however, disagree.[79] They did not explicitly state that they support a subspecies designation, but they say that recognizing the wolves as a subspecies or a set of genetically distinct populations is important for conservation. My team and I responded, noting the subjectivity of subspecies and the need to manage populations regardless of terminology.[80]

The wolf in southeast Alaska is important because the USFWS considered the *ligoni* subspecies for listing as an endangered species under the ESA. Acknowledging the lack of consensus in the scientific literature, USFWS noted that *ligoni* may or may not be a subspecies, but considered *ligoni* to be a subspecies for their ESA decision. However, USFWS did not list it because they found that the wolf in southeast Alaska is neither threatened nor endangered with extinction.[81] The finding by Chambers and associates (who are USFWS biologists) that *ligoni* is not a legitimate subspecies, and the finding that it is a legitimate subspecies in the ESA ruling, is scientific inconsistency within the USFWS, however Chambers and associates' opinion was prior to the ruling of the ESA.[82]

Another interesting facet of the subspecies designations is the introduction of wolves from the range of the northern timber wolf subspecies (*Canis lupus occidentalis*) into the northern US Rocky Mountains including Yellowstone National Park, which is within the original range of the Great Plains wolf (*Canis lupus nubilus*). These wolves were introduced (and listed as a distinct population segment under the ESA), apparently without concern for the subspecies of origin. This situation may well become more complex if the wolves in the northern US Rockies expand southward and come into contact and interbreed with the Mexican wolf subspecies in New Mexico and Arizona. There are already some reports of wolves in Colorado that are from the Northern Rockies population of gray wolves.

The northern timber wolf subspecies that has been introduced into the Northern Rockies is a large wolf from western Canada and Alaska. Skull and body sizes indicate that the northern timber wolf is larger than the Great Plains wolf. For example, plains wolf males weigh ninety-five to ninety-nine pounds, and northern wolf males weigh eighty-five to 115 pounds, occasionally reaching

130 pounds. A male northern wolf caught in Alaska in 1939 weighed 175 pounds.

The data showing the northern timber wolf subspecies is generally larger than the Great Plains wolf subspecies is consistent with reports that the introduced wolves in the northern US Rockies are considerably larger than the native Great Plains wolves. In Yellowstone National Park, one male wolf has been caught that weighed at 143 pounds with a full stomach and another weighed 147 pounds with an empty stomach. However, it is important to recognize that we do not know if the size differences of wolves in these areas is due to heritable (i.e., genetic) or environmental (e.g., good nutrition) causes.

In summary, these observations indicate that wolf subspecies are uncertain. A simple solution for conservation and management is to designate populations by geographic area instead of typological and subjective subspecies.[83] This has practical applications for several ESA cases in addition to the wolf. Recent research indicates that several subspecies listed under the ESA are not supported by genetic data, including the coastal California gnatcatcher (*Polioptila californica californica*),[84] the southwestern willow flycatcher (*Empidonax traillii extimus*),[85] the Preble's meadow jumping mouse (*Zapus hudsonius preblei*),[86] the wood bison (*Bison bison athabascae*),[87] and the wolf subspecies we've already discussed. This indicates there is unwarranted taxonomic inflation of wildlife subspecies designations similar to unwarranted species designations.[88] For wildlife management the traditional use of geographic populations as management units is more appropriate than subspecies.

Ecotypes

Schweitzer and associates assessed genome variation in gray wolves to identify genes under selection and local adaptation in six geographic regions (called ecotypes).[89] Their analysis identified potential local adaptation in different populations related to seasonality of precipitation and vegetation. They further found that Arctic and High Arctic wolf ecotypes have several genes under selection, and discuss their conservation value in light of climate change.[90] This is a rigorous study, and is similar to studies of genes underlying performance traits in livestock. A key point, however, is that studying adaptive variation due to selection is not the basis of taxonomy and my focus in this chapter. The "ecotypes" reflect local adaptation (or environmental influence) and not necessarily ancestry. This can result in what is called homoplasy (or convergence) in taxonomy, in which populations share traits due to common selection pressure, not ancestry. This is a highly technical subject, but important to consider when classifying animals.

Wolf Populations and the ESA

As noted above, "population" is a general term, and wolf populations can be designated at any scale. For our discussion, the population category used in the ESA, distinct population segment (DPS), is relevant. A population must be discrete and significant to be considered a DPS. Discreteness can be simple geographic separation or a combination of factors, including genetic differentiation. Significance reflects the importance of the population to the "taxon to which it belongs" which can be the entire species or a subspecies. Significance is usually a subjective judgment by USFWS or NMFS. Despite Congress's order that DPS be used "sparingly" under the ESA, the USFWS and NMFS has listed many DPS including wolf populations.[91]

In the case of wolves, the introduced population in the northern US Rocky Mountains (Montana, Wyoming, and Idaho) and the population in the western Great Lakes region (Minnesota, Wisconsin, and Michigan) were each listed as a DPS under the ESA. The USFWS hopes to remove the gray wolf from the list of endangered species in the contiguous United States as it has already exceeded population numbers originally established to determine its establishment; however, there has been a continual series of courtroom battles about this.

Endangered Species Act protections for gray wolves in Montana, Wyoming, and Idaho were removed in 2011 (Figure 2). The wolves have spread into Washington, Oregon, California, Utah, and Colorado (Figure 3).

In September 2014, the Federal District Court for Washington, DC, vacated the delisting of wolves in Wyoming under the ESA. Therefore, these wolves were again listed as a nonessential experimental population in all of Wyoming. However, on December 19, 2014, following two court orders, the USFWS reinstated regulatory protections under the ESA for the gray wolf in Wyoming and the western Great Lakes on February 20, 2015.

On November 9, 2015, the Oregon Fish and Wildlife Commission delisted wolves from protection under the Oregon Endangered Species Act.

Wolves in Minnesota, Wisconsin, and Michigan—the Great Lakes population—have now spread into North and South Dakota, Iowa, Illinois, Nebraska, and Ontario. Wolves were delisted in the Great Lakes States in 2012. However, as a result of a federal court decision, wolves in the western Great Lakes area (including Michigan, Minnesota, and Wisconsin) were relisted under the Endangered Species Act, effective December 19, 2014.

This review indicates that the current state of North American wolf taxonomy (and wolf ESA listings) is very confusing and not definitive. However, the management of wolves in the United States, first under the federal ESA and then by States after ESA delisting, can be influenced by taxonomic assessments. Both populations in the northern US Rocky Mountains and western

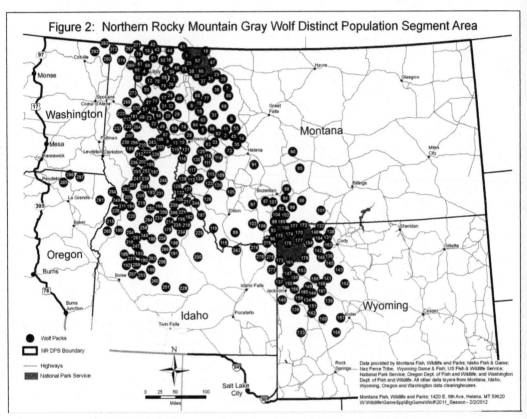

Figure 2. Wolf distribution in the northern US Rocky Mountains. From the US Fish and Wildlife Service website (15 August 2013), http://www.fws.gov/mountain-prairie/species/mammals/wolf/20120828MapPackDistributionFigure%202.jpg.

Gray Wolf - Western Great Lakes Distinct Population Segment

Figure 3. Wolf distribution in the Great Lakes States. From the US Fish and Wildlife Service website (15 August 2013), http://www.fws.gov/midwest/wolf/delisting/WGLDPS.html.

Great Lakes region have been managed with considerable population increases. However, both populations are of questionable taxonomic status. As noted above, the wolves in the northern US Rocky Mountains were transplants of the northern timber wolf subspecies into the historic range of the plains wolf subspecies.

The wolves in the western Great Lakes states are within the range of the plains wolf subspecies[92] but are also considered the eastern wolf species and there has been mixing of wolves and coyotes in this area.[93] There are also inconsistencies in which subspecies are considered important in some cases (e.g., the Mexican wolf), but subspecies designations are ignored in others (e.g., introducing the non-native northern timber wolf subspecies into the northern US Rocky Mountains).

A number of colleagues and I contend that for wildlife management it is most important to recognize that the wolves in the northern US Rocky Mountain and western Great Lakes regions fill the ecological role of a large predator, regardless of opinions regarding species, subspecies, and DPS. Extensive effort and resources can be wasted debating wolf species and subspecies status and saving "pure" species or subspecies, instead of focusing on practical wildlife management.[94]

It's perhaps most important that taxonomy should be practiced as a science by taxonomists and not by wildlife biologists, and the inherent uncertainty associated with subspecies, DPS, and some species designations should be openly and widely acknowledged. In particular, the USFWS should not favor one taxonomic opinion on species, subspecies, or DPS designation, and ignore equally valid differing opinions. Considering all of these points, my conclusion is that wolf management and wildlife management in general, including implementation of the ESA, should focus on populations in specific geographic areas, without being shackled by unresolvable debate over species, subspecies, and DPS taxonomy.

As my colleagues and I stated in our genetic study in 2015: "For wildlife management, the traditional use of geographic populations as management units . . . is more appropriate than subspecies. The scientific rigor of population genetics, systematics, and taxonomy, and their application to management and conservation, would be enhanced by implementation of this practice."[95]

Endnotes:

1.   NRC (National Research Council). 1995. Science and the Endangered Species Act. National Research Council, Committee on Scientific Issues in the Endangered Species Act. National Academy Press, Washington, D.C., USA.
2.   Darwin, C. 1859. *On the Origin of Species*. Murray, London.

3. Cronin, M.A. 2006. "A Proposal to eliminate redundant terminology for intra-species groups." Wildlife Society Bulletin 34:237–241. And Cronin, M. A., M. D. MacNeil, and J. C. Patton. 2005. "Variation in mitochondrial DNA and microsatellite DNA in caribou (*Rangifer tarandus*) in North America." *Journal of Mammalogy* 86:495–505.

4. Cronin, M.A. and L.D. Mech. 2009. "Problems with the claim of ecotype and taxon status of the wolf in the Great Lakes region." *Molecular Ecology* 18:4991–4993.

5. Avise, J.C. 2000. *Phylogeography, the History and Formation of Species.* Harvard University Press, Cambridge, Massachusetts, USA.

6. Mayr, 1970, *Populations, Species, and Evolution.* Belknap Press, Harvard University Press, Cambridge, MA, USA. And Mayr, E. 1982. *The Growth of Biological Thought.* Cambridge MA: Harvard University Press.

7. Chambers et al. 2012

8. Rutledge et al. 2015, vonHoldt et al. 2016

9. Mayr, E. 1970.OP CIT

10. Avise, J.C., and R.M. Ball Jr. 1990. "Principles of genealogical concordance in species concepts and biological taxonomy." *Oxford Survey of Evolutionary Biology* 7:45–67.

11. Mayr (1970), OP. CIT

12. Vanzolini, P. E. 1992. *Third world Museums and Biodiversity.* Pages 185–198 in N. Eldredge, editor. Systematics, ecology, and the biodiversity crisis. Columbia University Press, New York, New York, USA.

13. Futuyma, D. J. 1986. *Evolutionary Biology.* Sinauer Associates, Sunderland, Massachusetts, USA.

14. Ehrlich, P. R. 2000. *Human Natures.* Island Press, Shearwater Books, Washington, D.C., USA.

15. O'Gara, B.W. 2002. "Taxonomy" Chapter 1, pages 3–65 in: *North American Elk Ecology and Management.* Compiled and edited by D. E. Toweill and J. W. Thomas. Smithsonian Institution Press, Washington D.C. and London.

16. Zink, R.M. 2004. The role of subspecies in obscuring avian biological diversity and misleading conservation policy. Proceeding of the Royal Society of London B. 271:561–564.

17. Haig, S.M., E.A. Beever, S.M. Chambers, H.M. Draheim, B.D. Dugger, S. Dunham, E. Elliott-Smith, J.B. Fontaine, D.C. Kesler, B.J. Knaus, I.F. Lopes, P. Loschi, T.D. Mullins, and L.M. Sheffield. 2006. "Taxonomic considerations in listing subspecies under the US Endangered Species Act." *Conservation Biology* 20:1584–1594.

18. Cronin, M.A. 2007. "The Preble's meadow jumping mouse: subjective subspecies, advocacy and management." *Animal Conservation* 10:159–161.

19. Cronin 2006, OP. CIT.

20. vonHoldt, B.M., J.A. Cahill, Z. Fan, I. Gronau, J. Robinson, J.P. Pollinger, B. Shaprio, J. Wall, and R.K. Wayne. 2016. "Whole-genome sequence analysis shows that two endemic species of North American wolf are admixtures of the coyote and gray wolf." *Science Advances* 2:doi: 10.1126/sciadv.1501714. (July 27, 2016).

21. Wayne, R.K. and H.B.Schaffer. 2016. "Hybridization and endangered species protection in the molecular era." *Molecular Ecology* 25:2680–2689.

22. Zachos, R.E., M. Apollonio, E.V. Bärmann, M. Festa-Bianchet, U. Göhlich, J.C. Habel, E. Haring, L. Kruckenhauser, S. Lovari, A.D. McDevitt, C. Pertoldi, G.E. Rossner, M.R. Sanchez-Villagra, M. Scandura, and F. Suchentrunck. 2013. "Species inflation and taxonomic artefacts-A critical comment on recent trends in mammalian classification." *Mammalian Biology* 78:1–6.

23. Marris, E. 2007. "The species and the specious." *Nature*. 446:250- 253. and The *Economist*. 2007. Hail Linnaeus. Conservationists and polar bears should heed the lessons of economics.

24. Crandall, K.A. 2006. "Advocacy dressed up as scientific critique." *Animal Conservation* 9:250–251. and Cronin 2007, OP. CIT.

25. https://www.genome.gov/12511476/2004-advisory-dog-genome-assembled/.

26. Fan, Z., P. Silva, I. Gronau, S. Wang, A. Serres Armero, R. M. Schweizer, O. Ramirez, J. Pollinger, M. Galaverni, D. Ortega Del-Vecchyo, L. Du, W. Zhang, Z. Zhang, J. Xing, C. Vilà, T. Marques-Bonet, R. Godinho, B. Yue, and R. K. Wayne. 2016. "Worldwide patterns of genomic variation and admixture in gray wolves." *Genome Research* 26:163–173.

27. Ibid.

28. Nowak, R.M. 1995. "Another look at wolf taxonomy." Pages 375–397 in Carbyn LN, Fritts SH, Seip DR, editors. Proceedings of the second North American symposium on wolves. Edmonton, Alberta: Canadian Circumpolar Institute, University of Alberta.

29. Chambers, S.M, S.R. Fain, B. Fazio, and M. Amaral. 2012. "An account of the taxonomy of North American wolves from morphological and genetic analyses." *North American Fauna* 77:1–67. doi:10.3996/nafa.77.0001

30. Nowak 1995. OP CIT

31. reviewed by Chambers et al. 2012. OP CIT

32. vonHoldt et al. 2016. OP CIT.

33. Chambers et al. 2012. OP CIT

34. Rutledge, L.Y., P. J. Wilson, C. F.C. Klütsch, B. R. Patterson, and B.N. White. 2012. "Conservation genomics in perspective: A holistic approach to under-standing *Canis* evolution in North America." *Biological Conservation* 155:186–192.

35. Chambers et al. 2012. OP CIT

36. vonHoldt, B.M., J.P. Pollinger, D.A. Earl, J.C. Knowles, A.R. Boyko, H. Parker, E. Geffen, M. Pilot, W. Jedrzejewski, B. Jedrzejewski, V. Sidorovich, C. Creco, R. Ettore, M. Musiani, R. Kays, C.D. Bustamante, E.A. Ostrander, J. Novembre, and R.K. Wayne. 2011. "A genome-wide perspective on the evolutionary history of enigmatic wold-like canids." *Genome Research* 21:1294–1305.

37. Rutledge, L.Y., S. Devillard, J.Q. Boone, P.A. Hohenlohe, and B.N. White. 2015. RAD sequencing and genomic simulations resolve hybrid origins within North American *Canis*. Biology Letters 11:20150303. http://dx.doi.org/10.1098/rsbl.2015.0303.

38. vonHoldt et al. 2016. OP CIT

39. Cronin, M.A., A. Cánovas, A. Islas-Trejo, D.L. Bannasch, A.M. Oberbauer, and J.F. Medrano. 2015a. "Single nucleotide polymorphism (SNP) variation of wolves (*Canis lupus*) in Southeast Alaska and comparison with wolves, dogs, and coyotes in North America." *The Journal of Heredity*. 106:26–36.

40. Cronin, M.A. 1993. Mitochondrial DNA in wildlife taxonomy and conservation biology: cautionary notes. Wildlife Society Bulletin 21:339–348.

41. Cronin 1993. OP CIT (Cronin, 2006. OP CIT Zink 2004).

42. Goldman, E.A. 1944. Classification of wolves: part II. Pages 389–636 in Young SP, Goldman EA, editors. "The wolves of North America. Washington, D.C.: The American Wildlife Institute." And Hall, E.R. 1981. *The Mammals of North America*, 2nd. Ed. John Wiley and Sons, New York.

43. Table 2 of Chambers et al. 2012. OP CIT.

44. Nowak 1995, 2002. OP CIT

45. Chambers et. al. 2012 OP CIT

46. vonHolt et. al. 2016 OP CIT

47. Goldman 1944 OP CIT

48. Nowak, 1995, 2002 OP CIT

49. Wayne, R.K. and C. Vilá. 2003. "Molecular genetic studies of wolves." In: Mech LD, Boitani L, editors. *Wolves: Behavior, Ecology, and Conservation*. Chicago: University of Chicago Press. p. 218–238

50. Cronin et al. 2015a OP CIT

51. vonHoldt et al. 2011 OP CIT

52. Leonard, J.A., C. Villa, and R.K. Wayne. 2005. "Legacy lost: genetic variability and population size of extirpated US grey wolves (*Canis lupus*)." *Molecular Ecology* 14:9–17.; and Hailer, F. and J.A. Leonard. 2008. "Hybridization among three native North American *Canis* species in a region of natural sympatry." *PLOS ONE*. 3(10), e333. doi: 10.1371/\ journal.pone.0003333.

53. García-Moreno, J., M.D. Matocq, R.S. Roy, E. Geffen, and R.K. Wayne. 1996. "Relationships and genetic purity of the endangered Mexican wolf based on analysis of microsatellite loci." *Conservation Biology* 10:376–389. And Hedrick, P.W., P.S. Miller, E. Geffen, and R.K. Wayne. 1997. "Genetic evaluation of the three captive Mexican wolf lineages." *Zoo Biology* 16:47–69.

54. Fan et al. 2016. OP CIT

55. Cronin et al. 2015a OP CIT

56. Fredrickson, R., P. Hedrick, R.K. Wayne, B.M. vonHoldt, and M. Phillips. 2015. "Mexican wolves are a valid subspecies and an appropriate conservation target." *Journal of Heredity* 106: 415–416.

57. Cronin, M.A., A. Cánovas, A. Islas-Trejo, D.L. Bannasch, A.M. Oberbauer, and J.F. Medrano. 2015b. "Wolf Subspecies: Reply to Weckworth et al. and Fredrickson et al." *The Journal of Heredity* 106:417–419.

58. Cronin et al. 2015a OP CIT

59. Fan et. al. 2016 OP CIT

60. Fredrickson et al. (2015. OP CIT

61. Cronin et al. 2015a, 2015b, OP CIT

62. Ibid

63. Hendricks, S.A., P.C. Charruau, J.P. Pollinger, R. Callas, P.J. Figura, and R.K. Wayne. 2015. "Polyphyletic ancestry of historic gray wolves inhabiting US Pacific states." *Conservation Genetics* 16:759–764. And Hendricks, S.A., P.R. Sesink Clee, R.J. Harrigan, J.P. Pollinger, A.H. Freedman, R. Callas, P.J. Figura, and R.K. Wayne. 2016. "Re-defining historical geographic range in species with sparse

records: Implications for the Mexican wolf reintroduction program." *Biological Conservation* 194:48–57.

64. Mayr 1982, OP CIT

65. Patten, M.A. and K.F. Campbell. 2000. "Typological thinking and the conservation of subspecies: the case of the San Clemente Island loggerhead shrike." *Diversity and Distributions* 6:177–188.

66. Hendricks et al. 2015, 2016. OP CIT

67. Cronin and Mech 2009. OP CIT

68. Wayne and Schaffer 2016 OP CIT

69. Fan et al. 2016. OP CIT and Fredrickson et al 2015. OP CIT,

70. Hendricks et al. 2016. OP CIT, (Wayne and Schaffer 2016), OP CIT

71. Hendricks et al. 2016. OP CIT

72. Leonard et al. 2005. OP CIT

73. Vilá, C, I.R. Amorim, J.A. Leonard, D. Posada, J. Castroviejo, F. Petrucci-Fonseca, K.A. Crandall, H. Ellegren, and R.K. Wayne. 1999. "Mitochondrial DNA phylogeography and population history of the gray wolf *Canis lupus*." *Molecular Ecology* 8:2089–2103. And Fredrickson et al. 2015, Hendricks et al. 2016, Wayne and Schaffer 2016). OP CIT

74. Wilson and Brown 1953; Cronin 2006; Cronin and Mech 2009). OP CIT

75. Goldman 1944, and Hall 1981. OP CIT

76. Nowak 1995, 2002, and Chambers et al. 2012. OP CIT

77. Cronin et al 2015a OP CIT

78. Chambers et al. 2012. OP CIT

79. Weckworth, B, N. Dawson, S. Talbot, and J. Cook. 2015. "Genetic distinctiveness of Alexander Archipelago wolves (*Canis lupus ligoni*), reply to Cronin et al. (2015)." *Journal of Heredity* 106:412–414.

80. Cronin et. al. (2015b) OP CIT

81. FWS. 2016. Endangered and Threatened Wildlife and Plants; 12-Month Finding on a Petition to List the Alexander Archipelago Wolf as an Endangered or Threatened Species. Fish and Wildlife Service. Federal Register Vol. 81, No. 3 Wednesday, January 6, 2016 Proposed Rules pages 435–458.

82. Chambers et al. 2012

83. Wilson and Brown 1953.

84. Zink, R.M., J.G. Groth, H. Vazquez-Miranda, and G.F. Barrowclough, 2013. "Phylogeography of the California gnatcatcher (*Polioptila californica*) using multilocus DNA sequences and ecological niche modeling: Implications for conservation." *The Auk* 130:449–458.

85. Zink, R.M. 2015. "Genetics, morphology, and ecological niche modeling do not support the subspecies status of the endangered the Southwestern Willow Flycatcher (*Empidonax traillii extimus*)." *The Condor Ornithological Applications* 117:76–86.

86. Malaney, J.L.and J.A. Cook. 2013. "Using biogeographical history to inform conservation: the case of the Preble's meadow jumping mouse." *Molecular Ecology* 22:6000–6017.

87. Cronin MA, MacNeil MD, Vu N, Leesburg V, Blackburn H, Derr J. 2013. "Genetic variation and differentiation of extant bison subspecies and comparison with cattle breeds and subspecies." *Journal of Heredity.* 104:500–509.

88. Zachos et al. 2013.

89. Schweizer, R.M., B.M. vonHoldt, R. Harrigan, J.C. Knowles, M. Musiani, D. Coltman, J. Novembre, and R.K. Wayne. 2016a. "Genetic subdivision and candidate genes under selection in North American gray wolves." *Molecular Ecology* 25:380–402. doi: 10.1111/ mec.13364. And Schweizer, R.M., J. Robinson, R. Harrigan, P. Silva, M. Galverni, Marco Musiani, R.E. Green, J. Novembre, and R. K. Wayne. 2016b. "Targeted capture and resequencing of 1040 genes reveal environmentally driven functional variation in grey wolves." *Molecular Ecology* 25:357–379.

90. Schweitzer et al. (2016b) OP CIT.

91. (Cronin 2006), OP CIT.

92. Nowak 1995. OP CIT and Nowak, R.M. 2002. "The original status of wolves in eastern North America." *Southeastern Naturalist* 1:95–130.

93. Chambers et al. 2012. OP CIT.

94. Cronin 1993, 2007, Cronin and Mech 2009, OP CIT; and Ramey, R.R. 2012. "On the origin of specious species." In: *Institutions and Incentives in Regulatory Science.* J. Johnston (ed.), Lexington Books.

95. Cronin et al. 2015a.

CHAPTER 19

. . .

Of What Value Are Imported Canadian Wolves?

By Rob Arnaud and Ted B. Lyon

Photo credit: Dennis Donahue/Shutterstock.com

"Wolves are elusive, difficult to see, and occur in low population densities—all characteristics that inevitably confound the ecotourist's desire to see or get close to a wild wolf."

—Matthew Wilson

One of the justifications of allowing relocated wolf populations to increase with little or no opposition is they have tremendous economic benefits. This chapter presents an evaluation of that argument.

Lost Revenue

Two economic techniques that can be used for quick evaluation of an environmental program or project are monetary and non–monetary.

Historically, hunters and fishermen generate annual (monetary) revenues used to maintain and manage America's prized big-game herds, game birds, and fish stocks. Contributions from sportsmen and outdoor organizations have added to the revenues generated by license sales. For decades, this money was used successfully to balance and manage America's wildlife, contributing billions directly into America's economy annually, including the following examples.[1]

When the wood duck population began to decline, duck hunters and their primary support organization, Ducks Unlimited, built predator-resistant wood duck boxes and placed them in swamps; that helped bring America's wood duck population back to close to three million, enough to support annual harvests in the hundreds of thousands.[2]

When hunters became aware that the number of wild turkeys was declining, they helped restore them to their original range. Today, wild turkeys are found in virtually every state in the nation.[3]

When Gulf Coast fishermen saw that the redfish were being decimated by netters, they formed the Gulf Coast Conservation Association and stopped it by legislation.[4] (Incidentally, Ted Lyon helped pass the first legislation in the nation banning netting of redfish on Texas coasts.)

There now are more rare African antelope preserved on Texas ranches than there are in the African wild.[5] Hunters, sportsmen, property owners, ranchers, farmers, community businesses, outdoor organizations, and state wildlife agencies alike have successfully managed and maintained America's abundant wildlife for decades. When Canadian wolves were imported by the US Fish and Wildlife Service (USFWS) and released in Yellowstone National Park in 1995 and 1996, those wolves have multiplied and destroyed America's prized big-game herds, decimating moose, elk, mule deer, and other wildlife—along with thousands of head of livestock throughout the West, forcing outfitters and other industry-related businesses to close or to lay off employees and bring to an end what thousands of citizens have worked for decades to establish.

With the disappearance of this wildlife have gone huge sums of revenue generated historically by hunters to maintain these herds of game animals and the businesses connected to them—outfitters, motels, cafes, gas stations, sporting goods stores, and groceries. State wildlife agency budgets where these big-game herds were traditionally located have experienced plummeting hunting license sales. For example, since 2008, Montana, Idaho, and Wyoming non-resident hunting license sales have significantly declined, resulting in budget shortfalls since license sales represent a large portion of those agencies' budgets.

As of 2014, the total undersold licenses in Montana amount to more than $3.3 million in potential annual revenue for Fish Wildlife and Parks. The agency is now facing a $5.75 million shortfall starting in fiscal year 2017. Idaho non-resident big-game tags have fallen short, costing the department

$3.5 million. Wyoming Game and Fish, after being denied a 20 percent license fee boost by its legislature, has been forced to cut its budget by $4.6 million a year.[6]

Ranchers and farmers whose cattle, sheep, horses, and dogs have been slaughtered like the nation's wildlife are likewise losing hundreds of thousands of dollars' worth of livestock to these wolves. To be reimbursed for these losses, ranchers must be sure that each head of livestock killed by wolves is independently verified by the government. According to T. R. Mader, research director of the Abundant Wildlife Society of North America, as many as 90 percent of livestock wolf kills could be classified as "unconfirmed" based on Animal Damage Control Officer reports.[7] He has pointed out that:

> Virtually all of the carcass is usually consumed by the wolves, scavengers and decay, especially in hot weather, rapidly eliminates evidence confirming cause of death. Coyotes, eagles, fox, skunks, crows, ravens, and magpies—all are usually waiting to feed on what's left of a carcass once the wolves finish.
>
> And dense undergrowth and vegetation, especially in heavy timber, often make it difficult to even find a kill once a rancher notices livestock are missing.

Ranchers or farmers have to make up these losses out-of-pocket, which cuts deep into their annual operating revenue and profit margins and directly affects their livelihood and ability to stay in business. Rocky Barker with the *Idaho Statesman* captured this financial reality in an article he wrote following the slaughter of 176 of a rancher's prize rams caused by wolves:

> 176 sheep died on the west slope of the Tetons in Idaho when wolves panicked them and they ran over a ridgeline and were trampled or suffocated. The sheep were part of a 2,400-head band owned by the Siddoway Sheep Co, grazing in the area of Fogg Hill near Tetonia. Wildlife Services confirmed that wolves were the culprits, based on eyewitnesses, bite marks, and tracks. At about $200 a head (for reimbursement), the Siddoway's loss is $35,000.[8]

It doesn't take many slaughters like this to put a ranch or farm family out of business.

Anyone familiar with wolves—never in danger of biological extinction— had to know what would happen once their numbers exploded. Benjamin Corbin, a trapper and hunter, gave as good a first-hand description of what thousands of wolves running rampant across America's western states looked like in 1900 (while pitching his bosses for a raise) and the tens of thousands of livestock they were killing.[9]

This domestic livestock and wildlife tragedy began in 1995–96 with the USFWS introducing into Montana and Idaho thirty-one Canadian wolves, which have multiplied quickly into thousands. Their slaughter of America's prized big-game animals, if not stopped, has the potential to rank alongside the carrier pigeon and buffalo exterminations.

Wolves as Revenue Generators

Supporters of the introduction of the Canadian wolves into the United States have tried to put a counter "value" on these predators, saying they generate interstate commerce and significant revenue through:

- "Wolf howlings" (monetary),
- "Wolf watchings" (monetary), or
- The notion that wolves are "priceless" and that you cannot put a monetary value on them.

Wolf Howlings

The "wolf howlings" gambit to establish "commercial value" of wolves first appeared in the 1970s when the USFWS trapped what red wolves they could find roaming between Texas and Louisiana, put them in a captive breeding program, and declared the species "extinct in the wild." When they checked the red wolves' DNA, however, it was found that feral dogs and coyotes had polluted the gene pool, so technically, the species may have already been "extinct" when they were captured and, thus, ineligible for any kind of "protection" under the ESA.[10]

Time passed, the red wolves multiplied in captivity, and the USFWS proposed introducing them in Kentucky and Tennessee. State officials, stockmen, and others in those states met them at the gate and told them they didn't want the wolves. So the USFWS shifted its wolf release to federal land in the Alligator River National Wildlife Refuge in North Carolina and nearby Pocosin Lake Refuge, assuring the public that, if one of their collared wolves strayed off federal land, the agency would simply "recapture and return the animal to the refuge."

The wolves quickly exited these forests and multiplied, and new generations of collarless wolves began scouring private land for meals, some more than a hundred miles away. A landowner who shot one threatening his livestock was fined and required to feed captive red wolves for a year. When North Carolina's Hyde and Washington counties began having similar problems with the wolves, they passed local ordinances making it legal for ranchers and farmers to kill wolves they believed a threat to people or livestock on their land. (Today there are forty-five to sixty red wolves in the wild in North Carolina.)

A lawsuit—*Gibbs vs. Babbitt*—challenged the USFWS's authority to prevent landowners from killing wolves that were harming their livestock.[11] This suit, filed in Federal District Court for the Eastern District of North Carolina, argued that citizens had the right under state law to protect their property from marauding wolves, the ESA notwithstanding. In court, USFWS lawyers argued that ranchers and farmers who killed wolves that killed livestock and wildlife on their property were "interfering with interstate commerce and thus violating the Commerce Clause."

Establishing a "fact" that wolves generated interstate commerce was critical for USFWS's long-range plans for the wolves because the Commerce Clause prohibited states from interfering with interstate commerce (the wolves), which was another shield, like the Endangered Species Act (ESA), for the USFWS to use to justify the release of imported Canadian wolves in Montana and Idaho. So, what kind of interstate commerce would wolves generate? The kind—argued contract lawyers for USFWS—that tourists generated when they came from places like New York to North Carolina to "hear the wolves howl."

Gibbs's lawyers examined the logbook at the federal forest where the so-called "wolf howlings" were held and found approximately two hundred people had attended these events over the previous decade. And no one knew if they came to hear the wolves howl or just happened to be camping there when the "howlings" were held.

Other Types of Commerce

What other type of "commerce" could be harmed if a red wolf was shot? Growth of a valuable wolf pelt industry might be disrupted. Gibbs lawyers again checked the historical records and could find no value placed on wolf pelts in the last century in North Carolina. They did find a bounty on wolves in the late 1700s, however.

What other type of "commerce" might be affected? Researchers who crossed state lines to study red wolves would be deprived of a resource if red wolves were shot. The Federal District Court for the Eastern District of North Carolina agreed with USFWS lawyers, and the case was referred to the 4th US Circuit Court of Appeals which, in a split decision, agreed with the lower court. The federal court's logic refuting the landowners' claim that they had a right to kill wolves killing their livestock went something like this:

Wolves were "things of interstate commerce" because they moved across state lines and their movement was followed by "tourists, academics, and scientists."

That the killing of the wolves implicated this "variety of commercial activities associated with interstate commerce" and thus was constitutional under the Commerce Clause.

- That landowners killing wolves on private lands would reduce the total number of wolves, which in turn "could reduce the number of wolves on federal land."
- That the ESA regulation allowing these wolves to roam unchecked was an integral part of an overall federal scheme to "conserve valuable wildlife resources important to the welfare of the country," and that, since the landowners were killing wolves to protect their livestock and since livestock were clearly part of interstate commerce, Congress had the authority to regulate the wolves because the law affected interstate commerce and was, therefore, allowed under the Commerce Clause, even if that effect was a negative one.

The Competitive Enterprise Institute subsequently filed a petition for certiorari with the Supreme Court, but on February 20, 2001, the Supreme Court refused to hear the case. Wolf advocates were worried about *Gibbs* from the beginning. Had the Supreme Court heard the case and ruled in favor of Gibbs, it would have prevented the USFWS from using the Commerce Clause to let wolves wreak havoc on domestic livestock and private property throughout America, and it stopped the USFWS from coming onto private property to charge stockmen and property owners with interfering with interstate commerce for killing wolves once states passed laws to deal with red wolf depredations.

Wolf Watchings

Washington, DC, and USFWS shifted from "wolf howlings" to "wolf watchings" as the next interstate commerce *raison d'être*. Yellowstone National Park became ground zero for the introduction of Canadian wolves and for "wolf watchings," which the USFWS said would substantially increase tourism (interstate commerce) at the Park. Early economic forecasting by the USFWS predicted people hoping to see a wolf in Yellowstone would add an estimated twenty-three million dollars annually to the greater area.

An often-quoted 2005 study of the value of Yellowstone's wolf ecotourism—written by John Duffield, a University of Montana economist, and funded by the Yellowstone Park Foundation, an official fundraising partner of the Park—estimated that at least thirty-five million dollars annually in economic benefits would flow from wolf ecotourism into Yellowstone and could even double as the "multiplier effect" of wolf-ecotourism money circulated through the local economy.[12]

Duffield began his study in 2004 with random samplings taken at Park entrances and from the Lamar Valley parking areas where wolves were most likely to be seen. About two-thirds of these visitors (64 percent) responded. Based on the survey, it was estimated that approximately 10 percent of Yellowstone's visitors came to see a wolf; this was when the wolf population was at its highest in the Park. Since then, Yellowstone's wolf population has decreased 50 percent. Of their favorite food sources—moose and elk—the moose herd that historically maintained itself at approximately one thousand animals has been decimated, the majority of which were eaten by wolves along with 80 percent of the Park's historic elk herd. The *Associated Press* reported on March 13, 2013, that:

> Scientists from the Park and the Montana Department of Fish, Wildlife and Parks said the Northern Yellowstone elk herd is down 6 percent this winter to 3,915 animals. The herd peaked at about 20,000 animals in 1992. That was just a few years before gray wolves were reintroduced to the Yellowstone area from Canada after being absent from the region for decades. . .[13]

Lack of adequate food, deadly territorial battles between wolf packs for dominance, and wolves contracting distemper and mange have accounted for declining wolf numbers in Yellowstone in recent years.

When the Wyoming Sierra Club also surveyed visitors to Yellowstone in 2005 and asked them why they visited, their number one reason for coming to the Park was to view the scenery, and the number two reason was to watch the wildlife. Only 3.5 percent said they would not come if there was no opportunity to see a wolf.[14] Elk and moose were their two most popular ungulates, along with the grizzly.

Additional reports have found that the presence of large numbers of wolves in Yellowstone has driven the surviving elk away from frequenting open fields and into thicker timber and cover to have a better chance of evading wolves. As a result, the chance of seeing elk in Yellowstone is declining, and moose have virtually disappeared. All this has occurred since wolves follow their prey. So as the number of elk and moose either declined, or they fled into cover farther from the roads, wolf watching opportunities have declined, unless the wolves become habituated and are attracted to roads for roadkills, garbage, and handouts.

Outfitters that attempted to react to the disappearance of the big-game herds in the Rockies by offering wolf-watching tours reported that, by 2005, tourists were more interested in seeing grizzlies and geysers than wolves. The 1994 federal environmental impact statement (EIS) of the Canadian wolf introduction into the United States predicted that the wolves' presence would result in a 5 to 10 percent increase in annual visitation to Yellowstone and generate twenty million dollars annually for Idaho, Montana, and Wyoming.[15] Instead, from 1995 through 2009, Yellowstone visitation declined in eleven years and remained essentially unchanged in the others.[16]

People who study eco-tourism and wolf watching say that opportunities to see wolves without professional assistance are rare and limited to areas of open terrain.

> Wolves are elusive, difficult to see, and occur in low population densi-
> ties—all characteristics that inevitably confound the ecotourist's desire
> to see or get close to a wild wolf. . . . Only with the aid of science, tech-
> nology, and adaptations in recreational strategies have current oppor-
> tunities for viewing and hearing wolves in the wild become possible.[17]

In the upper Midwest, wolves are even more difficult to see due to dense for-
est cover and different predation patterns. "Wolf watching" in the Midwest is focused at the International Wolf Center in Ely, Minnesota (which reports that it draws about fifty thousand visitors a year) and at Wolf Park in Battle Ground, Indiana, where captive wolves in enclosures ensure that people get to look at a wolf.[18]

In the Southwest, more than twenty-six million dollars has been spent trying to reintroduce Mexican wolves, excluding the additional costs associated with wolves in the mountain states. It's estimated that only about a hundred of these animals exist in the wild after a thirty-year attempt to reintroduce them. USFWS appears to be using a similar reintroduction strategy tried with the red wolves with exception: they are either inadvertently or deliberately opening the door to a Mexican wolf-Canadian wolf cross as the Canadian wolves drift into New Mexico's northern mountains and points south.[19] Such a large area with so few animals whose genes may already be mixed with those of feral dogs and coyotes generates the following question: What is a "genuine" Mexican wolf?

The probability of people generating "wolf revenue" or "wolf dollars" by driving to Chihuahuan or Sonoran deserts for Mexican howlings" or "watchings" is negligible. The greatest chance to view Mexican wolves currently is if they are habituated to being fed by offerings from people—roadkill and carcasses of slaughtered domestic animals—which is how they are fed in captivity.

Conclusion

"Wolf watchings" and "wolf howlings" will never generate the revenue it takes to cover the minimum costs required to manage these predators. And unlike elk, moose, deer, and bighorn sheep hunters, who return annually, "wolf watchers" who see a wolf generally are satisfied with a one-time sighting for their "bucket list" and move onto their next quest. "Wolf watchers" returning year after year in the same way that traditional hunters do exists only in the mind of wolf promoters.

Hunting license sales to take a wolf will make no difference in replacing the revenues from disappearing hunters, who no longer make their annual trips to Western states because the traditional big-game populations there have either disappeared or declined to the point that it's no longer worth the trip for most. Attempts by the Western states to raise the price of hunting licenses only exacerbates the declining hunter shortage because there is no game to hunt, and increasing the cost of hunting licenses cannot make up revenues lost from lack of hunters, most of which historically came from out of state.

When wolf hunting was permitted in Montana in 2009, license sales from wolf licenses brought in $325,916, an amount that did not come close to covering the million dollars a year it costs the state to manage them. Nor did it even move the needle on Montana's fifty-seven million dollars annual wildlife budget, 66 percent of which comes from the sale of the state's traditional hunting and fishing licenses. Wolves cannot pay their way like the herds they decimated did.

Today, these imported wolves have spilled out of the Northern Rockies into Washington, Oregon, Utah, California, and Colorado in the West. In New England, wolf numbers are growing. Michigan, Minnesota, and Wisconsin have an estimated six thousand in their area, while the USFWS continues trying to reintroduce the Mexican "wolf" in New Mexico and Arizona and debates whether a significant self-sustaining red wolf population can be maintained in North Carolina.

Some researchers now suggest that the number of Canadian wolves in North America is in the tens of thousands, none of which were wanted or needed for economic or environmental reasons. In their wake lie tens of millions of dollars in unnecessary and unwarranted expenses generated by the USFWS using taxpayer dollars, along with decimated wildlife populations and dead cattle, sheep, horses, guard dogs, and other domestic livestock and pets.

Of what value then are these imported Canadian wolves?

Endnotes:
1. "Hunting and Fishing: A Force As Big As All Outdoors," http://www.nssf.org/ PDF/research/bright%20stars%20of%20the%20economy.pdf
2. *Wood Duck Report.* USDA Natural Resources Conservation Service, ftp://ftpfc:sc. egov.usda.gov/WHMI/WEB/pdg/woodduck%281%29.pdf
3. "A Dedication to Wild Turkey Conservation," http://www.nwtf. org/conservation/?utm _source=nwtf.org&utm_medium=conserve-article&utm_campaign=spring-hunting

4. Plugger, Wade, *Fishing the Gulf Coast: Rudy Grigar, W.R. McAffee*, Texas Tech University Press, 1997; http://www.amazon.com/Plugger-Wade-Fishing-Gulf-Coast/dp/0896725103

5. "Texas Ranchers Fight To Breed, Hunt Endangered Antelope," http://articles.latimes.com/2012/apr/03/nation/la-na-nn-texas-antelope20120403

6. http://www.gohunt.com/read/montana-license-demand-slows

7. "Abundant Wildlife Society of North America Fact Sheet, T.R. Mader, Research Director," http://www.aws.vcn.com/fact.html

8. "Rocky Barker: After wolf attacks, rancher regrets . . . -Idaho Statesman," www.Idahostatesman.com/2013/09/02/2738613/after-wolf-att

9. "Benjamin Corbin's advice, The Wolf Hunter's Guide. A firsthand picture of tens of thousands of wolves running rampant in the Mountain and Western States in 1900," Internet Archive, archive.org, Ebook and Texts Archive, The Library of Congress, http://archive.org/details/ corbinsadviceorw00corb

10. *Canid News*, Vol. 3, 1995, "Hybridization: The Double-Edged Threat," by Ron Nowak, Canid Specialist Group, www.canids.org/publicat/cndnews3/hybridiz.htm

11. *Gibbs v. Babbitt*, 214 F.3d 483 (4th Cir. 2000).

12. Duffield, John, Chris Neher, and David Patterson, "Wolves and People in Yellowstone: Impacts on the Regional Economy," September 2006, http:// www.defenders.org/publication/wolves-and-people-yellowstone

13. "Major Yellowstone elk herd continues steep decline," *Missoulian, Associated Press*, March 8, 2013

14. http://missoulian.com/news/state- andregional/major-yellowstone-elk-herd-continues-steep-decline/article_ea10e4cc-880a-11e2-9ef3-0019bb2963f4.html Retrieved from http://wolfsource.blogspot.com/; http://www.yellowstonepark.com/MoreToKnow/Show NewsDetails.aspx?newsid=182

15. "Wolf restoration worth millions of dollars to the economies of Idaho," www.forwolves.org/ralph/wolf-economic-impact.htm

16. Yellowstone National Park Visitor Statistics Page; http:www.yellowstone. co/stats.htm

17. "The Human Dimensions of Wolf Ecotourism in North America," by Matthew A. Wilson, Departments of Sociology and Rural Sociology, University of Wisconsin-Madison; http://www.wolf.org/wolves/learn/wow/regions/Canada_subpages/wolf_human_interactions2.asp

18. Schaller, David T., "The Ecocenter as a Tourist Attraction: Ely and the International Wolf Center," Ely, MN: IWC, 1999.

19. "Feds Make Mexican Wolf Deals," by Center for Biological Diversity, August 29, 2013: http:pinedaleonline.com/news/2013/08/FedsmakeMexicanWolfd.htm

Photo credit: ericlafrancais/Shutterstock.com

CHAPTER 20

· · ·

I Rest My Case

By Ted B. Lyon

May it please the court, ladies and gentlemen of the jury of public opinion. I want to first of all thank you for taking the time to read this book. The case this book presents is concerned with addressing the environmental, economic, and political crisis caused by wolf proliferation across North America and the myths that have been perpetuated by those who say that they believe it's a good thing to dramatically increase the population of wolves back into the lower forty-eight states. We, the authors and contributors, are not anti-wolf, but are seeking a way to manage wolf populations so that there is minimal conflict with people, reasonable contact with game species, and perseveration of the genetic stability of wolves so the species will continue to exist as it has for thousands of years. We hope that by now you agree with us or at least we have given things to think about.

Like many other people, when I started on this journey in 2007, I didn't understand that wolves are some of the most destructive animals on Earth. But working on this book has shown me that through scientific research which goes back well over one hundred years and continues on today, wolves devastate elk, caribou, deer, and moose wherever they exist together unless wolf populations are dramatically controlled.

From Russia to Canada, from Montana to Idaho, the same story repeats itself. Where wolf populations expand, big-game herds disappear or are severely impacted.

Teddy Roosevelt, perhaps our greatest conservation president, once said that wolves were "beasts of waste and desolation." The tragedy that has happened in Yellowstone National Park to the elk and moose and to other large ungulate herds in Montana and Idaho is beyond comprehension. It is unbelievable that the elk herd in Yellowstone, which was at nineteen thousand when wolves were introduced in 1995, now numbers fewer than four thousand elk in 2016. The moose count in the same area has gone from at least twelve hundred in 1995 to now almost zero.

What has been shocking to me is that wolves, while causing all of this destruction, have been elevated to an almost exalted status by the Federal Government and some in the environmental movement. I am shocked that no one at the top of the Yellowstone National Park hierarchy has told the truth about what was happening to the elk, the mule deer, and the moose. The only statements that have come out of the park about wolves are from the chief wolf biologist for Yellowstone, who is an admitted wolf proponent. No one with authority has come forward publicly to defend the elk, the mule deer, the wild sheep, or the moose that are disappearing from the Park. Instead, the wolf has become a cash cow for groups that espouse environmental conservation, but actually are using wolves to raise money so they can hinder conservation efforts by state and federal agencies by forcing them to spend precious time and money to defend themselves from wolf advocates.

This book hopefully has proven to the readers that wolves are not an economic boom to the economy as has been represented by many environmental groups when one considers the costs of growing wolf populations to farmers, ranchers, guides, and hunters, as well as increased potential for negative encounters between wolves and people. And, after the newness of wolves has worn off, wildlife watchers say that they actually are more interested in seeing bears than wolves.

When wolves cause huge losses of wild game there is a resultant loss of license revenues and expenditures by hunters that is dramatic, as has been seen in Idaho and Montana, where millions of dollars in revenue have been lost due to fewer hunters coming in to hunt big game. Many years of raising funds to support conservation efforts for game animals have been quickly cancelled by wolf populations that are artificially created and justified by research that is heavily biased toward wolf populations higher than can be sustained for any extended period of time.

The wolf that was introduced into Yellowstone and Idaho is here to stay according to every wolf expert that has been quoted in popular print. We need to learn to live with wolves, and to regulate them so that areas where wolves are found are managed to minimize conflicts between people and wolves and support a healthy wildlife population that is consistent with protecting livestock, pets, and people, as well as respecting wolves.

One of the most significant economic costs of increasing wolf populations is that wolves also cause dramatic damages to ranchers who have no way of recovering from their losses due to programs that require them to prove beyond a reasonable doubt that wolves caused a kill. Studies cited in this book show that for every cow that is identified as a wolf kill, there are at least seven other cattle that are killed. Many are not even found because a wolf pack eats almost everything and leaves little or nothing to prove or disprove what killed the animal, and following a wolf kill, many other predators and scavengers can and do feed on wolf kills to obliterate all proof that wolves are responsible for the kill.

The United States Department of Agriculture reported that wolves killed over eight thousand head of cattle in 2010 alone. At an average cost of one thousand dollars per cow in 2013, that's a direct economic loss of approximately eight million dollars to the cattle industry in states that harbor wolves.

That loss, however, doesn't even touch the indirect losses suffered by ranchers whose cattle are continually harassed by wolves. Studies done in Idaho, Oregon, and New Mexico demonstrate much greater costs because of the stress that cattle suffer, which result in loss of weight and the failure to get pregnant.

What has also been shown here is that Congress should require each federal agency that pays attorney fees for the enforcement of the Equal Access to Justice Act to keep an accounting of those payments, what they are for, and to send that to the General Accounting Office, which shall send a report to Congress each year detailing these costs. We also feel that these payments should be made public, so the public at large can appreciate the environmental and economic impacts of wolf advocates.

In addition, each federal district court should be required by statute to send a copy of any judgment awarding attorney fees under the act to the Government Accounting Office as well.

Far beyond the costs of relocating some Canadian wolves into the United States, it's been estimated that since the 1970s the Federal Government and the States have spent close to two hundred million dollars to reintroduce Mexican wolves in New Mexico and Arizona and Canadian gray wolves into Idaho, Montana, and Wyoming. Assuming that the government's estimated numbers of wolves are correct, there are roughly ninety-seven Mexican wolves in New Mexico and Arizona and a minimum of roughly eighteen hundred wolves in Montana, Idaho, and Wyoming. Assuming that the numbers in the Northern Rockies are wrong (and many people who live there believe that they are undercounted) and instead using the four thousand total wolves estimated, then the cost to reintroduce wolves is fifty thousand dollars per wolf ($200,000,000 ÷ 4,000 = $50,000).

A reasonable person when thinking about governmental policy and what is good for all wildlife and the economy would have to come up with the conclusion that further federal expenditures to propagate a species that causes such huge economic losses to people living in the areas affected is not rational.

The chapters by Laura Scheberger and Jess Carey, who have been intimately involved with the Mexican Gray Wolf Recovery Program, call into question whether or not Congress should allow for any further expenditures of this program, which has been going on for over thirty-four years at a cost of over twenty million dollars. According to the US Fish and Wildlife Service, there are currently ninety-seven Mexican wolves in the wild. Divide fifty-eight into twenty-six million dollars which is what has been spent on the Mexican Gray Wolf Recovery Program, and the cost of each Mexican wolf is over four hundred thousand dollars per wolf. That's not rational under any sense of the word when one considers that there are fifty thousand to one hundred thousand wolves in Canada, another eighteen hundred to four thousand and perhaps more in Montana, Idaho, and Wyoming and that wolves have expanded into Oregon, Washington, and many other states. There are also several thousand wolves in Minnesota, Michigan, and Wisconsin. Continued expenditures of federal or state tax dollars to promote wolves is not sound governmental policy. Additionally, all across North America there are at least three hundred thousand and maybe five hundred thousand wolves and wolf-dogs in captivity. The wolf species is here to stay. And I note here that as wolves can and do breed with coyotes and dogs, as well as wolves, that wolves in the wild, especially in the Lower 48, are increasingly hybrids, and not pure wolves at all.

I also think people who are part of the environmental movement need to take a close look at the groups that use the wolf as a rallying cry to raise money and file endless lawsuits over wolf hunts. If you are an environmentalist and believe in the preservation of our wild species, why would you support an organization that continually tries to expand wolves when, in fact, you are destroying other wildlife in order to do so. Isn't this exploitation of wolves? And just what are these groups doing with the millions of dollars that they bring in by your donation, grants, and litigation fees they collect? To what extent are their continual threats and litigations weakening the state and federal wildlife management agencies that are charged with managing wildlife? Some environmental organizations work out ways that they can work with and support agencies rather than constantly being adversaries.

Hunting, fishing, and sportsmen groups across the United States should also understand that when they work together and can put aside their differences, they are as powerful as any group in the country and can truly impact public policy. When farm and ranch groups combine with organized sportsmen groups, then they are even more powerful.

By the time people in Oregon and Washington fully understand what the wolves have done to their big-game herds and to their ranching industry, it may be too late. Hopefully people from those states will read this book and understand the political template for success in changing the laws in those states to allow for the control of wolves before they destroy their big-game hunting industry and severely affect ranchers and cattlemen.

I did not pick this fight nor this issue to write a book about, but after six years of study I think that I am on the right side of the preservation of our wild game.

Others may disagree, which is their right, but if they look at all the research that's been done on wolves for over one hundred years, I think they will conclude, as the experts in this book and I have, that it's clear that wolves need to be stringently controlled if we are to have abundant game herds, as well as allow people to co-exist with wolves on some level.

We also have clearly shown that wolves can and do attack pets and people. Research from all around the world shows that where they are not managed, habituated wolves can and will enter human settlements, where they will eat garbage, roadkill, livestock, pets, and ultimately attack people. Peer-reviewed studies in North America, Europe, and Asia, support the habituation model presented in this book. As we have also pointed out, a primary reason why wolves in North America have not attacked and killed as many people as they have in Eurasia is that we have more firearms to protect ourselves, and wolves quickly learn this and retreat when they are not habituated.

There is also a serious need to recognize that wolves may be vectors for diseases that can harm livestock, wildlife, pets, and people.

I told my wife that I expected this book to be widely praised by some groups and widely condemned by others who may use sophisticated methodology, even misinformation, to attack every facet of the book. As they say in Texas, "Bring it on."

"No one seriously advocates more than a small sprinkling of wolves. When they reach a certain level they will certainly have to be held down to it . . . . In thickly settled counties, we cannot have wolves, but in parts of the north we can and should."

—Aldo Leopold

APPENDIX A: LETTERS

· · ·

Will N. Graves 900 Hillen Drive
Millersville, MD 21108
3 Oct 1993

ATTN : Ed Banks, US Fish & Wildlife Service Project Leader

Dear Mr. Banks,

Thank you for the opportunity to comment on the DEIS regarding "The Reintroduction of Gray Wolves to Yellowstone National Park & Central Idaho." I support Alternative 3, the No Wolf Alternative.

1. Diseases, Worms, and Parasites. I was surprised that the DEIS did not make a detailed study on the impact issue of diseases, worms, and parasites (page 9). I believe an EIS is not complete without a detailed study covering the diseases, worms and parasites that wolves would carry, harbor, and spread around in YNP and in Idaho. The study should cover the potential negative impact of these diseases on wild and domestic animals, and on humans. I believe the potential negative impact of these diseases is a valid reason not to reintroduce wolves into YNP and to Idaho.

Countless articles about the diseases, worms, and parasites carried, harbored, and spread around by wide ranging wolves have been published in a magazine sponsored by the former Soviet Ministry of Agriculture. For example, a Soviet biologist reported that gray wolves are carriers of a number of types of worms and parasites which are dangerous for animals and for humans. According to this biologist, the main <Rff!'" is cestoda. Over approximately a ten year period, the Soviets conducted a controlled study on this subject. They made the following observations. When and where wolves were almost eliminated in a given rese.arch area, (where almost all wolves were killed by each spring and new wolves moved into the controlled area only in the fall) infec-tions of taenia hydatiqena in moose and boar did not occur in more than 30 to 35% of the animals. The rate of infections were 3 to 5 examples in each animal. When and where wolves were not killed in the controlled areas in the spring, and where there were 1 or 2 litters of wolf cubs, the infections in moose and boar of taenia hydatiqena reached 100% and up to 30 to 40 examples of infection (infestation) were in each moose and boar. Each year the Soviets studied 20 moose and 50 boar. The research was documented and proved that even in the presence of foxes, racoons, and domestic dogs, ONLY THE WOLF was the basic source of the infections in the moose and boar. Examinations of 9 wolves showed that each one was infected with taenia hydatiqena with an intensity of 5 to 127 examples. This confirmed the soviet conclusions. The damage done by taenia hydatiqena to cloven footed game animals is documented by

Soviet veterinarians. My concern is that if gray wolves in the former USSR carried and spread to game animals dangerous parasites, then there must be danger that gray wolves in YNP and in Idaho would also spread parasites. Why should we subject our game animals, and possibly our domestic animals to such danger?

If wolves are planted in YNP and in Idaho, I believe the wolves will undoubtedly play a role in the epizootiology and epidemiology of rabies. The wolf has played an important role, or perhaps a major role, as a source of rabies for humans in Russia, Asia, and the former USSR. From 1976 to 1980 a wolf bite was the cause of rabies in 3.5% of human cases in the Uzbek, Kazakh, and Georgian SSRs and in several areas of the RSFSR. Thirty cases of wolf rabies and 36 attacks on humans by wolves were registered in 1975-78 only in the European area of the RSFSR. In the Ukraine, wolf rabies constituted .8% of all cases of rabies in wildlife in 1964 to 1978. The incidence of wolf rabies increased six fold between 1977 & 1979, The epizootic significance of the wolf has been shown in the Siberian part of the former USSR. Between 1950 and 1977 a total of 8.7% of rabies cases in the Eastern Baikal region were caused by wolf bites. In the Aktyubinsk Region of Kazakhstan, of 54 wolves examined from 1972 to 1978, 17 or 31.5% tested positive for rabies. During this period, 50 people were attacked by wolves and 33 suffered bites by rabid wolves. This shows that healthy wolves also attack and bite humans. Recent Russian research states that as the numbers of hybrid wolves increases, the likelihood of a healthy hybrid wolf attacking humans also increases. Russian wildlife specialists state that when there is no hunting of wolves, the possibility of wolves attacking humans also increases, as the wolves lose their fear of humans.

Wolves not only have and carry rabies, but also have carried foot and mouth disease and anthrax. Wolves in Russia are reported to carry over so types of worms and parasites, including echinococcus, cysticercus and the trichinellidae family.

Prior to planting wolves into YNP and Idaho, I respectfully request a detailed study be made on the potential impact wolves will have in regard to carry, harboring, and spreading diseases.

I believe bringing wolves into YNP and into Idaho will increase the cost of meat and animal products. It will take a lot of work and cost a lot of money for our ranchers to protect their animals from wolves. Why should we subject our ranchers to these increased costs?

2. Prey Animals Normally Killed by Wolves. I do not understand the wide discrepancy in the number of prey animals projected to be killed annually by one grey wolf in YNP/Idaho, and the numbers of prey animals killed annually by one grey wolf in Russia. The Soviets have professionally documented the number of prey animals "normally" killed by one wolf in one year. I realize there are many factors involved. However, Soviet literature states that one grey wolf will kill SO deer each year (about 1 per week), or up to 90 saiga (Saiga tatarica L.) or 50 to 80 boar, or about 8 to 10 moose. I understand that the US estimate is one wolf will kill one deer every 23 days. Why is there such a difference in these figures?

Russian and Soviet literature is filled with examples reporting that wolves kill extremely large numbers of game and domestic animals. In the Krasnoyarskij Region wolves kill 30 to 40,000 reindeer and 700 to 800 moose each year. In the Putorana Plato, just east of Norilsk, wolves kill about 20,000 reindeer each year.

3. Lustful Killing by Wolves. It is well documented that wolves are lustful killers. Look at some of the figures. Here are some examples from Soviet literature. Eight wolves killed 50 reindeer in about two days. In 1978 one wolf killed 39 reindeer in one attack, and another wold killed 29 in one attack. A pack of wolves killed over 200 sheep in a few short attacks. There are many, many examples of lustful killing of animals by wolves in the former USSR.

In Sweden in 1977 one wolf killed between 80 and 100 reindeer in 19 days. Is there any doubt that a high majority of Swedish reindeer owners (over 70%) reject efforts to protect wolves?

Does the FWS expect that wolves in the US will not carry out lustful killing attacks, especially during periods of crusted or heavy snow? What would these attacks have on the populations of small game herds? Do the US estimates on the number of prey expected to be taken by wolves account for any lustful attacks?

4. Wolf Attacks on Humans in the Former USSR. In the former USSR, it was not unusual for wolves to attack or threaten humans. (Details are available.) Some of these attacks were done by non rabid wolves. There are remote villages in Siberia that have been under siege by wolves. The villagers would not dare go out of their houses at night for fear of wolf attacks. I believe bringing wolves into the US will create a threat to humans that is just not necessary. There is extensive information in the world about the behavior of wolves. Gray wolves attack and kill humans in Asia, so I do not support planting. them anywhere in the US.

5. . . . Wolves Affect Structure of Prey Populations. Wolves not only reduce the size of prey populations, but also have a marked effect on the structure of the prey population. Detailed research by the Soviets showed that in areas inhabited by wolves, about 50% of the moose and deer calves do not reach the age of six months. In areas where the wolves had been exterminated, the loss of calves of moose and deer was only 7 to 9%. The negative effects of wolves on populations of ungulates is especially apparent in years when there is heavy snow or crusted snow. In just one severe winter, wolves can almost completely destroy the ungulate population in the area. In these conditions, wolves pay little attention to the animals they killed, because it is so easy for the wolves to kill the almost helpless animals. Documenting these lustful killings is almost impossible due to the conditions.

6. Cost. I do not support spending 6 million dollars to reintroduce an experimental wolf population. In this high deficit period, we need to cut all unnecessary spending, and this is a good area to cut. During the years of the USSR, wolves cost the Soviets about 45 million dollars per year. The US need to keep our "wolf costs" to a minimum.

7. Reduction in Harvest of Female Elk. Why should we let wolves kill female elk rather than let US hunters bag them? Let us keep the wolves out, and if the ungulate population becomes to large for food and cover, then we should let hunter bag the excess game. Part, or even all of the meat could be given to the poor. I believe wildlife managers should use hunters to control excess populations of game, and not wolves.

8. States Manage Own Land & Resources. I believe eadh state should have primary authority and rights to manage its own land and resources.

9. The Wolf is the Most Dangerous & Damaging of all Predators of Fauna. Am eminent Soviet professional wrote that if all the predators of fauna were placed in a list according to the degree of damage and danger to humans, the wolf, without a doubt, would be in first place at the very top of the .list, and would far outdistance its closest competitors.

I believe it is time to stop idealizing the role of the wolf in nature. The wolf is a powerful, vicious, dangerous, damaging and lustful predator. The wolf is an almost perfectly designed killing machine. Some Soviet researchers state that US wolf specialists generally overestimate the selectivity role of the wolf in nature. Wolves do kill weak and deformed animals, but wolves kill many perfectly fit, healthy animals. Wolves select prey animals based on many circumstances. It is a fact that wolves prefer females that are in late stages of pregnancy and young animals. There are many people who, try to emphasize that wolves select only old, weak, sick or deformed animals for prey; the facts do not support this belief.

I recently read a book for children about wolves. The book showed a white tail deer about two years old, and stated that this is a healthy deer and thus does not have to be afraid of wolves since a healthy deer can run faster than a wolf. If you accept this as fact, then how do you explain that thousands upon thousands of perfectly healthy, fit deer are killed each year by wolves. It is time for us to teach the truth about wolves. I still meet people who believe that wolves in the northers regions of Canada do not kill reindeer for food - that the wolves prey only on lemmings. Then they say that a "specialist" on nature proved that by a detailed study about wolves which he published in a book.

Wolves cause some prey to suffer terribly. Some Soviet technical literature about wolves describes in detail how wolves kill their prey. It is not a pleasant subject. It may take several days for a few wolves to kill a large, male moose. During this time the wolves are actually eating the moose alive. Wolves often eat sheep when they are still alive.

After reading Russian and Soviet literature about wolves for so many years, and talking to Russians who have had experiences with wolves, I have come to the conclusion that many American wildlife biologists have become enamoured/infatuated with the wolf. To these American wolf experts, it appears that "the wolf can do no wrong." Although I am not a biologist, I have learned a lot about wolves and their behaviour in Czarist Russia and in the USSR. What I have learned has convinced me that it would be a mistake to plant wolves in the YNP and in central Idaho.

Signed
Will Graves
mailed on 30 Oct '93
to mr. Bangs

THE STATE OF ARIZONA

GAME AND FISH DEPARTMENT

5000 W. CAREFREE HIGHWAY PHOENIX,
AZ 85086-5000
WWW.AZGFD .GOV

GOVERNOR JANICE
K. BREWER

COMMISSIONERS
CHAIRMAN, J.W. HARR IS, LLJCSON
ROBERT E. MANSELL, WINSLOW
KURT R. DAVIS, PHOENIX
EDWARD "PAT" MADDEN, FLAGSTAFF

DIRECTOR
LARRY D. VOYLES

DEPUTY DIRECTOR
IY E. GRAY

September 9, 2013

The Honorable Sally Jewell Secretary
United States Department of the Interior
1849 C Street NW
Washington, DC 20240

The Honorable Dan Ashe Director
United States Fish and Wildlife Service
1849 C Street NW
Washington, DC 20240

Dear Secretary Jewell and Director Ashe:

It is my duty as Chairman of the Arizona Game and Fish Commission to forward the attached Commission Resolution passed by this body at our September 6 Commission meeting in Show Low, Arizona, to express our disappointment at the refusal to hold public scoping hearings in Arizona regarding the proposed expansion of Mexican wolf conservation, an expansion that will almost certainly deliver a preponderance of the program's social, financial, and biological costs to the citizens of Arizona.

As your partners in this effort since the program's birth, the passion of the resolution's language reflects our regret at what seems like an opportunity being missed by the Department of Interior and the U.S. Fish and Wildlife Service to show their respect for those who share the working lands of Arizona with this species and who will live with the consequences of this program, potentially in perpetuity.

The best meetings in Sacramento, Albuquerque, or even the nation's capital, will not suffice as substitutes for meeting on the affected land with the affected parties.

Sincerely,

The Arizona Game and Fish Commission

A RESOLUTION OF THE ARIZONA GAME AND FISH COMMISSION CONCERNING SCOPING HEARINGS ON THE EXPANSION OF MEXICAN WOLF CONSERVATION

WHEREAS, the Arizona Game and Fish Commission and Arizona Game and Fish Department (AGFD) have been partners with the United States Fish and Wildlife Service (USFWS) in Mexican wolf conservation since the effort's inception; and

WHEREAS, the Endangered Species Act (ESA) envisions a significant role for state fish and wildlife agencies in endangered species conservation and recovery efforts; and

WHEREAS, since the Mexican Wolf conservation effort promulgated under section 1O(j) of the ESA, Arizona has been the only state in which primary releases of Mexican wolves have been allowed; and

WHEREAS, the primary capacity for the Interagency Field Team has been supplied by personnel from AGFD, and

WHEREAS, the congressional record clearly reflects congressional intent that rules promulgated under section 1O(j) of the ESA will reflect an agreement between the USFWS and the affected states; and

WHEREAS, the National Environmental Policy Act requires adequate public notice and opportunity to comment which should involve public hearings near the location of the proposed action; and

WHEREAS, the USFWS has received written requests to hold public scoping hearings in Arizona directly from the AGFD, Graham County, Greenlee County, Navajo County, Apache County, Gila County, and specifically from US Congressman Paul Gosar, US Congresswoman Ann Kirkpatrick, and Senator Jeff Flake; and

WHEREAS, the USFWS has publicly announced its decision to hold scoping hearings in Washington, DC, where there are no plans to conserve Mexican wolves; and to hold public scoping hearings in Sacramento, CA where there are no plans to conserve Mexican wolves; and

WHEREAS, the USFWS has elected to hold a public scoping hearing in Albuquerque, NM as the only state in which Mexican wolf conservation would be involved; and

WHEREAS, the USFWS has specifically not elected to hold any public scoping hearings in Arizona; and

WHEREAS, the citizens of Arizona will unarguably be the most affected by the decisions to be made;

NOW, THEREFORE, BE IT RESOLVED that the Arizona Game and Fish Commission expresses its extreme concern that the USFWS has refused to hold public scoping hearings in Arizona regarding proposals to expand Mexican Wolf conservation efforts;

BE IT FURTHER RESOLVED that the Arizona Game and Fish Commission demands that the Secretary of the Interior and Director of the USFWS reconsider this decision and schedule public scoping hearings in the State of Arizona with at least one meeting to occur in a community geographically located within the proposed I O(j) area.

ADOPTED on the 61 day of September, 2013 by the Arizona Game and Fish Commission.

John W. Harris
Chairman
Arizona Game and Fish Commission

APPENDIX B

* * *

Global Socioeconomic Impact of Cystic Echinococcosis

Christine M. Budke,* Peter Deplazes,* and Paul R. Torgerson*

CYSTIC ECHINOCOCCOSIS (CE) IS AN emerging zoonotic parasitic disease throughout the world. Human incidence and livestock prevalence data of CE were gathered from published literature and the Office International des Epizooties databases. Disability-adjusted life years (DALYs) and monetary losses, resulting from human and livestock CE, were calculated from recorded human and livestock cases. Alternative values, assuming substantial underreporting, are also reported. When no underreporting is assumed, the estimated human burden of disease is 285,407 (95% confidence interval [CI] 218,515-366,133) DALYs or an annual loss of US $193,529,740 (95% CI $171,567,331-$217,773,513). When underreporting is accounted for, this amount rises to 1,009,662 (95% CI 862,119-1,175,654) DALYs or US $763,980,979 (95% CI $676,048,731-$857,982,275). An annual livestock production loss of at least US $141,605,195 (95% CI $101,011,553-$ 183,422,465) and possibly up to US $2,190,132,464 (95% CI $1,572,373,055-$2,951,409,989) is also estimated. This initial valuation demonstrates the necessity for increased monitoring and global control of CE.

Cystic echinococcosis (CE) is a condition of livestock and humans that arises from eating infective eggs of the cestode *Echinococcus granulosus*. Dogs are the primary definitive hosts for this parasite, with livestock acting as intermediate hosts and humans as aberrant intermediate hosts. The outcome of infection in livestock and humans is cyst development in the liver, lungs, or other organ system. The distribution of E. *granulosus* is considered worldwide, with only a few areas such as Iceland, Ireland, and Greenland believed to be free of autochthonous human CE. However, CE is not evenly distributed geographically (Figure 1) (1). For example, the United States has few cases in livestock and most human cases are imported. The same is true for regions of Western and Central Europe. In many parts of the world, however, CE is considered an University of Zürich, Zürich, Switzerland emerging disease. For example, in the former Soviet Union and Eastern Europe, the number of observed cases has dramatically increased in recent years (2-4). Additionally, in other regions of the world, such as parts of China, the geographic distribution and extent of CE are greater than previously believed (5).

---

* University of Zürich, Zürich, Switzerland

CE not only causes severe disease and possible death in humans, but also results in economic losses from treatment costs, lost wages, and livestock-associated production losses. To date, no global estimates exist of CE burden (total health, socioeconomic, and financial cost of a given disease to society) in humans or livestock. Such an estimate is imperative since it can be used as a tool to prioritize control measures for CE, which is essentially a preventable disease.

Two methods previously used to assess disease burden are disability adjusted life years (DALYs) and the calculation of monetary losses (6). DALYs were first developed in the 1990s and were used in the Global Burden of Disease (GBD) Study to determine the worldwide burden of disease due to both communicable and noncommunicable causes (7). Although the application of DALYs is becoming more commonplace, the use of DALYs and the methods behind the creation of this measure remain debatable (8). The GBD Study was an extensive undertaking; however, echinococcosis was not among the conditions studied. Nevertheless, DALYs have been applied to cystic echinococcosis and alveolar echinococcosis, caused by *E. multilocularis,* on a small scale in western China (9). Likewise, monetary evaluations have been applied to CE infections in humans and livestock only at a local level *(10-14).* Global burden indicators not only give an idea of the scope of the disease under study, but can also be used to direct limited financial resources to sites where they can be most effective. Because of the magnitude of applying burden of disease measurements on a global scale, this study must be considered a preliminary estimate. Nevertheless, this report should increase awareness of the global impact of CE by both the public health and livestock sectors.

Materials and Methods

CE Incidence in Humans

Data on country-specific annual reported human CE cases were obtained from the Office International des Epizooties (OIE), World Health Organization Handistatus II database for the years 1996-2003 *(15).* This information was then merged with published case reports from numerous countries and logged into an Excel spreadsheet

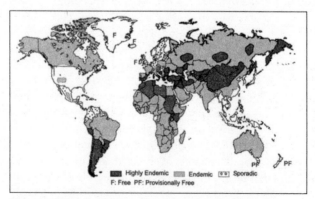

Figure 1. Global distribution of zoonotic strains of *Echinococcus granulosus.* Adapted from Eckert and Deplazes, 2004 [1]. Copyright Institute für Parasitologie, Universität Zürich); used with permission.

(Microsoft Corp., Redmond, WA, USA). Type and quality of incidence data varied by country or region; however, most data consisted of annual numbers of detected cases per susceptible population or was converted into this form for analysis purposes. If both an OIE-reported and a literature-based value were available, the larger of the 2 was used. However, if the higher value appeared to be from a survey that evaluated a highly disease-endemic region and was, therefore, not applicable to the entire country, a corresponding adjustment was made. In addition, we assumed that «10% of annual cases are not officially diagnosed, and those patients do not receive medical attention because of their socioeconomic status or the subclinical nature of the illness. Based on past studies, this estimate is most likely conservative (12,14). For example, in China, mass ultrasound screening in remote areas has shown high prevalence rates of CE (9). A number of these patients have advanced clinical disease but would not normally have access to treatment because of poverty and distance from medical facilities. Human cases of CE are also systematically underreported by the healthcare establishment, with up to 75% of clinic or hospital-diagnosed cases never recorded in local or national databases or published reports (16,17). Therefore, adjustments were made to account for the substantial underreporting of known treated cases.

CE Prevalence in Livestock

Numbers of annual reported CE cases in slaughtered livestock (sheep, goats, cattle, camels, and swine) for the years 1996-2003 were obtained from the OIE-Handistatus II database (15). This information was merged with abattoir studies performed in numerous countries. If data from both sources were available, the larger of the 2 estimates was used. However, if the higher value appeared to be from a region that was highly disease-endemic and was not appropriate for a countrywide estimate, an adjustment was made. Prevalence per species, for each country, was applied to the estimated number of slaughtered animals per year, with 2004 livestock numbers obtained from the FAO-STAT database (18). The assumption was made that approximately one fourth of sheep and goat populations, one sixth of cattle and camel populations, and the entire swine population would be slaughtered annually, based on estimated average species' lifespan (e.g., approximately one fourth of a country's sheep population would be slaughtered annually, with a typical animal life expectancy of 4 years). Such a general estimate was used because of the large amount of variation in animal production practices between and within countries. As with the human incidence data, the true number that were positive for *E. granulosus* at slaughter is substantially higher than reported. Therefore, a correction factor was used to estimate true prevalence.

Application of DALYs to Human Incidence Data

The DALY formula (shown below) was applied to global human incidence data.

$$-\left[\frac{DCe^{-\beta a}}{(\beta + r)^2} \left[e^{-(\beta + r)(L)} \left(1 + (\beta + r)(L + a)\right) - \left(1 + (\beta + r)a\right) \right] \right]$$

In this equation, D is a disability weight, P is an age-weighting function parameter, C is an age-weighting correction constant, r is a discount rate, a is age at clinical onset,

Table 1. Outcome of surgery for cystic echinococcosis in humans

Country (y)	No. patients	Cure (%)	Morbidity (%)	Relapse (%)	Death (%)	Reference
Greece (1984–1990)	56	40 (72)	13 (23)	3 (5)	0	(9)
Italy (1950–1987)	298	244 (82)	27 (9)	15 (5)	12 (4)	(20)
Turkey (1992–1999)	95	32 (34)	38 (40)	24 (25)	1 (1)	(21)
Turkey (1990–1995)	108	88 (81)	19 (18)	0	1 (1)	(22)
Greece (1985–1990)	67	59 (86)	4 (6)	3 (6)	1 (2)	(23)
Italy (1982–1994)	89	70 (79)	17 (19)	1 (1)	1 (1)	(24)
Total	**713**	**533 (75)**	**118 (17)**	**46 (6)**	**16 (2)**	

and L is the duration of disability or time lost because of death (7). Disability weight for CE was assigned a multinomial distribution based on numerous retrospective studies evaluating postoperative outcome (Table 1) *(19-24)*. The percentage of patients projected to improve after surgery was assigned a disability weight of 0.200 (Dutch weight for clinically disease-free cancer) for 1 year, the percentage of patients projected to have substantial postsurgical conditions was assigned a disability of 0.239 (GBD weight for preterminal liver cancer) for 5 years, the percentage of patients projected to have recurrent disease was assigned a disability of 0.809 (GBD weight for terminal liver cancer) for 5 years, and the percentage of patients projected to die postoperatively were assigned a disability of 1 (indicating death) for the remainder of their predicted lifespan *(7,25)*. An assumption was also made that ≈10% of cases are not reported and do not receive medical treatment. These cases were assigned a disability weight of 0.200 (Dutch weight for clinically disease-free cancer) for 10 years *(25)*. For the GBD Study, a standardized life table was used for L (7).

Economic Evaluation of Human-associated Losses

Overall cost per human surgical case was based on findings from previous international studies (Table 2) *(11,13,14,26,27)*. Expenses taken into consideration included diagnostic costs, surgical cost, hospitalization, and postoperative costs. The average cost per surgical patient was shown to be significantly correlated ($R^2 = 0.898$, $p < 0.05$), with the country-specific per capita gross national income (per capita GNI) (Atlas Method) (Table 2). Therefore, the linear regression coefficient was used as a predictor of treatment costs for each disease-endemic country. In addition to medical costs and single-year wage losses, past studies have indicated an average 2.2% postoperative death rate for surgical patients (Table 1). Approximately 6.5% of cases also are assumed to relapse and require a prolonged recovery time (Table 1) (11). Therefore, these outcomes were also taken into account. We assumed that, in addition to surgical cases, «10% of cases are not officially diagnosed each year, and those patients never receive treatment. Wage losses for this group were thus taken into consideration. Economic losses in humans were also evaluated, taking into account the nearly 4-fold degree of underreporting of patients who received treatment.

Economic Evaluation of Livestock-associated Losses

Production-based losses attributable to infected sheep, goats, cattle, camels, and pigs were estimated. Losses from liver condemnation, defined as the action of preventing

Table 2. Average cost per surgical case of cystic echinococcosis

Country	Years	Average cost per case (US $)	% of real per capita GNI* per patient	Refernce
Jordon	2002	701.50	40	(26)
Spain	1987–2001	10,915.00	76	(27)
Tunisia	2000	1,481.00	71	(11)
Uruguay	2000	6,721.00	110	(14)
Wales, UK	2000	13,600.30	54	(13)

*World Bank Atlas method for converting data in national currency to US dollors; GNI, gross national income.

the sale of livers deemed unfit for human consumption (sheep, goats, cattle, pigs, camels), reduction in carcass weight (sheep, goats, cattle), decrease in hide value (sheep, cattle), decrease in milk production (sheep, goats, cattle), and decreased fecundity (sheep, goats, cattle) were taken into account. Only liver-associated losses in camels and pigs are presented since few studies have evaluated production losses from echinococcosis in these species *(28)*. Losses from liver condemnation are assumed to occur since hepatic pathology is associated with infection in swine and camels *(29)*. Losses from liver condemnation were presumed proportional to those used for the analysis of the economic impact of CE in Jordan *(12)*. Decrease in hide value (20%) and decrease in fecundity (11%) were presumed proportional to values suggested by numerous Soviet studies conducted from the 1950s through the 1980s (28). Reductions in carcass weight (2.5%) and milk production (2.5%) were also based on previous reports *(30)*.

Analysis

Spreadsheet models were constructed in Excel to estimate global impact of CE in terms of DALYs and monetary losses. Total disease effects, in DALYs lost or monetary costs, was calculated by summing all of the constituent components. Uncertainty in parameter estimates was modeled by using Monte Carlo techniques (6). Briefly, all parameters were assigned a probability distribution based on the quantity and quality of reported data. Macros were written in Excel to sample across these distributions, with 10,000 iterations of each model calculated. Mean and 95% confidence intervals (CIs) for losses were then determined from these iterations.

Reported global human incidence was assigned a normal distribution, with a standard deviation of 5%. Adjustments were then made to account for the nearly 4fold degree of underreporting of treated cases believed to occur *(16,17)*. In addition, cases that would not be officially acknowledged had to be accounted for, i.e., cases in persons who never receive treatment in a hospital. We therefore assumed that «10% (uniform distribution of 8% to 12%) of cases would not be detected. This estimate is conservative compared to other country-specific estimates *(12,14)*.

The DALY formula was applied to worldwide CE cases in a stochastic manner similar to that used to apply DALYs to echinococcosis cases in a region of western China *(10)*. Mean age of clinical onset (a) was allocated a uniform distribution of 30 to 40 years, established on the basis of various studies (Table 3) *(4,9,21,31-34)*. Numerous and varying reports have indicated the sex of CE-positive persons with women tending to be infected at a higher rate than men. Based on these reports, we assigned a uniform

Table 3. Average age at ultrasound detection or surgery

Country	Years	Average cost at onset/detection (y)	Refernce
China	2001–2003	35*	(9)
Jordon	1994–2000	31–45†	(31)
Kenys (Turkana)	1979–1982	21–30*	(32)
Kyrgyzstan	1991–2000	22†	(4)
Morocco	2000–2001	32*	(33)
Turkey	1992–1999	44†	(21)
Uruguay	1991–1992	45*	(34)

*Age at time of ultrasound detection
†Age at surgery

distribution of 50% to 60% of infected persons as female (4,35). Number of DALYs lost, using incidence values corrected and uncorrected for underreporting of surgical incidence, was determined.

Human-associated economic losses were applied in a stochastic manner similar to that used for a region of western China (10). Variability in surgical treatment costs, due to CE, was modeled by using a uniform distribution of 50% to 90% of per capita GNI per country and was weighted by each country's contribution to global human CE incidence (36). Lower income, higher unemployment, or both has been associated with a diagnosis of CE (4,10). Consequently, a decrease in wages earned was assumed, at least for the year of initial diagnosis and treatment. Therefore, all patients were assigned a uniform loss of 50% to 90% of country-specific per capita GNI for 1 year (36). Approximately 6.5% of patients were also assigned a 50%-90% wage loss for 4 additional years because of relapse and prolonged recovery time. In addition, 2.2% of patients were assigned a 100% wage loss until the expected retirement age of 65 due to postsurgical death. A standard 3% discounting rate was applied to all income losses (7). In addition to surgical cases, «10% of cases (uniform distribution of 8% to 12%) annually were assumed to not be officially diagnosed. A 25% wage loss for 5 years was consequently assigned to this population. This estimate is conservative and does not take into account income losses attributable to undiagnosed cases with fatal outcomes. Projections were made that assumed the absence and presence of underreporting of surgical incidence (16,17). In addition to using real per capita GNI (Atlas Method), calculations were also performed by using purchasing power parity (ppp) adjusted per capita GNI.

As with human-associated economic losses, livestock- associated losses were applied in a stochastic manner (10). Livestock prices were given uniform distributions of US $30-$60 for sheep, US $15-$30 for goats, US $150-$350 for cattle, US $300-$600 for camels, and US $55-$75 for pigs. Uniform distributions were used because of the large regional variations in prices and assigned in accordance with baseline prices for most affected countries. Production losses were assumed to follow a log-normal distribution; most affected animals were lightly infected, and only a small proportion of animals had severe losses. As with human cases, substantial underreporting of livestock infection was recognized, since official reporting is not mandatory in most countries. Therefore, a uniform correction factor of 1.5 to 2 was used to approximate true economic losses. A large uniform distribution was used because of the lack of information concerning true global prevalence of CE in livestock. This lack will, therefore, be represented in the wide confidence limits obtained.

Results

DALYs

Regional findings for predicted global burden of CE in terms of DALYs lost, with 95% CIs, can be found in Table 4. The most conservative estimate of number of global DALYs lost is 285,407 (95% CI 218,515-366,133), with no consideration for disease underreporting. Estimated number of global DALYs lost, taking into consideration nonreported surgical cases, is 1,009,662 (95% CI 862,119-1,175,654).

Human-associated Economic Losses

Findings for predicted regional burden of human CE in economic terms, with 95% CI, can be found in Table 5. Global losses, assuming no underreporting, are estimated at US $193,529,740 (95% CI $171,567,331-$217,773,513). Losses, adjusted for underreporting, are estimated at US $763,980,979 (95% CI $676,048,731-$857,982,275). When ppp adjusted per capita GNI is used instead of real per capita GNI, estimated annual overall losses, without correction for underreporting, are US $484,878,359 (95% CI $432,898,134-US $542,048,125). When corrected for underreporting, annual losses are estimated at US $1,918,318,955 (95% CI $1,700,574,632-$2,142,268,992) (Table 5).

Table 4. Estimated lobal impact of cystic echinococcosis in terms of DALYs lost

Region*	Total unadjusted DALYs lost (95% CI †)	Total adjusted DALYs lost (95% CI)
Western Europe, USA, Canada, Australia New Zealand	11,842 (8,977–15,722)	41,891 (30,949-55,014)
Middle Eastern Crescent	104,503 (79,291–135,722)	370,056 (275,228–486,353)
Formerly socialist economies of Europe and Russia	17,317 (13,129–22,371)	61,369 (45,800–ll0,077)
China	112,451 (85,001–145,898)	398,015 (295,922–521,879)
Other Asia and Islands	1,130 (851–1, 462)	4,003 (2,971–5,256)
Sub-Saharan Africa	2,639 (1,926–3,518)	9,314 (6,664-12,623)
Latin America and the Caribbean	14,834 (11,252–19,241)	52,693 (38,787–09,380)
India	20,691 (15,666-26,822)	73,364 (54,518–96,263)
World	285,407 (218,515-366, 133)	1,009,662 (862, 119–1,175,654)

"Regional breakdown of disability-associated life years (DALYs) lost is based on that used in the Global Burden of Disease study (7) †CL, confidence interval.

Table 5. Global annual cystic echinococcosis-associated economic losses to humans

Region*	Total unadjusted economic losses (95% CI)* (US $)	Total adjusted economic losses (95% CI)† (US $)
Western Europe, USA, Canada, Australia, New Zealand	$309,983,585 ($244,256,327–$383,371,741)	$354,460,281 ($277,178,852–$440,438,597)
Middle Eastern Crescent	$197,276, 106 ($158,870,204–$240,282, 181)	$564,496,304 ($454,402,304–$690,682,060)
Formerly socialist economies of Europe and Russia	$46,896,902 ($37,750,210-$57,355,873)	$143,921,865($114,323,294–$176,555,114)
China	$146, 129,578 ($114,279,187–$181,937,463)	$663,712, 150 ($516,048,103–$826,353,341)
Other Asia and Islands	$1,535,990 ($1,159,946–$1,946,632)	$2,412,386 ($1,826,342–$3,074,240)
Sub-Saharan Africa	$832,295 ($649,915–$1,035,681)	$5, 176,229 ($3,710,869–$6,969,680)
Latin America and the Caribbean	$48,396,449 ($38,408,001–$59,672, 173)	$120,717,047 ($95,789,339–$146,939,896)
India	$12,930,073 ($9,674,489–$16,499,072)	$63,422,693 ($47,576,673–$80,430,630)
World	$763,980,979 ($676,048,731–$857,982,275)	$1,918,318,955 ($1,700,574,632–$2, 142,268,992)

* Income losses based on per capita gross national income (GNI) (Atlas method); Cl, confidence interval.
† Income losses based on purchasing power parity–adjusted per capita GNI.

Livestock-associated Economic Losses

Estimated livestock-associated losses, with 95% CI, can be found in Table 6. Minimal annual losses, assuming liver condemnation alone with no correction for underreporting, is estimated at US $141,605,195 (95% CI $101,011,553-$183,422,465). However, when losses from additional production factors (decreased carcass weight, decreased milk production, decreased hide value, decreased fecundity) are taken into account, losses range from US $1,249,866,660 (95% CI $942,356,157-$1,622, 045,957), not taking into account underreporting, up to US $2,190,132,464 (95% CI $1,572,373,055-$2,951,409, 989), when underreporting is considered.

Discussion

Even without correcting for the underreporting of human and livestock cases, CE has a substantial global disease impact in terms of DALYs and monetary losses. The importance of using both indicators is illustrated by the proportional difference in DALYS lost versus economic losses per region (Tables 4 and 5). If only monetary losses were evaluated, the severity of the situation in poorer regions would be underestimated because of the decreased income and economic value of livestock products relative to more economically prosperous regions. For example, China is responsible for 40% of the world's CE DALYs but only 19% of human-associated economic losses. However, losses based on ppp-adjusted per capita GNI give a better picture of the relative distribution of

Table 6. Global annual cystic echinococcosis-associated livestock production losses

Category	Economic losses (95% CI) (US $)*
Liver condemnation†	$141,605,195 ($101,011,553-$183,422,465)
Decreased carcass weight†	$241,525,979 ($100,335,764–$518,035,773)
Decreased hide value‡	$34,871, 148 ($23,965,776–$46,162,828)
Decreased milk production§	$378,722, 717 ($279,048,143–$495,682,356)
Decreased fecundity§	$453, 141,617 ($278,287,046–$67 1,424,319)
Overall cost (no correction factor)	$1,249,866,660 ($942,356,157--$1,622,045,957)
Overall cost (adjusted for underreporting)	$2,190,132,464 ($1,572,373,055–$2,95 1,409,989)

*CI, confidence interval.
†Sheep, goats, cattle, camels, pigs.
‡Sheep, cattle.
§Sheep, qoats, cattle.

disease impact (Table 5). When the number of DALYs lost, taking into account the recognized underreporting of human cases, is compared with those of other parasitic conditions evaluated by the World Health Organization (WHO), worldwide losses due to CE are slightly less than those caused by African trypanosomiasis (1,525,000) and more than those caused by onchocerciasis (484,000) or Chagas disease (667,000) *(37)*. Even though estimated number of DALYs lost from CE is greater than estimated losses from multiple members of the tropical disease cluster, CE continues to be excluded from funding associated with conditions related to low socioeconomic status. This exclusion best illustrated by evaluating research and training funding provided by the United Nations Children's Fund (UNICEF)/United Nations Development Programme (UNCP)/World Bank/ WHO-supported Special Programme for Research and Training in Tropical Diseases (TDR). If funding for CE were placed on the same scale as TDR-supported diseases, based on estimated DALYs lost, CE should receive approximately US $1,200,000 annually (Figure 2) *(38)*. For now, however, CE continues to be widely underappreciated by most international agencies. These findings emphasize the need for CE to be taken seriously as a global public health condition, regardless of its economic implications. What makes this disease exceptional, however, is that it is not only a substantial human health problem but also has a considerable economic effect on the livestock industries of some of the most socioeconomically fragile countries.

In addition to reporting the estimated global burden of CE, this study has shown the need for more accurate reporting of infected humans and livestock. Very few country-specific estimations of the true incidence of CE in humans have been made and no studies, to the authors' knowledge, that estimate its true prevalence in livestock *(16,17)*. Presentation of the substantial economic losses for both the public health and agricultural sectors will also, we hope, encourage countries and international organizations to more closely examine potential control programs and cost-sharing methods between the 2 affected sectors *(10)*.

The values presented in this paper are not definitive but instead estimates of the severity of the global situation from human- and livestock-associated CE. Considerable

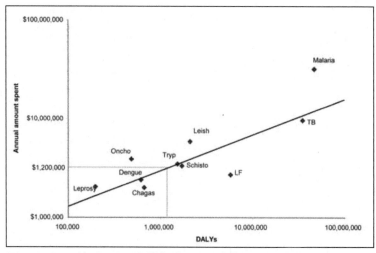

Figure 2. Annual budget (in US $) for diseases included in the United Nations Children's Fund/UNDP/World Bank/World Health Organization-supported Special Programme for Research and Training in Tropical Diseases (TDR) compared to their estimated global disability-associated life years (DALYs). The thinner lines indicate estimated DALYs lost because of cystic echinococcosis (CE) and the recommended funding level based on the TDR 2004–2005 approved program budget (Oncho, onchocerciasis; Tryp, trypanosomiasis; Schisto, schistosomiasis; Leish, leishmaniasis; LF, lymphatic filariasis; TB, tuberculosis). This figure does not take into account the substantial regional variability in both the estimates of DALYs lost and the annual budget for the diseases illustrated.

sums of money have been invested in the investigation and control of such parasitic conditions as lymphatic filariasis and onchocerciasis. Although these conditions can result in severe human disease, unlike CE they do not have severe secondary economic implications, such as massive livestock production losses *(39,40)*. In addition, regional control programs that have been implemented and recommended thus far for CE, based on combinations of dog deworming, stray dog culling, sheep and goat vaccination, and education programs, have been shown to be very cost effective *(10,27)*. CE is, therefore, a worthy condition for research and control program implementation, with substantial anticipated return on invested funding.

Acknowledgments

We thank F.-X. Meslin of WHO for encouraging us to undertake this study.

The authors received financial support from the University of Zurich, an Ecology of Infectious Diseases program grant from the US National Institutes of Health, and the National Science Foundation (TWO 1565-02), and the International Association for the Promotion of Co-operation with Scientists from the New Independent States of the Former Soviet Union (INTAS 01-500, INTAS 03-51-5661).

Dr. Budke was a research scientist at the Institute of Parasitology, University of Zurich, Switzerland, where she studied the transmission and economic effects of echinococcosis. She is currently an assistant professor of epidemiology at the College of Veterinary Medicine and Biomedical Sciences, Texas A&M University. Her research interests include the epidemiology of emerging and zoonotic diseases.

References

1. Eckert J, Deplazes P. Biological, epidemiological, and clinical aspects of echinococcosis, a zoonosis of increasing concern. Clin Microbiol Rev. 2004;17:107–25.
2. Torgerson PR, Shaikenov BS, Baitursinov KK, Abdybekova AM. The emerging epidemic of echinococcosis in Kazakzhstan. Trans R Soc Trop Med Hyg. 2002;96:124–8.
3. Todorov T, Boeva V Human echinococcosis in Bulgaria: a comparative epidemiological analysis. Bull World Health Organ. 1999;77:110–8.
4. Torgerson PR, Karaeva RR, Corkeri N, Abdyjaparov TA, Kuttubaev OT, Shaikenov BS. Human cystic echinococcosis in Kyrgystan: an epidemiological study. Acta Trop. 2003;85:51–61.
5. Chai JJ. Epidemiological studies on cystic echinococcosis in China—a review. Biomed Environ Sci. 1995;8:122–36.
6. Carabin H, Budke CM, Cowan LD, Willingham III AL, Torgerson PR. Methods for assessing the burden of parasitic zoonoses: cysticer- cosis and echinococcosis. Trends Parasitol. 2005;21:327–33.
7. Murray CJL, Lopez AD. The global burden of disease: a comprehensive assessment of mortality and disability from disease, injuries, and risk factors in 1990 and projected to 2020. Cambridge: Harvard University Press; 1996.
8. Anand S, Hanson K. Disability-adjusted life years: a critical evaluation. J Health Econ. 1997;16:685–702.
9. Budke CM, Qiu J, Wang Q, Zinsstag J, Torgerson PR. Utilization of DALYs in the estimation of disease burden for a high endemic region of the Tibetan plateau. Am J Trop Med Hyg. 2004;71:56–64.
10. Budke CM, Qiu J, Wang Q, Torgerson PR. Economic effects of echinococcosis on a highly endemic region of the Tibetan plateau. Am J Trop Med Hyg. 2005;73:2–10.
11. Majorowski MM, Carabin H, Kilani M, Bensalah A. Echinococcosis in Tunisia: a cost analysis. Trans R Soc Trop Med Hyg. 2005;99:268–78.
12. Torgerson PR, Dowling PM, Abo-Shehada MN. Estimating the economic effects of cystic echinococcosis. Part 3: Jordan, a developing country with lower-middle income. Ann Trop Med Hyg. 2001;95:595–603.
13. Torgerson PR, Dowling PM. Estimating the economic effects of cystic echinococcosis. Part 2: an endemic region in the United Kingdom, a wealthy, industrialized economy. Ann Trop Med Hyg. 2001;95:177–85.
14. Torgerson PR, Carmona C, Bonifacino R. Estimating the economic effects of cystic echinococcosis: Uruguay, a developing country with upper-middle income. Ann Trop Med Hyg. 2000;94:703–13.
15. OIE-Handistatus II, Office International des Epizooties, Paris; 2005. [cited 4 Jan 2006] Available from http://www.oie.int/hs2/report.asp.
16. Serra I, Garcia V, Pizarro A, Luzoro A, Cavada G, Lopez J. A universal method to correct underreporting of communicable diseases. Real incidence of hydatidosis in Chile, 1985–1994. Rev Med Chil. 1999;127:485–92.
17. Nazirov FG, Ilkhamov IL, Ambekov NC. Echinococcosis in Uzbekistan: types of problems and methods to improve treatment [article in Russian]. Medical Journal of Uzbekistan. 2002;2/3:2–5.

18. FAOSTAT. FAO statistical databases, February 2004 ed. Food and Agricultural Organization of the United Nations, Rome; 2004. [cited 4 Jan 2006] Available from http://www.apps.fao.org/.

19. Gogas J, Papachristodoulou A, Zografos G, Papastratis G, Gardikis S, Markopoulos C, et al. Experience with surgical therapy of hepatic echinococcosis [article in German]. Zentralbl Chir. 1997; 122: 339–43.

20. Cirenei A, Bertoldi I. Evolution of surgery for liver hydatidosis from 1950 to today: analysis of a personal experience. World J Surg. 2001;25:87–92.

21. Yorganci K, Sayek I. Surgical treatment of hydatid cysts of the liver in the era of percutaneous treatment. Am J Surg. 2002;185:63–9.

22. Ozacmak ID, Ekiz F, Ozmen V, Isik A. Management of residual cavity after partial cystectomy for hepatic hydatidosis: comparison of omentoplasty with external drainage. Eur J Surg. 2000;166:696–9.

23. Vagianos CE, Karavia DD, Kakkos SK, Vagenas CA, Androulakis JA. Conservative surgery in the treatment of hepatic hydatidosis. Eur J Surg. 1995;161:415–20.

24. Altieri S, Doglietto GB, Pacelli F, Costamagna G, Carriero C, Murigani M, et al. Radical surgery for liver hydatid disease: a study of 89 consecutive patients. Hepatogastroenterology. 1997;44:4 96–500.

25. Stouthard MEA, Essink-Bot ML, Bonsel GJ. Disability weights for diseases: a modified protocol and results for a Western European region. Eur J Public Health. 2000;10:24–30.

26. Nasrieh MA, Abdel-Hafez SK, Kamhawi SA, Craig PS, Schantz PM. Cystic echinococcosis in Jordan: socioeconomic evaluation and risk factors. Parasitol Res. 2003;90:456–66.

27. Jimenez S, Perez A, Gil H, Schantz PM, Ramalle E, Juste RA. Progress in control of cystic echinococcosis in La Rioja, Spain: decline in infection prevalences in human and animal hosts and economic costs and benefits. Acta Trop. 2002;83:213–21.

28. Romazanov VT. Evaluation of economic losses due to echinococcosis. In: Lysendo A, editor. Zoonosis control: collection of teaching aids for international training course vol. II. Moscow: Centre of International Projects GKNT;1983. p. 283–85.

29. Njoroge EM, Mbithi PM, Gathuma JM, Wachira TM, Gathura PB, Magambo JK, et al. Astudy of cystic echinococcosis in slaughter animals in three selected areas of northern Turkana, Kenya. Vet Parasitol. 2002;104:85–91.

30. Polydorou K. Animal health and economics. Case study: echinococcosis with a reference to Cyprus. Bull Off Int Epizoot. 1981;93:981–92.

31. Al-Qaoud KM, Craig PS, Abdel-Hafez SK. Retrospective surgical incidence and case distribution of cystic echinococcosis in Jordan between 1994 and 2000. Acta Trop. 2003;87:207–14.

32. Macpherson CNL. An active intermediate host role for man in the life cycle of *Echinococcus granulosus* in Turkana, Kenya. Am J Trop Med Hyg. 1983;32:397–404.

33. Macpherson CNL, Kachani M, Lyagoubi M, Berrada M, Bouslikhane M, Shepherd M, et al. Cystic echinococcosis in the Berber on the Mid Atlas mountains, Morocco: new insights into the natural history of the disease in humans. Ann Trop Med Parasitol. 2004;98:481–90.

34. Cohen H, Paolillo E, Bonifacino R, Botta B, Parada L, Cabrera P, et al. Human cystic echinococcosis in a Uruguayan community: a sonographic, serologic, and epidemiologic study. Am J Trop Med Hyg. 1998;59:620–7.

35. Schantz PM, Wang H, Qiu J, Liu FJ, Saito E, Emshoff A, et al. Echinococcosis on the Tibetan plateau: prevalence and risk factors for cystic and alveolar echinococcosis in Tibetan populations in Qinghai Province, China. Parasitology. 2003;127:S109-20.

36. World Bank data and statistics. World Bank Group; 2005. [cited 4 Jan 2006] Available from http://www.worldbank.org/data/.

37. World Health Organization. The world health report. Geneva: the Organization; 2004. Available from http://www.who.int/whr/2004.
38. United Nations Children's Fund/United Nations Development Programme/World Bank/World Health Organization Special Programme for Research and Training in Tropical Diseases. Approved Programmed Budget 2004–2005. 2003. TDR/PB/04- 05/Rev. 1. [cited 4 Jan 2006] Available from http://www.who.int/ tdr/ublications/publications/budget_04.htm.
39. Ramaiah KD, Pradeep KD. Mass drug administration to eliminated lymphatic filariasis in India. Trends Parasitol. 2004;20:449–502.
40. Remme JH. Research for control: the onchocerciasis experience. Trop Med Int Health. 2004;9:243–54.

Address for correspondence: Paul R. Torgerson, WHO Collaborating Centre for Parasitic Zoonoses, Institute of Parsitology, University of Zurich, Winterthurerstrasse 266A, 8057 Zurich, Switzerland; fax: 41-4463-58907; email: paul.torgerson@access.unizh.ch

APPENDIX C

. . .

European Echinococcosis Registry: Human Alveolar Echinococcosis, Europe, 1982–2000

Petra Kern,* Karine Bardonnet.† Elisabeth Renner,‡ Herbert Auer, ¶Zbigniew
Pawlowski,‡ Rudolf W. Ammann,‡ Dominique A. Vuitton,# Peter Kern,** and the
European Echinococcosis Registry[1]

Surveillance for alveolar echinococcosis in central Europe was initiated in 1998. On a voluntary basis, 559 patients were reported to the registry. Most cases originated from rural communities in regions from eastern France to western Austria; single cases were reported far away from the disease-"endemic" zone throughout central Europe. Of 210 patients, 61.4% were involved in vocational or part-time farming, gardening, forestry, or hunting. Patients were diagnosed at a mean age of 52.5 years; 78% had symptoms. Alveolar echinococcosis primarily manifested as a liver disease. Of the 559 patients, 190 (34%) were already affected by spread of the parasitic larval tissue. Of 408 (73%) patients alive in 2000, 4.9% were cured. The increasing prevalence of *Echinococcus multilocularis* in foxes in rural and urban areas of central Europe and the occurrence of cases outside the alveolar echinococcosis-endemic regions suggest that this disease deserves increased attention.

Human alveolar echinococcosis, caused by the metacestode of the fox tapeworm *Echinococcus multilocularis*, is considered to be the most pathogenic zoonosis in temperate and arctic regions of the Northern Hemisphere. Transmission to humans occurs when eggs of the tapeworm, excreted by the final hosts (usually foxes), are accidentally ingested. The larva's primary target organ is the liver, where it proliferates slowly,

---

* 'University of Ulm, Ulm, Germany; †University Hospital, Besançon, France; ‡University Hospital Zürich, Zürich, Switzerland; §University of Vienna, Vienna, Austria; ¶University of Medical Sciences, Poznan, Poland; #University of Franche Comté, Besançon, France; and

** University Hospital Ulm, Ulm, Germany

1 The European Echinococcosis Registry (EurEchinoReg) is a surveil-lance network for human alveolar echinococcosis coordinated by D.A.Vuitton (France) and P. Kern (Germany). Registration of human casesis organized by H. Auer (Austria), Y. Carlier (Belgium), L. Kolarova(Czech Republic), K. Bardonnet (France), P. Kern (Germany), P.S.Craig (Great Britain), I. Prousalidis (Greece), A. Siracusano (Italy), J.van der Giessen (Netherlands), Z. Pawlowski (Poland), E. Renner, R.W. Ammann (Switzerland), and N. Altintas (Turkey).

but the larva also spreads into extrahepatic structures and even metastasizes to distant organs. In earlier untreated cohorts, the fatality rate exceeded 90% within 10 years (1). The introduction of benzimidazoles for alveolar echinococcosis treatment in 1976 has considerably improved the prognosis (2,3). Long-term follow-up of 117 patients showed that the 5-year actuarial survival rate increased to 88% with this improved management (4). As chemotherapy is parasitostatic only, long-term administration is mandatory for most patients (5,6). Radical surgical excision, the only curative treatment, is feasible in a few select cases (7).

In Europe, previous assessments of human cases did not cover all alveolar echinococcosis-endemic areas at comparable periods. In Switzerland, where laboratory-diagnosed alveolar echinococcosis was a reportable disease until 1997, the annual incidence ranged from 7.2 to 10.4 (0.10–0.18/100,000) and did not markedly vary during a 36-year period (8). In Austria, an average incidence of 2.5 cases per year corresponded to an incidence of 0.034/100,000 from 1985 to 1999 (9). These low numbers of human infections throughout a whole country failed to alarm public health authorities. However, two findings are beginning to attract more attention: 1) high annual incidence rates occurring regularly in particular regions, e.g., the Swiss Jura (0.74/100,000) (10); and 2) a presumed range extension of the parasite in its sylvatic life cycle.

In Europe, the Red Fox *(Vulpes vulpes)* is the most important final host for *E. multilocularis*. Reviews based on the data collected during the past decade have shown that the natural range of the parasite extends farther to the east and north in Europe than previously thought (11,12). Defined rural areas have been monitored regularly for many years, and increasing parasite prevalence rates in foxes have been recorded (13). Clusters of high endemicity (60% to 75%) have been found (14). Increasing fox populations have been reported from several European countries (13). Foxes migrating to urban areas are also causing concerns: *E. multilocularis* prevalence rates of 20% and 48% have been recorded in Stuttgart, Germany (11), and Zürich, Switzerland, respectively (15).

Knowledge of the parasite's range and prevalence in animal hosts has thus grown during recent years. However, comprehensive assessments of human alveolar echinococcosis covering the known risk areas across European countries have not been performed. To provide baseline data for future risk calculations and to establish a prospective case retrieval, the European Echinococcosis Registry (EurEchinoReg) created a network of reporting centers in 11 countries of western and central Europe and Turkey. This report provides the status of reporting, origin, and clinical and epidemiologic data of such patients reported to the registry up to the year 2000.

Methods

Case Retrieval
Case detection and data collection have been organized by each participating country according to the existing infrastructure of the national health systems and the availability of data sources. In the EurEchinoReg, experts from universities (research units and hospitals) and public health authorities cooperate in eight countries of the European Union (Austria, Belgium, France, Germany, Greece, Great Britain, Italy, and

the Netherlands), and in Switzerland, Poland, the Czech Republic, and Turkey. Patient data are stripped of identifiers and sent to two subregistries (the University of Franche-Comté, Besançon, France and the University of Ulm, Ulm, Germany), where they are controlled and approved for electronic recording.

Case Definition and Period of Inclusion

Diagnosis of alveolar echinococcosis is confirmed by 1) positive histopathology, if available and/or 2) typical liver lesion morphology identified by imaging techniques (ultrasound scan, computed tomography, and magnetic resonance imaging) with or without the detection of serum antibodies (serology). Positive serologic results without suggestive imaging findings or positive histopathology does not qualify for a case definition.

The period of inclusion began in 1982, when benzimida-zoles, ultrasound, and other imaging techniques (which facilitated diagnosis, treatment, and follow-up) were introduced. The registry includes all confirmed new cases from January 1982 to December 2000, as well as cases with a diagnosis from earlier periods, provided the patients were alive in 1982 and their diagnosis was confirmed with the appropriate techniques in 1982 or later.

Case Report Form and Completeness of Registration

Patients were asked to allow their nominal registration at their national center, in conformity with the national legislation for data privacy to avoid double registration and to facilitate follow-up. Two questionnaires are used: an epidemiologic part to be answered by the patient and a clinical part to be completed by the reporting physician. In addition to demographic baseline data, we gathered information on year of diagnosis, disease manifestation at the time of the diagnosis, co-existing conditions; diagnostic and therapeutic measures, year of death, presumed cause of death, places of residence, and occupation in agriculture, forestry, and gardening.

Data files from the study groups in Austria, France, Germany, and Switzerland were the basis for the European patient registry; additional case files were collected by active case finding and with the help of physicians from hospitals and private practices. Completeness of registration can be assumed: 1) in France, since access to patient files is facilitated by a centralized distribution of albendazole by a few university hospitals; 2) in Austria, since laboratory diagnosis is made in a single institution; and 3) in Switzerland, where alveolar echinococcosis was a reportable disease until 1997; case reports are thus complete from the 1970s until 1997. Underreporting is likely in Germany, where reporting relies entirely on the cooperation of family physicians and clinicians. In Belgium, Greece, and Poland, alveolar echinococcosis seems to be newly emerging, and cases are discussed in the medical community; the cases reported to the registry should reflect the true prevalence in these countries.

Data Analysis

The combined data sets for all European patients are kept in an Access database (Microsoft Corp., Redmond, WA). Descriptive analyses were made with SAS software

V8 (SAS Institute, Inc., Cary, NC). The regional distribution of alveolar echinococcosis cases was mapped with the software package RegioGraph 5.1 (GfK MACON AG, Waghausel, Germany).

Results

Epidemiology
The total number of verified alveolar echinococcosis cases reported to the registry was 559; 42.0% were diagnosed in France, 23.6% in Germany, and 21.1% in Switzerland (Table 1). Fifteen patients acquired the infection outside their reporting country, 7 of these cases originated from one of the neighboring countries, 8 were of non-European origin.

During the reporting period, the number of new cases varied from year to year. From 1981 to 2000, a peak incidence of 36 was noted in 1988; aside from this 1 year, reports ranged from 15 to 27 patients. A total of 258 patients were male (46.2%) and 301 female (53.8%) (gender ratio 1:1.2). The median age at first diagnosis was 56 (mean 52.5, range 5–86 years) and was almost equal in men and women (Figure 1). The proportion of patients <20 years old was 2.1% (12/559); 88 (15.7%) were >69 years of age. For four patients (0.7%), the year of birth was missing. Three of the four children in this case series, ages 5 and 7 when diagnosed, had severe organ damage; two were immunocompromised.

Information on potential risk factors was available for 210 (37.6%) patients from Austria, Germany, Greece, and France (Table 2), including 97 men and 113 women. Of these, 21.9% were farmers. In addition, of all the patients engaged in other professions (including housewives and students), 46.2% regularly farmed, gardened, or performed related activities as a pastime. Of all pensioners and unemployed patients, 62.2% also gardened, farmed, or the like. Most patients (70.5%) owned or formerly kept dogs and cats. Among these pet owners, 105 persons also actively farmed or gardened. Only 15 patients (7.1%) did not farm, garden, or own pets.

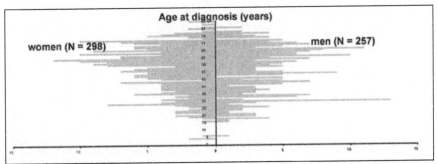

Figure 1. Patients with alveolar echinococcosis reported to the European Registry. Age at first diagnosis by gender (N=555, year of birth missing for 4 patients).

Table 1. Number of patients with alveolar echinococcosis, Europe

Yr of first diagnosis	Reporting country									
	Austria	Belgium	France	Germany	Great Britain	Netherlands	Switzerland	Poland	Greece	Total[a]
Until 1980	8	0	23	30	0	0	43	0	0	104
1981–1985	12	0	60	11	0	0	29	0	0	112
1986–1990	11	0	80	17	0	0	17	2	0	127
1991–1995	13	0	40	26	0	0	16	6	0	101
1996–2000	10	3	32	48	1	1	13	6	1	115
Total no. of patients[b]	54 (1)	3	235	132(6)	1(1)	1(1)	118(6)	14	1	559

[a]Status of notification to the European Echinococcosis Registry as of August 2001.

[b]Includes 15 non-autochthonous cases (in parentheses); 7 of them originated from neighboring countries, 3 from Turkey, 3 from the Newly Independent States, 1 from Kazakhstan, and 1 from Afghanistan.

Geographic Distribution

Figure 2 gives the residence at the time of diagnosis or at the time of the last medical report for 532 alveolar echinococcosis patients; cases were autochthonous from the countries represented on the map. The patient from Greece lived in Macedonia. Data were unavailable for 18 patients.

Most residences were clustered in defined regions: central France, French Jura and Savoy, Swiss Jura and northeastern Switzerland, southern Germany, and western Austria. Single cases were identified in Belgium, the northern regions of France, Germany, and Poland, and northeastern Austria. For the period 1980–1999, a total of

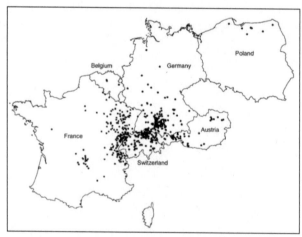

Figure 2. Regional distribution of autochthonous alveolar echinococcosis in Europe, from 532 diagnoses ascertained from 1982 to 2000. Dots represent place of residence (at time of diagnosis or last medical record) of 1–5 patients. In Austria, Belgium, Germany, and Poland, administrative units for locating patients are the municipality; in France and Switzerland, dots are placed at random in larger units ("Arrondissement" for France, "Kanton" for Switzerland). Source: European Echinococcosis Registry, Ulm, Besançon, 2001. Used with permission.

Table 2. Possible exposure risks assessed for 210 patients with alveolar echinococcosis

Occupation	N (%)	Activity in agriculture, gardening, forestry, hunting			Ownership of dogs, cats, or both		
		Yes	No	Missing	Yes	No	Missing
Farmers	46 (21.9)	46	0	0	39	2	5
Nonfarmers[a]	119 (56.7)	55	56	8	80	13	26
Occupation not specified, including unemployed and pensioners	45 (21.4)	28	13	4	29	6	10
Total	210 (100.0)	129	69	12	148	21	41

[a]For example, tailors, hairdressers, cooks, nurses, drivers, teachers, students, and housewives.

201 cases were reported from Turkey; all originated from the Asian part of the country, mostly from eastern Anatolia. However, the aggregated data (reviewed by Altintas et al. [16]) could not be combined with the detailed datasets from western and central Europe. No autochthonous cases were reported from the Netherlands, the Czech Republic, the Slovak Republic, Italy, or the U.K.

Clinical Data

Table 3 lists the main diagnostic procedures, conducted within a time span of 6 months after initial examination, which led to the diagnosis of alveolar echinococcosis. A total of 53.5% of diagnoses were definitely confirmed by positive histopathology; 38.5% were ascertained by imaging techniques combined with serology, or imaging alone, when obtaining tissue specimens for analysis was not possible. Information on diagnostic procedures was missing for 7.7% of the patients.

Table 3. Diagnostic procedures to ascertain the diagnosis of alveolar echinococcosis[a]

Histopathology[b]	Imaging[c]	Serology[d]	No. of patients (%)
+	+	+	176 (31.5)
+	+	−	48 (8.6)
+	−	+	19 (3.4)
+	−	−	56 (10.0)
Subtotal			299 (53.5)
-	+	+	192 (34.3)
-	+	−	25 (4.5)
Subtotal			217 (38.8)
Data not available			43 (7.7)
Total			559 (100.0)

All documented techniques, applied during 6 months after initial examination. +, positive result in the respective tests/examinations; −, negative result, tests/examinations not done, or data not available.
[b]Examination of liver tissue samples carried out on material removed by surgery, diagnostic laparoscopy, or, in rare instances, by fine needle biopsy.
[c]Comprised one or several examinations, i.e., ultrasound, computed tomography (CT), or magnetic resonance imagry (MRI) of the abdomen. In some cases, X-ray, CT, or MRI were available on brain, chest, or other organs.
[d]Included screening methods using different crude antigen preparations in indirect hemagglutination or enzyme-linked immunosorbent assays (ELISA). In addition, purified and recombinant antigen preparations such as Em2+, Em10 or Em18 were used in ELISA, Western blot, or both.

Table 4. Location of the primary lesions at first diagnosis in alveolar echinococcosis

Primary infection site	No. of patients
Liver	541 (96.8%)
Spleen, peritoneum, lung, vertebra, brain, kidneys, heart	13 (2.3%)
Data not available	5 (0.9%)
Total	559

In 397 (71.0%) of the 559 cases, the diagnosis was made after the patients reported symptoms; 66 (11.8%) cases were disclosed by chance in the course of a general medical checkup or an examination related to other diseases; and 18 (3.2%) cases were found during studies that screened for alveolar echinococcosis. Data on these circumstances were not available for 78 (14%) of 559 patients.

The primary infection site was the liver for almost all patients, and primary extrahepatic lesions without any involvement of the liver were diagnosed in 13 patients (Table 4). At first diagnosis, the liver was the only affected organ for 351 (62.8%) of 559 patients. Damage to the liver included single or multiple lesions in one or more segments of one or both liver lobes, the hilus region, the intrahepatic portal vein, hepatic vein, or bile duct. Approximately one third (34%) of the patients (190/559) were already affected by a spread of the larval tissue either in continuum into neighboring organs, by the formation of distant metastases, or both. Specific details of organ damage were available for 178 of 190 patients. The organs most frequently affected by continuous growth were the diaphragm (59 patients), kidneys or adrenal glands (26 patients), and lungs and pleura (15 patients). Metastases occurred mainly in the lungs (39 patients), brain (17 patients), and spleen (10 patients).

At the time of reporting, 267 (47.8%) of the 559 patients had undergone surgery and received benzimidazoles; 200 (35.8%) were treated with these drugs alone, and 48 (8.6%) by surgery alone. A total of 29 patients (5.2%) underwent liver transplantation. Twenty-two patients (3.9%) had not received any treatment during the time between diagnosis and notification; another 13 (2.3%) had apparently had inadequate treatment. For 9 patients (1.6%) the chosen treatment options were not specified.

By December 2000, 73.0% of the patients were alive, 21.3% had died, and 5.7% were lost to follow-up (Table 5). Of the patients still alive, disease activity was assessed at their last clinical examination as follows: cured (20 patients, 4.9%); stable or regressive (226 patients, 55.4%); or progression, sequelae, or complications caused by larval growth or occurring after intervention (43 patients, 10.5%). An assessment was not provided for 119 patients (29.2%); many of them had been diagnosed recently, and treatment had just begun. Death was definitely associated with alveolar echinococcosis in 13 (10.9%) of the 119 cases; in 15 cases (12.6%) death was probably related to this disease. In 20 patients (16.8%), death was definitely independent of the diagnosis of alveolar echinococcosis. No assessment was available for 71 patients (59.7%).

Discussion

In 1998, the EurEchinoReg network initiated the assessment of human alveolar echinococcosis across European borders. The reasons for promoting concerted efforts to survey a disease thought to be rare in Europe were as follows: 1) the disease is one of the most aggressive chronic liver diseases, 2) comprehensive assessments of human cases covering all known risk areas were not available, 3) the routes of transmission to humans are still hypothetical, and 4) the range of the parasite in its life cycle seems to have extended, posing threats in previously unaffected areas.

The assessment included patients from former clinical studies and cases identified by active case finding. Nine European countries reported on 559 patients; cases were autochthonous from seven of these countries. The median numbers per year did not vary during two decades (24 in 1980s; 22 in 1990s). Underreporting from previous years was responsible for a seemingly increasing incidence in Germany; underreporting since 1998 explains a decline in Switzerland. High numbers in France in the mid-1980s could be an effect of mass screenings performed at that time in alveolar echino-coccosis-endemic areas, which may have raised awareness of the disease. In the past, the number of verified and published cases from Europe (Austria, France, Germany, and Switzerland) amounted to 844 cases or 10.6 cases per year (published between 1900 and 1980) (17). The patient numbers from our report reflect what is probably an opti-mal detection rate owing to improved technology. Thus, low but constant incidence is characteristic of the occurrence of human alveolar echinococcosis in Europe today.

In this parasitic infection, a long incubation period seems to precede diagnosis. Albeit difficult to prove, the initial asymptomatic period is assumed to last 5–15 years (1). (This conclusion is derived from the small proportion of patients <20 years old at diagnosis [2.1% in this report]). Determining the time and place of infection is diffi-cult. Assuming that in humans, who are unsuitable hosts for *E. multilocularis,* repeated or long-term exposure is required before an infection becomes established, these condi-tions are more likely to be met by outdoor activities close to the place of residence than by travels to alveolar echinococcosis-endemic areas. We therefore assume that for most cases, the place of residence is most likely the area of infection. A complete documen-tation of all the places where the patients had lived during their lives was available for approximately 30%. Mobility of this patient subgroup was low, in conjunction with long-term farming.

The distribution of alveolar echinococcosis in Europe shows a core area with a high density of cases and border areas with clusters of a few patients or single cases. The core area covers large parts of the classic alveolar echinococcosis-endemic regions in Austria, France, Germany, and Switzerland, including those where the index cases from each of these countries have been identified since 1855 (17). In these areas, recent screening studies have detected not only a small number of manifest diseases but also self-cured infections (aborted lesions, first described by Rausch et al. [18]), and sero-positivi- ty rates of up to 2% (14,19,20). Fifteen persons with aborted hepatic lesions (lesions with characteristic calcification) and positive serologic results were reported to the registry but were excluded from this analysis, since a definite diagnosis based on histopathologic or molecular findings had not been provided. Together with a persis-tent *E. multilocularis* seroprevalence, such reports point to a manifest infection pressure in the core area.

In the core area, a consistently high prevalence of *E. multilocularis* in foxes has been reported, e.g., >50% in southwestern Germany (11), 44% in western Bavaria (21), 65% in eastern France (13), and 35% in western Austria (9). In the border area with less frequent and more dispersed human cases, fewer investigations have been undertaken to establish parasite prevalence, and the figures determined rely on low numbers of examined foxes. The prevalence was generally low, e.g., 13% in northern Germany (11), 8% to 21% in eastern Bavaria (21), and 10% in eastern Austria (9). In Belgium, the first three patients with autochthonous infections lived in areas with low parasite prevalence (final report to the European Commission, Directorate General V (EurEchinoReg, unpub. data, 1999). No prevalence data are available for northern France. In Poland, parasite prevalence was initially investigated in 1994. All registered patients live in the northeastern districts with the highest prevalence (20% and 36%) (22). In Greece, sporadic cases had been reported previously (23).

Whether low parasite prevalence exerts an infection pressure relevant for transmission to humans remains questionable. Recent investigations have shown that foci of high prevalence can persist, even for long periods, in regions where the overall infection rates in foxes are low (e.g., foci of 25% in areas with 5% in eastern Germany) (24). Similar foci may exist in other regions but are undetected to date. Thus, human infection can probably occur in regions with low overall parasite prevalence, and we regard case reports from areas remote from the core area as strong hints of new areas at risk. Therefore, threshold findings at diagnosis should not be rated as incompatible with the disease when the patient lived in an area where human alveolar echinococcosis was unknown before. According to Eckert et al. (12), all regions with a proven occurrence of *E. multilocularis* in Red Foxes indicate a "potential risk area," irrespective of the magnitude of prevalence rates. This view is the basis for the current concept of a continuous distribution of the parasite in Europe from central France to Poland. Future studies should, therefore, address the redefinition of risk areas for alveolar echinococcosis and the population at risk.

Transmission of the parasite to humans occurs rarely, and individual risk factors for human disease are not well understood. In Europe, only one case-control study has been published; this study included 21 patients and 84 controls from Austria (25). A high association of the disease was found with cat ownership and hunting, but because of the low case number the study was of limited power. Farming did not seem to have an impact on infection risk. In China, a population-based study showed that farming was the most important risk factor (26). In Alaska, dog ownership was found to be associated with the disease (19 patients, 38 controls) (27). None of these studies found an association of the disease with a history of picking and eating wild berries and mushrooms or raw produce from unfenced gardens. Also, neither fox hunters in China (26) nor trappers in South Dakota, United States, are affected by the disease (28). The records of 210 patients from the European registry data show that 21.9% were farmers; another 39.5% were engaged in farming, gardening, hunting, or working in forestry as a pastime; 70.5% of all patients kept dogs or cats. These data point to a high frequency of putative exposure, but the lack of a comparison group does not allow an evaluation of the risk potential of these activities. These activities may be characteristic of most people in rural communities in Europe. For Europe, the questions of how risk behavior can be defined and how exposure can best be prevented are, therefore, still unanswered.

Within the last 20 years, major improvements have been made in the diagnosis and treatment of alveolar echinococcosis. Definite diagnosis by histopathology was available in 53.3% of this case series; the remaining cases were ascertained by imaging with or without specific serology. At diagnosis, 34% of the patients were already

Table 5. Vital status of patients with alveolar echinococcosis, as of December 2000

Yr of first diagnosis	No. of patients			Interval between diagnosis and death (yrs)
	Alive	Lost to follow-up	Deceased	
Until 1980	63	4	37	4-29
1981-1985	73	11	28	<1-14
1986-1990	84	9	34	<1-10
1991-1995	85	6	10	1-4
1996-2000	103	2	10	<1-1
Total	408	32	119	

affected by advanced larval growth; when the parasitic tissue does not affect important organs or vessels, it may go unnoticed for prolonged periods. This fact may also explain a diagnosis late in a patient's life. In Europe, the mean age at diagnosis was slightly higher than in non-Caucasian populations, i.e., in Hokkaido, Japan (48.7 years) (29), or in China (35.7 years) (30).

Immunodeficiencies, e.g. HIV infection (31) or immunosuppressive therapy after liver transplantation (32), may possibly accelerate the manifestation of alveolar echinococcosis. Chemotherapy with benzimidazoles was the only treatment for one third of the registered patients; complete cure after surgery and adjuvant chemotherapy was achieved in only 4.9%. For most patients, stability can be achieved with long-term chemotherapy, with or without surgery or interventional radiology, but the disorder remains chronic.

This report is the first collection of data on human alveolar echinococcosis in Europe. Our study confirms that an infection with this parasite is still dangerous. A low annual incidence persists in the previously known foci. However, case reports from regions remote from the core area indicate that the disease is spreading. We therefore recommend that the occurrence of this potentially reemerging zoonosis should be continuously monitored in western and central Europe.

Acknowledgments
We thank the hospital and clinic physicians who informed the registry about alveolar echinococcosis cases; students who searched for and extracted data from case report forms; and Alain Gérard, Jacques Beytout, Jérome Watelet, and Thierry Saurin, who made the patients' files available in Nancy, Clermont-Ferrand, and Lyon, France.

The pilot phase received financial support of the European Commission, Directorate General V (SOC 97 20239805F01). In Germany, the registry work was additionally financed by the University of Ulm, the Paul-Ehrlich-Gesellschaft e.V., and GlaxoSmithKline GmbH&Co. KG, Munich. In Switzerland, the registry was financed in part by the Swiss National Science Foundation, Janssen Pharmaceutica, Berse, Belgium, the Echinococcosis Foundation in Zürich, and a number of private sponsors. In France, the study received support from the Conseil Régional de Franche Comté.

Dr. Kern is a research assistant at the Department of Biometry and Medical Documentation at the University of Ulm, Germany. She is responsible for the data collection of human cases of alveolar echinococcosis, data control, and analysis in the European Echinococcosis Registry.

References

1. Ammann RW, Eckert J. Cestodes. Echinococcus. Gastroenterol Clin North Am 1996;25:655–89.
2. Schantz PM, Van den Bossche H, Eckert J. Chemotherapy for larval echinococcosis in animals and humans: report of a workshop. Zeitschrift für Parasitenkunde 1982;67:5–26.
3. Wilson JF, Rausch RL, McMahon BJ, Schantz PM. Parasiticidal effect of chemotherapy in alveolar hydatid disease: review of experience with mebendazole and albendazole in Alaskan Eskimos. Clin Infect Dis 1992;15:234–49.
4. Bresson-Hadni S, Vuitton DA, Bartholomot B, Heyd B, Godart D, Meyer JP, et al. A twenty-year history of alveolar echinococcosis: analysis of a series of 117 patients from eastern France. Eur J Gastroenterol Hepatol 2000;12:327–36.
5. Ammann RW, Ilitsch N, Marincek B, Freiburghaus AU. Effect of chemotherapy on the larval mass and the long-term course of alveolar echinococcosis. Swiss Echinococcosis Study Group. Hepatology 1994;19:735–42.
6. Reuter S, Jensen B, Buttenschoen K, Kratzer W, Kern P. Benzimidazoles in the treatment of alveolar echinococcosis: a comparative study and review of the literature. J Antimicrob Chemother 2000;46:451–6.
7. Ishizu H, Uchino J, Sato N, Aoki S, Suzuki K, Kuribayashi H. Effect of albendazole on recurrent and residual alveolar echinococcosis of the liver after surgery. Hepatology 1997;25:528–31.
8. Eckert J, Deplazes P. Alveolar echinococcosis in humans: the current situation in Central Europe and the need for countermeasures. Parasitol Today 1999;15:315–9.
9. Auer H, Aspöck H. Human alveolar echinococcosis and cystic echinococcosis in Austria: the recent epidemiological situation. Helminthologia 2001;38:3–14.
10. Ammann RW, Fleiner Hoffmann A, Eckert J, Schweizerische Echinokokkose-Studiengruppe (SESG). Schweizerische Studie für Chemotherapie der alveolären Echinokokkose—Rückblick auf ein 20jähriges klinisches Forschungsprojekt. Schweiz Med Wochenschr 1999;129:323–32.
11. Romig T, Bilger B, Dinkel A, Merli M, Mackenstedt U. *Echinococcus multilocularis* in animal hosts: new data from western Europe. Helminthologia 1999;36:185–91.
12. Eckert J, Conraths FJ, Tackmann K. Echinococcosis: an emerging or re-emerging zoonosis? Int J Parasitol 2000;30:1283–94.
13. Giraudoux P, Raoul F, Bardonnet K, Vuillaume P, Tourneux F, Cliquet F, et al. Alveolar echinococcosis: characteristics of a possible emergence and new perspectives in epidemiosurveillance. Médicine des Maladies Infectieuses 2001;31:247–56.
14. Romig T, Kratzer W, Kimmig P, Frosch M, Gaus W, Flegel WA, et al. An epidemiologic survey of human alveolar echinococcosis in southwestern Germany. Römerstein Study Group. Am J Trop Med Hyg 1999;61:566–73.
15. Hegglin D, Bontadina F, Gloor S. From the alpine to the urban fox—adaptive behavior of the red fox (Vulpes vulpes). Advances in Ethology 1998;33:119.
16. Altintas N. Cystic and alveolar echinococcosis in Turkey. Ann Trop Med Parasitol 1998;92:637–42.
17. Fesseler M. Vergleich der Endemiegebiete von *Echinococcus multilocularis* und Tollwut in Mitteleuropa [dissertation]. Zürich: Universität Zürich; 1990.

18. Rausch RL, Wilson JF, Schantz PM, McMahon BJ. Spontaneous death of *Echinococcus multilocularis*: cases diagnosed serologically (by Em2 ELISA) and clinical significance. Am J Trop Med Hyg 1987;36:576–85.

19. Gottstein B, Saucy F, Deplazes P, Reichen J, Demierre G, Busato A, et al. Is high prevalence of *Echinococcus multilocularis* in wild and domestic animals associated with disease incidence in humans? Emerg Infect Dis 2001;7:408–12.

20. Bresson-Hadni S, Laplante JJ, Lenys D, Rohmer P, Gottstein B, Jacquier P, et al. Seroepidemiologic screening of *Echinococcus multilocularis* infection in a European area endemic for alveolar echinococcosis. Am J Trop Med Hyg 1994;51:837–46.

21. Nothdurft HD, Jelinek T, Mai A, Sigl B, von Sonnenburg F, Löscher T. Epidemiology of alveolar echinococcosis in southern Germany (Bavaria). Infection 1995;23:85–8.

22. Malczewski A. CE and alveolar echinococcosis in eastern Europe. In: Craig P, Pawlowski Z, editors. Cestode zoonoses: echinococcosis and cys- ticercosis. Amsterdam: IOS Press; 2002. p. 81–9.

23. Theodoropoulos G, Kolitsopoulos A, Archimandritis A, Melissinos K. Echinococcose alvéolaire hépatique: trois observations en Grèce. La Nouvelle Presse Médicale 1978;7:3056.

24. Tackmann K, Löschner U, Mix H, Staubach C, Thulke HH, Conraths FJ. Spatial distribution patterns of *Echinococcus multilocularis* (Leuckart 1863) (Cestoda: Cyclophyllidea: Talveolar echinococcosisniidalveolar echinococcosis) among red foxes in an endemic focus in Brandenburg, Germany. Epidemiol Infect 1998;120:101–9.

25. Kreidl P, Allersberger F, Judmaier G, Auer H, Aspöck H, Hall AJ. Domestic pets as risk factors for alveolar hydatid disease in Austria. Am J Epidemiol 1998;147:978–81.

26. Craig PS, Giraudoux P, Shi D, Bartholomot B, Barnish G, Delattre P, et al. An epidemiological and ecological study of human alveolar echinococcosis transmission in south Gansu, China. Acta Trop 2000;77:167–77.

27. Stehr-Green JK, Stehr-Green P, Schantz PM, Wilson JF, Lanier A. Risk factors for infection with *Echinococcus multilocularis* in Alaska. Am J Trop Med Hyg 1988;38:380–5.

28. Hildreth MB, Sriram S, Gottstein B, Wilson M, Schantz PM. Failure to identify alveolar echinococcosis in trappers from South Dakota in spite of high prevalence of *Echinococcus multilocularis* in wild canids. J Parasitol 2000;86:75–7.

29. Aoki S, Uchino J, Sato N, Takahashi M, Shimamura T, Misawa K. Clinic op atho logical study on alveolar echinococcosis of the liver. In: Uchino J, Sato N, editors. Alveolar echinococcosis: strategy for eradication of alveolar echinococcosis of the liver. Sapporo, Japan: Fuji Shoin Sapporo; 1996. p. 101–7.

30. Zhou HX, Chai SX, Craig PS, Delattre P, Quere JP, Raoul F, et al. Epidemiology of alveolar echinococcosis in Xinjiang Uygur autonomous region, China: a preliminary analysis. Ann Trop Med Parasitol 2000;94:715–29.

31. Sailer M, Soelder B, Allerberger F, Zaknun D, Feichtinger H, Gottstein B. Alveolar echinococcosis of the liver in a six-year-old girl with acquired immunodeficiency syndrome. J Pediatr 1997;130:320–3.

32. Bresson-Hadni S, Koch S, Beurton I, Vuitton DA, Bartholomot B, Hrusovsky S, et al. Primary disease recurrence after liver transplantation for alveolar echinococcosis: long-term evaluation in 15 patients. Hepatology 1999;30:857–64.

Address for correspondence: Petra Kern, Dept. for Biometry and Medical Documentation, University of Ulm, Schwabstr. 13, D-89075 Ulm, Germany; fax: (0049) 731–5026902; e-mail: echinoreg@medizin.uni-ulm.de

EMERGING
INFECTIOUS DISEASES

A Peer-Reviewed Journal Tracking and Analyzing Disease Trends   Vol.7, No.3, May–Jun 2001

Search past issues of EID at www.cdc.gov/eid

EID
Online
www.cdc.gov/eid

Seasonal Cycles and
Susceptibility to Infections

Chapter Authors

Ted B. Lyon

An attorney specializing in complex litigation with over forty years of experience,
Ted Lyon has represented clients in more
than one hundred fifty jury trials and
was named one of the top one hundred
lawyers in America by the American Trial
Lawyers Association 2007 through 2016.
Ted served in both the Texas House of
Representatives (1979–1983) and the
Texas State Senate (1983–1993). He has
also been a police officer, a licensed fishing and hunting guide, and a teacher. Ted

has received numerous prestigious and meaningful awards including the 2012
Teddy Roosevelt Conservationist of the Year award, Wildlife Recovery Award
by Montana Sportsmen for Fish and Wildlife in 2014, Legends of the Bar
Recipient in 2016 from the Dallas Bar Association, and was featured in D
CEO Magazine as one of the Most Powerful Business Leaders in Dallas-Fort
Worth in 2016.

Will N. Graves

Will Graves began his professional career
with the US Department of Agriculture
as the leader of a team working on controlling hoof and mouth disease's spread
to the US. One of the primary vectors was coyotes. With the outbreak of
the Korean War, Will volunteered for
the US Air Force and was trained as a
Russian linguist. In order to accelerate and develop his skills in Russian, he

started to read Russian wildlife magazines and books. Wolves were often discussed, and soon his interest became focused on wolves in Russia. He asked every native Russian he met, including scientists, if they had any knowledge of wolves and began to record data and resources. Graves's interest in wolves grew into a serious hobby that continued after the war. For more information, visit: wolfeducationinternational.com

Valerius Geist

Valerius Geist was born in Ukraine in 1938 and immigrated to Canada in 1953. Geist earned a bachelor's degree in zoology from the University of British Columbia in 1960. His wildlife studies began with feral goats and continued with moose, Stone sheep, bighorn sheep, and Dall sheep. In 1966, Geist earned a doctorate in ethology, the study of animal behavior. He returned to Germany for a year of studying with Konrad Lorenz, and wrote his first award-winning book during that time. Geist joined the faculty of the University of Calgary in 1968, where he was a professor and Department Chairman of environmental science and biology until 1995. Geist has shared his research, knowledge, and ideas through fifteen books, seven wildlife policy reports, more than 120 scholarly papers, book chapters, and commentaries, as well as numerous magazine articles and contributions to documentary films covering an array of wildlife topics and species. Particularly relevant to this book, Geist has served as expert witness in wildlife law enforcement, environmental policy, native treaties, and animal behavior cases in both the US and Canada, including the death of Kenton Carnegie by wolves.

Don Peay

Don Peay earned his bachelor of science in chemical Engineering and his master's degree from BYU. After working at an aerospace firm, Don started his own engineering consulting firm, Petroleum Environmental Management Inc. With deer herd populations crashing, ranchers wanting to greatly reduce Utah's elk herds, and a fish and game agency saying the future was non-hunting programs, Don founded Sportsmen for Fish and Wildlife in 1994. Since then, Don has received numerous conservation awards in Utah and throughout the

West, including *Outdoor Life* naming him one of the Top 25 Conservationists in North America in 2008. In 2009, the Utah Legislature passed a unanimous resolution thanking Don for his work to help military veterans, to protect wildlife and their habitats, and to increase jobs, tourism, and quality hunting opportunities.

Arthur T. Bergerud

Dr. Arthur T. Bergerud has been a population ecologist involved in research on caribou populations in North America since 1955 and is considered the world's foremost authority on woodland caribou. His thirty-year study (1974 to 2004) of two caribou populations, one in Pukaskwa National Park (PNP) and the other on the Slate Islands in Ontario, is considered the most comprehensive study of caribou ever done. Along with Stuart Luttich and Lodewijk Camps, Bergerud authored *The Return of the Caribou to Ungava* (2008), which is the story of the George River caribou herd that increased from fifteen thousand animals in 1958 to seven hundred thousand in 1988, becoming the largest herd in the world at the time. For over twenty years, Bergerud, the former chief biologist with Newfoundland and Labrador's Wildlife Division, and his associates studied the George River herd all across Canada's tundra and taiga.

Heather Smith-Thomas

Heather Smith-Thomas grew up on a cattle ranch near Salmon, Idaho, and started writing about horses and cattle in high school. She graduated from the University of Puget Sound in 1966 with a BA in English and history. Heather, who writes regularly for more than twenty-five farm and livestock magazines and thirty horse publications, has written twenty books on such topics as horse care and cattle raising. She and her husband, Lynn Thomas, have been raising beef cattle and horses since 1967.

Rob Arnaud

Rob Arnaud is a fourth-generation Montanan, who grew up in Manhattan, Montana. His love of the outdoors and hunting led him to study animal science at Montana State University where he graduated with a BS in 1980. He has been the contracted wildlife manager and outfitter on several large ranches in the West and has served on a private wildlife board. He is a licensed outfitter in Montana, Wyoming, and California. In total, he provides trophy hunting opportunities on sixteen different ranches. He is also the president of the Montana Outfitters and Guides Association and is very active in lobbying for hunter rights and responsible wildlife management. Rob is fortunate to have hunted many of the North American animals.

Laura Schneberger

Laura Schneberger ranches with her husband and children in southwestern New Mexico on a historic cattle ranch located in the Black Range of the Gila Forest. She spends as much time as possible educating the public and writing articles on the realities of wolf reintroductions and their impacts on families and small communities that depend on livestock production for an economic base. Living in the middle of wolf and mountain lion territory has made her a much sought-after expert of wildlife activity in the Southwest.

Jess Carey

Jess Carey, Catron County wolf inspector, has lived in Reserve, New Mexico, for thirty-five years. Trained in woods skills by his father over fifty years ago, following a tour of duty in the US Marine Corps, initially, he made his living as a trapper. For the last ten years, prior to becoming a wolf inspector, Carey served as the elected sheriff, under-sheriff, and investigator for the 7th judicial district attorney's office in Catron County.

Karen Budd-Flaen

Karen Budd-Falen is an attorney, and with her husband Frank Falen, is the owner of the Budd-Falen Law Offices, L.L.C. located in Cheyenne, Wyoming. Karen served for three years in the Reagan Administration, US Department of the Interior, Washington, DC, as a Special Assistant to the Assistant Secretary for Land and Minerals Management. She later served as a law clerk to the

Assistant Solicitor for Water and Power, and has also worked as an attorney at Mountain States Legal Foundation. Karen has been featured in *Newsweek*'s "Who's Who: 20 for the Future" for her work on property rights issues, and was awarded Wyoming's Outstanding Ag Citizen in 2001, the "Always There Helping" award from the New Mexico Stock Growers Association in 2003, and the "Bud's Contract" award from the New Mexico Public Lands Council in 2006. Karen grew up as a fifth-generation rancher on a family-owned ranch in Big Piney, Wyoming. She received her undergraduate degrees and her law degree from the University of Wyoming. She is admitted to practice law before the State Court of Wyoming; US District Courts for Wyoming, Colorado, Nebraska, and the District of Columbia; US Courts of Appeals for the Seventh, Eighth, Ninth, Tenth, District of Columbia and Federal Circuits; US Court of Federal Claims; and US Supreme Court.

Joshua Tolin

Joshua Tolin is an associate attorney with Budd-Falen Law Offices, LLC, in Cheyenne, Wyoming, where he practices natural resources law, environmental law, administrative law, property law, and criminal law. Prior to joining BFLO, Joshua served as a judicial law clerk to the Honorable Judge Alan B. Johnson in the United States District Court for the District of Wyoming. Joshua also interned with the United States Attorney's Office for the District of Wyoming, where he practiced both civil and criminal law in the District of Wyoming and the Tenth Circuit Court of Appeals under their respective student-practice rules. Joshua earned his juris doctor with honors from the University of Wyoming in Laramie, where he was awarded the Archie G. McClintock Outstanding Law Student Award. Joshua served both as an editor for the Wyoming Law Review and a Teaching Assistant for Legal Research and Writing. Prior to law school, Joshua earned an MBA from the University of Nebraska Graduate College and a bachelor of science in political science from the University of Nebraska at Kearney.

Matthew A. Cronin

Matthew A. Cronin received a PhD in biology from Yale University in 1989, an MS in biology from Montana State University in 1986, and a BS in Forest Biology from State University New York, College of Environmental Science and Forestry in 1976. Cronin was a US Coast Guard officer from 1981 to 1984 and worked for the US Fish and Wildlife Service as a geneticist from 1989 to 1992. From 1992 to 2004, he worked in the private sector with a focus on wildlife research and impact assessments for the oil, timber, and mining industries. From 2004 to 2016, Cronin was a research professor of animal genetics at the University  of Alaska Fairbanks, School of Natural Resources and Agricultural Sciences, where he taught and focused his research on population genetics of wildlife and livestock. In 2016, he returned to the private sector with the Northwest Biotechnology Company consulting company in Anchorage and Bozeman, Montana. Cronin's work emphasizes proper application of science to natural resource management, and his experience with private industry, academia, and government provides insights for achieving multiple use objectives.

Index